Student Workbook

Third Edition

Intermediate Microeconomics
Theory and Applications

Heinz Kohler
Amherst College

SCOTT, FORESMAN/LITTLE, BROWN HIGHER EDUCATION
A Division of Scott, Foresman and Company
Glenview, Illinois   London, England

ISBN 0-673-46208-0

Copyright © 1990, 1986, 1982 Scott, Foresman and Company.
All Rights Reserved.
Printed in the United States of America.

1 2 3 4 5 6—VPI—94 93 92 91 90

PREFACE

This *Student Workbook* is designed to accompany the third edition of *Intermediate Microeconomics: Theory and Applications*. It is not a substitute for the text or for class discussion. Nor is it a summary of the text. Rather, it is a self-contained, self-help device, entirely devoted to making successful study easier for the student.

The *Student Workbook* is completely integrated with the text and should be used with the text chapters as they are studied. Answers to *all* questions and problems in any chapter of this guide can be found at the end of that chapter. Thus, students can examine and grade themselves during and after study of the text and before taking an actual class test.

The following order of study is particularly recommended for each chapter:

1. Read the text chapter, taking notes as you like.

2. Look at the chapter Summary in the text and ask yourself whether you could easily elaborate upon each of the Summary points. If necessary, review your notes or reread the relevant section of the chapter.

3. Look at the list of Key Terms at the end of the text chapter and ask yourself whether you could easily define them. If necessary, look at the Glossary in the back of the text or find the relevant term printed in boldface in the body of the chapter.

4. Answer the multiple-choice questions in the *Workbook*. There is one answer per question. Then check your answers against those at the end of the *Workbook* chapter.

5. Answer the true-false questions in this *Workbook* and then check the answers.

6. If you did not do well, it is time for a second and more thorough reading of the text. Find out *why* you made the mistakes you did when answering the two types of objective questions.

7. Now give yourself a more difficult test by turning to the Problems in this *Workbook*. Once more, detailed answers are available at the end of the chapter.

8. Finally, turn to the end-of-chapter *text* questions and problems. You can check to see if you're on the right track by comparing your odd-numbered answers with those in the Answer section at the end of the text.

9. Those students who wish to go beyond the text will find in the 31 biographies and the 4 appendices of this *Workbook* a number of fascinating starting points (see the Contents for a detailed listing).

Naturally, the above procedure is only a suggestion. After trying this procedure you may finally select another one that better suits your tastes and needs. I wish you success.

Heinz Kohler
Amherst College

THE KOHLER-3 PERSONAL COMPUTER DISKETTES

A set of programs for IBM personal computers and compatible machines has been specifically designed to accompany this text. The computer must have graphics capability. Although it is not required, a color display is preferable. The 20 programs are available either on two 5.25″ floppy disks or on a single 3.5″ rigid diskette.

KOHLER-3 covers all aspects of the text. It contains:

1. About 450 multiple-choice questions (along with nearly 200 graphs). For each question, the correct answer is identified and elaborate comments are provided on incorrect responses.

2. Some 200 challenging problems (with solutions when appropriate). These problems involve one of the following:
    a. Thought Experiments. You think about the effects of specified events on graphs or tables, then use the computer to check on your mental solution. The graphs and tables change in response to your prompts. Examples: the separation of substitution and income effects (Chapter 3), the 'random walk' of futures prices (Chapter 10), income redistribution and incentives (Chapter 14), externalities and their remedies (Chapter 16).
    b. Graphical Exercises. You interpret graphical information. Examples: measuring price elasticities (Chapter 3), isoquant analysis (Chapter 5), monopoly profit maximization or government price fixing (Chapter 7).
    c. Tabular Exercises. You interpret tabular information. Examples: the production function (Chapter 4), concepts of cost (Chapter 5), marginal revenue (Chapter 7), decision theory and game theory (Chapter 9).
    d. Solving Equations Algebraically. You learn to solve linear simultaneous equations algebraically. Examples: supply and demand in markets for goods and resources (Chapters 6 and 11), the Cournot equilibrium or the dominant-firm model (Chapter 8), general equilibrium (Chapter 13).
    e. Other Algebraic Exercises. You learn to use various special programs. Examples: maximization under constraints via linear programming (Chapter 5), determining fair and unfair insurance and gambles (via an expected-monetary-value program, Chapter 10), benefit-cost analysis (via a cash-flow program, Chapter 16).

3. A kit of tools that will enable you to solve all kinds of problems of your own. Most of these tools are found in Appendices A–D; some are built into chapters. They allow you to solve simultaneous equations, do linear programming, work compounding or discounting problems, find expected monetary values, perform matrix operations, and much more.

Initial Start-Up Procedure

The KOHLER-3 diskettes contain detailed operating instructions. For first-time-ever users, however, the following will be helpful:

1. Prior to loading KOHLER-3, in response to the DOS prompt (such as C>), type BASICA and press [ENTER]. (Failure to load BASICA and loading BASIC instead will prevent you from seeing graphs on the screen.)

2. At this point, place your diskette into the drive of your choice (for example, A) and, in response to the BASICA prompt (OK), type RUN"A:KOHLER" then press [ENTER]. (Naturally, you replace *A* by another letter if you are not using drive A.)

3. Some program features are automatic. Do not press any key unless told.

4. You can escape the program at any time by pressing [Ctrl] plus [Break]. If you also wish to return to DOS, type SYSTEM and press [ENTER].

5. You can always restart the program by the procedure noted in (1) and (2) if you see the DOS symbol. Just press the [F2] key if you see the OK symbol.

Hard Disk Installation Instructions

If you have the two floppy diskettes but your machine has a hard disk, you can install both floppies on it, along with a unified menu, by the procedure below. It is assumed here that the floppy disk is placed in drive A, that drive A is not a high-density drive, and that the hard disk is called C. If this is not the case, you must change the drive names accordingly.

1. Get the DOS prompt C> and place disk #1 in drive A. Then type COPY A:*.* C: (with a space before A and before C) and press [ENTER].

2. Repeat step 1 for disk #2.

3. Get the DOS prompt C> and type ERASE C:KOHLER.BAS (with a space before C) and press [ENTER].

4. Get the DOS prompt C> and type
RENAME C:KOHLER-H.BAS KOHLER. BAS
(with two single spaces as shown) and press [ENTER].

CONTENTS

1. Scarcity, Choice, and Optimizing 1
   Biography 1.1 Milton Friedman 11
   Appendix 1A Differential Calculus: A Review of the Basics 12
2. The Preferences of the Consumer 23
   Biography 2.1 Jeremy Bentham 36; 2.2 William S. Jevons 36;
   2.3 Francis Y. Edgeworth 37
3. The Demand for Goods 39
   Biography 3.1 Evgeny E. Slutsky 50; 3.2 John R. Hicks 50
4. The Technology of the Firm 51
   Biography 4.1 Johann H. von Thünen 63; 4.2 Harvey Leibenstein 65
5. Costs and the Supply of Goods 66
   Biography 5.1 Jacob Viner 79; 5A.1 Leonid V. Kantorovich 93
   Appendix 5A Linear Programming 80
6. Perfect Competition 95
   Biography 6.1 Alfred Marshall 111
7. Monopoly and Cartels 113
   Biography 7.1 Antoine A. Cournot 130
8. Oligopoly and Monopolistic Competition 132
   Biography 8.1 Edward H. Chamberlin 143
9. Decision Theory and Game Theory 145
   Biography 9.1 Frank H. Knight 154; 9.2 Oskar Morgenstern 154
10. Insurance and Gambling, Search and Futures Markets 156
    Biography 10.1 Joseph A. Schumpeter 168
11. Labor and Wages 170
    Biography 11.1 Joan V. Robinson 190
12. Capital and Interest 192
    Biography 12.1 Eugen von Böhm-Bawerk 204; 12.2 Irving Fisher 205;
    12.3 Theodore W. Schultz 207
    Appendix 12A Markets for Bonds and Stocks 208
13. General Equilibrium 214
    Biography 13.1 Adam Smith 224; 13.2 Marie Esprit Léon Walras 226;
    13.3 Wassily W. Leontief 227
    Appendix 13A Deriving the Leontief Inverse 229
14. Efficiency and Equity 233
    Biography 14.1 Vilfredo Pareto 258; 14.2 Friedrich A. von Hayek 259;
    14.3 Karl Marx 261
15. Property Rights, Antitrust, and Regulation 263
    Biography 15.1 George J. Stigler 271
16. Externalities and Public Goods 273
    Biography 16.1 Arthur C. Pigou 292; 16.2 Ronald H. Coase 292;
    16.3 Paul A. Samuelson 293; 16.4 James M. Buchanan 294

CHAPTER 1

Scarcity, Choice, and Optimizing

MULTIPLE-CHOICE QUESTIONS

Circle the letter of the *one* answer that you think is correct or closest to correct.

1. *Goods* are best described as
   a. physical objects, like bread, milk, or cars.
   b. means to satisfy material wants.
   c. intangible services.
   d. tangible commodities.

2. People's *desire for goods* is best measured by
   a. their demand for goods in the market.
   b. the quantities of goods they would take if all goods were offered at zero prices.
   c. the amount of purchasing power that backs up people's wishful thinking regarding goods.
   d. either (a) or (c).

3. *Resources* are best described as
   a. ingredients that can be used to make goods.
   b. human capital, land, and technology.
   c. labor, land, and capital (human plus financial).
   d. people able and willing to participate in the process of production.

4. Domesticated animals, irrigated farm land, and oil in storage tanks are perfect examples of what economists mean by
   a. natural resources.
   b. human resources.
   c. capital resources.
   d. human capital.

5. Which of the following, without question, are classified as *natural resources*?
   a. 100 acres of pasture land with 10,000 head of cattle.
   b. A waterfall, dam, and electric power plant.
   c. The wind, the tides, and an iron ore deposit.
   d. Both (a) and (c).

6. Which of the following might well be classified as (real) *capital resources*?
    a. Factory buildings, schools, airport control towers.
    b. Bridges, computers, typewriters.
    c. Bottles of ketchup, pasture land, sheep.
    d. All of the above.

7. The *process of production* of shoes includes
    a. their physical manufacture as well as their transportation over space and time until these shoes reach their ultimate user.
    b. their physical manufacture only.
    c. their physical manufacture and transportation over space, but not their storage.
    d. their physical manufacture and "transportation" (or storage) over time, but not their transportation over space.

8. The basic economic problem of scarcity arises because
    a. most people harbor an immense desire for goods.
    b. resources exist in limited quantities only; hence the set of goods people can produce in a period is finite.
    c. imperfect technology limits the quantity of goods people can produce per unit of resources.
    d. the set of goods people can produce in a period is insufficient to satisfy their desire for goods in that period.

9. Over time, as more resources and better technology are put to work and output rises, the extent of scarcity
    a. is bound to be reduced.
    b. is bound to be unchanged.
    c. is bound to increase.
    d. may be reduced, may remain unchanged, or may increase.

10. The *production-possibilities frontier* shows all the alternative combinations of two goods (or types of goods) that
    a. are wanted by people.
    b. cannot be produced.
    c. can be produced by using given resources fully and in the best possible way.
    d. can be produced with given resources—whether used fully and in the best possible way or not.

11. If there were no scarcity, the combination of goods that would fulfill the unrestrained material wants of people would lie
    a. to the right and above the production-possibilities frontier.
    b. to the left and below the production-possibilities frontier.
    c. on the production-possibilities frontier.
    d. at a point described by either (b) or (c).

When answering questions 12 and 13, consider text Figure 1.2, "The Production-Possibilities Frontier," which is reproduced on page 3.

12. The people of a society operating at point *E* would be incurring an opportunity cost of
    a. *AE*
    b. 0*D*.
    c. *BD* (while unnecessarily wasting all the resources that could have produced 0*D*).
    d. 0*D* (while unnecessarily wasting all the resources that could have produced *BD*).

13. If the people of this society were to move from *E* to *C,* they would have to give up
    a. no goods at all.
    b. 0*E* of military goods.
    c. *AE* of military goods.
    d. *BD* of civilian goods.

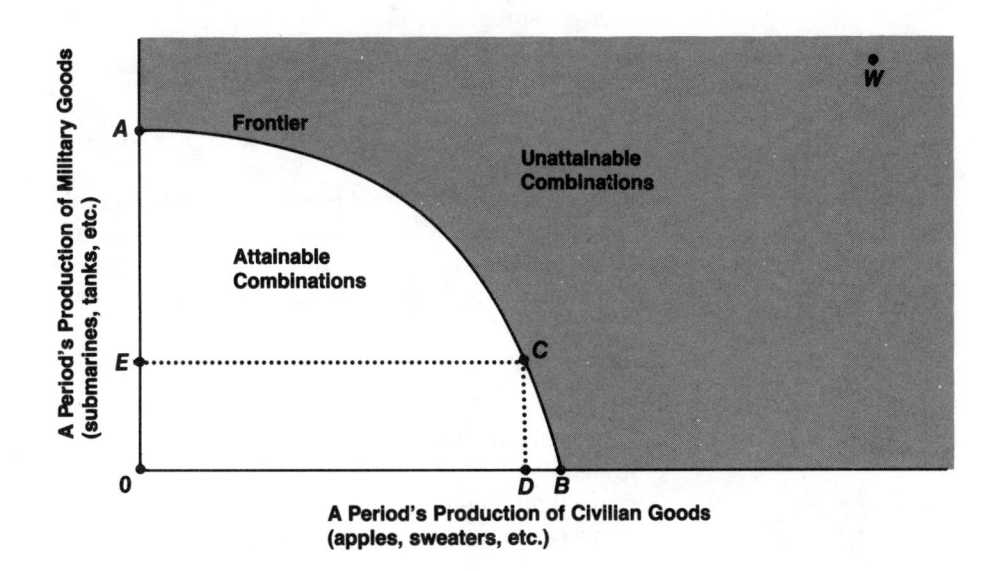

14. Hitler's guns-for-butter choice could be illustrated
    a. by a shift of the German production-possibilities frontier.
    b. by a move to the outside of the German production-possibilities frontier.
    c. by a move along the German production-possibilities frontier.
    d. by a move to the inside of the German production-possibilities frontier.

15. In the context of scarcity, every choice brings with it
    a. an advantage or benefit.
    b. a disadvantage or cost.
    c. an opportunity lost or an opportunity cost.
    d. all of the above.

16. The successful economizing of resources requires
    a. the making of all-or-nothing decisions.
    b. marginalist thinking about the objective possibility and the subjective welfare implication of small changes in variables.
    c. that people value the marginal benefit of any activity less than the marginal cost.
    d. that each activity is carried to the point of satiation.

17. *Marginal cost* could be the term used to describe
    a. the decrease in an activity's overall benefit that is attributable to a unit decrease in the level of that activity.
    b. the decrease in an activity's overall cost that is attributable to a unit decrease in the level of that activity.
    c. the increase in an activity's overall cost that is attributable to a unit decrease in the level of that actvity.
    d. either (a) or (b).

18. The values placed on the marginal cost tend to be
    a. smaller at higher levels of an activity.
    b. larger at higher levels of an activity.
    c. larger at lower levels of an activity.
    d. equal to those placed on the marginal benefit.

19. The values placed on the marginal benefit tend to be
    a. smaller at higher levels of an activity.
    b. larger at higher levels of an activity.
    c. smaller at lower levels of an activity.
    d. equal to those placed on the marginal cost.

20. The *total* benefit associated with a given level of any activity equals
    a. its total cost (because the very resources used in this activity could have been used in an alternative activity instead).
    b. the marginal benefit, but only at the optimum point.
    c. the sum of marginal benefits for all activity units up to this given level.
    d. the marginal benefit minus the marginal cost.

21. *Satiation* implies
    a. zero marginal cost.
    b. maximum total benefit.
    c. maximum marginal benefit.
    d. maximum total cost.

22. The *optimization principle* involves
    a. a comparison of an activity's marginal benefit and marginal cost.
    b. a recommendation that people who care to maximize their welfare expand any activity if they value its marginal benefit more highly than its marginal cost.
    c. a recommendation that people who care to maximize their welfare contract any activity if they value its marginal benefit less than its marginal cost.
    d. all of the above.

23. Which of the following is *false*?
    a. As long as the expansion of an activity brings positive marginal benefits, the activity level cannot be optimal.
    b. As long as the contraction of an activity is associated with declining marginal costs, the activity level cannot be optimal.
    c. As long as the expansion of an activity brings positive marginal costs, the activity level cannot be optimal.
    d. All of the above.

24. The optimization principle can be applied to almost anything, including the degree to which resources are spent on
    a. accident prevention.
    b. crime control.
    c. saving human lives.
    d. any of the above.

25. *Theorizing* is
    a. abstracting from reality.
    b. a process akin to producing a geographic map.
    c. letting the facts speak for themselves.
    d. both (a) and (b).

26. A theory that is based on false assumptions
    a. can never generate successful predictions.
    b. is bad and must be discarded.
    c. may, nevertheless, generate good predictions.
    d. is well described by (a) and (b).

27. According to Milton Friedman, truly significant theories
    a. necessarily have assumptions that are wildly inaccurate as descriptive representations of reality.
    b. abstract crucial elements from the mass of complex and detailed circumstances surrounding the phenomena to be explained and permit valid predictions on the basis of them alone.
    c. must be validated by showing the conformity of their assumptions with reality.
    d. are accurately described by (a) and (b).

28. Normative statements can be characterized as
    a. the essence of science.
    b. explanations of what causes what.
    c. prescriptions akin to preaching.
    d. both (a) and (b).

TRUE-FALSE QUESTIONS

In each space below, write a *T* if the statement is true and an *F* if the statement is false.

_____ 1. The basic economic problem of scarcity is less likely to be present among urban dwellers in Paris and New York than among African Bushmen and Amazon Indians.

_____ 2. Productive ingredients made by people are called *human resources*.

_____ 3. The growth of resources or technology over time is bound to reduce scarcity.

_____ 4. Over time, if the number of people or their desires multiplied, the extent of scarcity could nevertheless be reduced.

_____ 5. The change in an activity's overall benefit, which is attributable to a unit change in the level of that activity, is called the activity's *marginal (opportunity) cost*.

_____ 6. Satiation implies zero marginal benefit.

_____ 7. The rising marginal costs of an expanding activity may well be identical to the foregone rising marginal benefits of another activity that is also expanding.

_____ 8. The falling marginal costs of a contracting activity may well be identical to the rising marginal benefits of another activity that is expanding.

_____ 9. As long as the expansion of an activity brings positive marginal benefits, the activity level cannot be optimal.

_____ 10. As long as the contraction of an activity is associated with declining marginal costs, the activity level cannot be optimal.

_____ 11. A theory, like a map, can be useful precisely because it is unrealistic.

_____ 12. A good theory is one that makes prescriptive statements.

PROBLEMS

1. Consider the indicated combinations of goods that a firm could produce in a day, given fixed quantities of resources asnd current technology.

Choice	Jackets	Sweaters
A	55	0
B	50	20
C	40	35
D	30	44
E	17	50
F	0	55

   a. In the following graph, draw the firm's production-possibilities frontier as a smooth line through the given data sets.

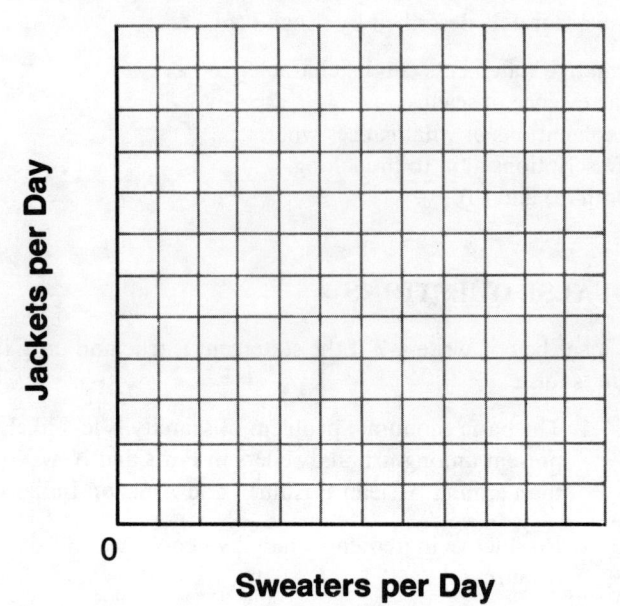

   b. In this graph, indicate the opportunity cost of producing 40 jackets.
   c. Indicate the opportunity cost of producing 20 sweaters.
   d. In the following graph, redraw the firm's production-possibilities frontier, indicating the overall benefits and costs associated with choice D.

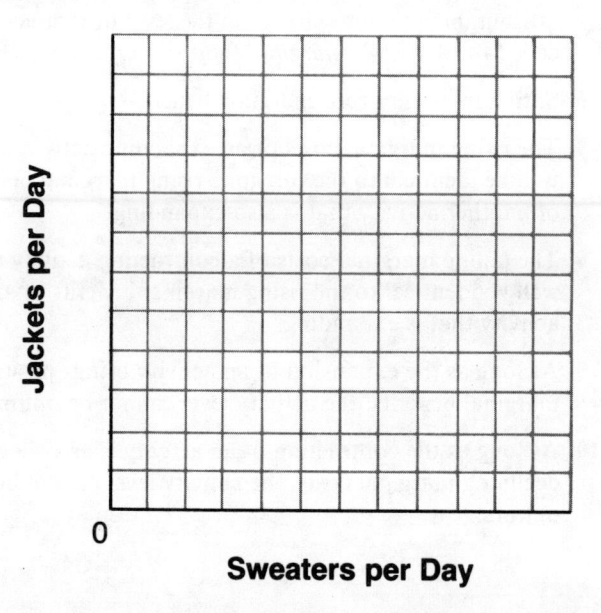

e. Redraw the production-possibilities frontier in the following graph, and indicate the marginal benefit and marginal cost of sweater production when replacing choice *D* by *E*.

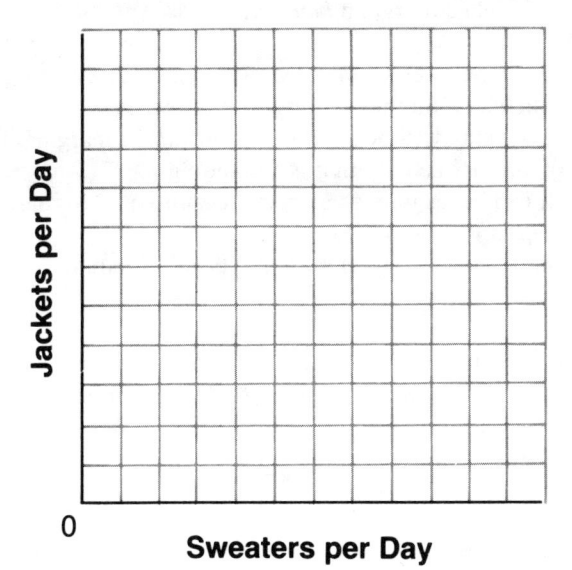

[Graph: Jackets per Day vs Sweaters per Day]

2. Imagine that the indicated values of total benefit and total cost were associated with the listed levels of an activity.
   a. Determine the activity's optimum level by focusing on the maximization of net benefit.
   b. Show how the same answer is obtained by focusing on marginal benefit and marginal cost. Note: Can you see how a graph of the data given here, along with your answers to (a) and (b), would look like text Figure 1.6, "Optimization"?

Activity Level	Total Benefit	Total Cost
1	100	100
2	200	110
3	280	130
4	340	160
5	380	200
6	400	250
7	400	310
8	380	380

3. A country faces a production-possibilities frontier that relates consumption goods, *C*, to investment goods *I*, and that is described by the equation $(I - 25)(C - 36) = 500$.
   a. What are the largest amounts of *C* and *I* that can be produced?
   b. If $I = 3$ is produced, what is the opportunity cost?

*4. If you know calculus (and a review of the basics is provided in Appendix 1A), determine the following from these equations of total benefit and total cost: $TB = 6Q - Q^2$ and $TC = .1Q^2$.
   a. The activity level, *Q*, that maximizes the total benefit.
   b. The activity level, *Q*, that maximizes the total net benefit.
   c. The size of the maximum total net benefit.

ANSWERS

Multiple-Choice Questions

1. b   2. b   3. a   4. c   5. c   6. d   7. a   8. d   9. d
10. c   11. d   12. c   13. a   14. c   15. d   16. b   17. b   18. b
19. a   20. c   21. b   22. d   23. d   24. d   25. d   26. c   27. d
28. c

True-False Questions

1. F (Even though those urban dwellers may have more goods, as long as their wants exceed their ability to satisfy them, they also suffer from scarcity.)

2. F (They are called *capital* resources.)
3. F (Wants can grow also.)
4. T (Output could grow even faster.)
5. F (It is called marginal *benefit*.)
6. T
7. F (They may well be identical to foregone rising marginal benefits of another activity that is *contracting*.)
8. F (They may well be identical to the *falling* marginal benefits of another activity that is expanding.)
9. F (It can be, as long as marginal cost is equal to marginal benefit.)
10. F (It can be, as long as marginal benefit is equal to marginal cost.)
11. T
12. F (Positive theory avoids such normative statements. A good theory makes accurate predictions. Prescriptive or normative statements do not qualify as theory at all.)

Problems

1.   a–c

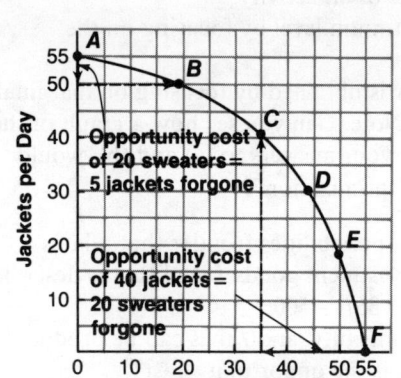

[Graph: Production possibilities frontier with points A(0,55), B(20,50), C(35,40), D(44,30), E(50,20), F(55,0). Opportunity cost of 20 sweaters = 5 jackets forgone. Opportunity cost of 40 jackets = 20 sweaters forgone.]

d

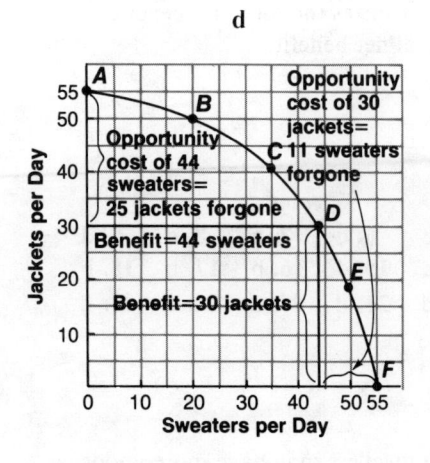

[Graph: Opportunity cost of 44 sweaters = 25 jackets forgone. Benefit = 44 sweaters. Opportunity cost of 30 jackets = 11 sweaters forgone. Benefit = 30 jackets.]

e

[Graph: Marginal cost = 13 more jackets forgone. Marginal benefit = 6 more sweaters. Point E at (44, 17).]

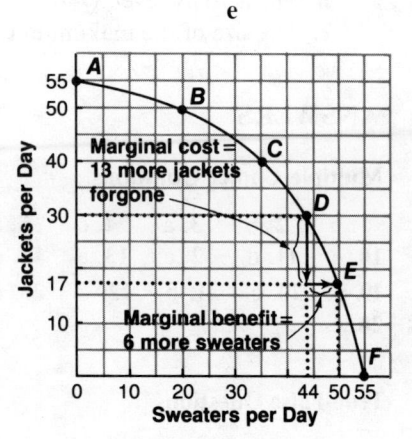

2. a. Net benefit can be calculated as the difference, in each row, between total benefit and total cost. The optimum (a maximum net benefit of 180 lies between 4 and 5 units of the activity.

Activity Level	Total Benefit − Total Cost = Net Benefit	
1	100 − 100 = 0	
2	200 − 110 = 90	
3	280 − 130 = 150	
4	340 − 160 = 180	optimum
5	380 − 200 = 180	optimum
6	400 − 250 = 150	
7	400 − 310 = 90	
8	380 − 380 = 0	

b. Marginal benefit and marginal cost can be calculated by taking the difference of adjacent numbers in the total benefit column and also in the total cost column (because these differences are in every case attributable to exactly a one-unit change in the activity level). The equality of (falling) marginal benefit and (rising) marginal cost (at 40) again points to an optimum between 4 and 5 units of the activity.

Activity Level	Total Benefit	Marginal Benefit	Total Cost	Marginal Cost	
1	100		100		
		+100		+10	
2	200		110		
		+80		+20	
3	280		130		
		+60		+30	
4	340		160		
		+40		+40	optimum
5	380		200		
		+20		+50	
6	400		250		
		0		+60	
7	400		310		
		−20		+70	
8	380		380		

3. a. If $I = 0$, we can find maximum C:

$$-25(C - 36) = 500$$
$$-25C + 900 = 500$$
$$400 = 25C$$
$$C = 16$$

The largest possible C is 16.

If $C = 0$, we can find maximum I:

$$(I - 25)(-36) = 500$$
$$-36I + 900 = 500$$
$$400 = 36I$$
$$I = 11.\overline{11}$$

The largest possible I is $11.\overline{11}$

b. If $I = 3$, it follows that

$$(3 - 25)(C - 36) = 500$$
$$-22C + 792 = 500$$
$$292 = 22C$$
$$C = 13.\overline{27}$$

Because maximum $C = 16$, producing $I = 3$ means forgoing $C = 16 - 13.\overline{27} = 2.\overline{72}$, which is the opportunity cost in question.

4. a. For maximum total benefit,

$$\frac{dTB}{dQ} = 6 - 2Q = 0. \text{ Hence } Q = 3.$$

$$\frac{d^2TB}{dQ^2} = -2 \text{ confirms a maximum.}$$

b. The total net benefit, $TNB = TB - TC = 6Q - 1.1Q^2$. For maximum,

$$\frac{dTNB}{dQ} = 6 - 2.2Q = 0. \text{ Hence } Q = 2.\overline{72}$$

$$\frac{d^2TNB}{dQ^2} = -2.2 \text{ confirms a maximum.}$$

Alternative solution by equating marginal benefit and marginal cost:

$$MB = \frac{dTB}{dQ} = 6 - 2Q$$

$$MC = \frac{dTC}{dQ} = .2Q$$

$$6 - 2Q = .2Q$$
$$6 = 2.2Q$$
$$Q = 2.\overline{72}$$

c. At $Q = 2.\overline{72}$, $TB = 6Q - Q^2 = 16.\overline{36} - 7.438 = 8.9256$

$$TC = .1Q^2 = .7438$$

$$TNB = TB - TC = 8.\overline{18}$$

Alternative solution:

$$TNB = 6Q - 1.1Q^2 = 16.\overline{36} - 8.\overline{18} = 8.\overline{18}.$$

BIOGRAPHY 1.1

Milton Friedman

Milton Friedman (1912–   ) was born in Brooklyn, New York, the son of immigrant parents. After studying at Rutgers and the University of Chicago, he worked for many years at the National Bureau of Economic Research and then received his doctoral degree at Columbia University in 1946. In 1951, he was honored with the Clark Medal, and in 1967 with the presidency of the American Economic Association; and in 1976, while serving as professor of economics at the University of Chicago, Friedman was awarded the Nobel Memorial Prize in Economic Science. He is now retired and a resident scholar at the Hoover Institution in Stanford, California.

Friedman's work covers a wide range of topics, but one underlying theme is the conviction that the market economy is the best economic system and works best when government performs only the minimal functions necessary to support it.

His most important work in macroeconomics concerns the role of money in the economy. *A Monetary History of the United States, 1867–1960* (written with Anna J. Schwartz, 1963) challenges the notions that the market economy without government intervention is inherently unstable and that monetary policy was powerless during the Great Depression. Instead, Friedman sees the Great Depression as a monument to the harm that can be done by a few who wield vast power over monetary policy.

In other areas, too, as Friedman sees it, the government is apt to do more harm than good when it intervenes in the operation of markets. As if led by an Invisible Hand, he says, government officials who seek to serve the public interest invariably end up serving a private interest instead. Much of this is discussed in his celebrated columns in *Newsweek* magazine, many of which are reprinted as *An Economist's Protest* (1972), and in his *Free to Choose: A Personal Statement* (1980), a book based on a series of PBS television shows of the same name.

Perhaps the most widely known of Friedman's books is *Capitalism and Freedom* (1962). He argues in this book that the kind of economic system that provides economic freedom—namely, competitive capitalism—also promotes political freedom. "Historical evidence speaks with one voice on the relation between political freedom and a free market," he says (p. 9). "I know of no example in time or place of a society that has been marked by a large measure of political freedom and that has not also used something comparable to a free market to organize the bulk of economic activity."

Finally, Friedman is the unchallenged leader of the "instrumentalists," who validate theories that prove to be good instruments for making predictions. Consider his own words:[1]

> The relevant question to ask about the "assumptions" of a theory is not whether they are descriptively "realistic," for they never are, but whether they are sufficiently good approximations for the purpose in hand. And this question can be answered only by seeing whether the theory works, which means whether it yields sufficiently accurate predictions. . . . A

---

[1] Milton Friedman, *Essays in Positive Economics* (Chicago: University of Chicago Press, 1953), pp. 15, 32, 40, and 41.

theory or its "assumptions" cannot possibly be thoroughly "realistic.". . . A completely "realistic" theory of the wheat market would have to include not only the conditions directly underlying the supply and demand for wheat but also the kind of coins or credit instruments used to make exchanges; the personal characteristics of wheat-traders such as the color of each trader's hair and eyes, his antecedents and education, the number of members of his family, their characteristics, antecedents, and education, etc.; the kind of soil on which the wheat was grown, its physical and chemical characteristics, the weather prevailing during the growing season; the personal characteristics of the farmers growing the wheat and of the consumers who will ultimately use it; and so on indefinitely. Any attempt to move very far in achieving this kind of "realism" is certain to render a theory utterly useless. . . . A meaningful scientific hypothesis or theory typically asserts that certain forces are, and other forces are not, important in understanding a particular class of phenomena. It is frequently convenient to present such a hypothesis by stating that the phenomena it is desired to predict behave in the world of observation *as if* they occurred in a hypothetical and highly simplified world containing only the forces that the hypothesis asserts to be important. . . . Such a theory cannot be tested by comparing its "assumptions" directly with "reality.". . . Yet the belief that a theory can be tested by the realism of its assumptions independently of the accuracy of its predictions is widespread and the source of much of the perennial criticism of economic theory as unrealistic. Such criticism is largely irrelevant.

APPENDIX 1A. DIFFERENTIAL CALCULUS: A REVIEW OF THE BASICS

The study of microeconomics inevitably deals with *relationships* in which one variable (called the *dependent variable*) is said to be related to, vary with, depend on, or "be a function of" one or more other variables (called *independent variables*). Consider how the quantity demanded (Q_D) or the quantity supplied (Q_S) is said to be related to price (P); how the quantity produced (Q) is said to vary with labor input (L) or capital input (K) or land input (T); how total revenue (TR) or total cost (TC) or total profit (Π) is said to depend on the quantity produced (Q); and how the utility (U) derived from good a is said to be a function of the quantity of good a consumed (Q_a).

Functional notation. Relationships such as these can be expressed symbolically. Thus the well-known relationship between quantity demanded and price that represents the *demand function* is written as $Q_D = f(P)$ and pronounced "Q sub D is a function of P." This does **not** mean that some number f, multiplied by some number P, equals Q_D. It does mean that there exists a relationsip between quantity demanded and price such that the value of Q_D is uniquely determined for any given value of P. The other relationships noted above can be similarly represented:

$Q_S = f(P)$	supply function
$Q = f(L)$	
$Q = f(K)$	production functions
$Q = f(T)$	
$TR = f(Q)$	revenue function
$TC = f(Q)$	cost function
$\Pi = f(Q)$	profit function
$U = f(Q_a)$	utility function

In an analogous fashion, one can represent relationships among more than 2 variables. Thus the quantity demanded of good *a* might be a function of the price of *a*, the price of *b*, the price of *c*, and income:

$$Q_{D_a} = f(P_a, P_b, P_c, I).$$

The quantity produced might be a joint function of labor, capital, and land input:

$$Q = f(L, K, T).$$

A person's utility might be derived from the consumption of good *a* as well as from the consumption of good *b*.

$$U = f(Q_a, Q_b).$$

In all cases, the $f(\ )$ symbol can be used to indicate the relationship in question. But note: If several functions are being discussed simultaneously, it is wise and customary to avoid confusion by distinguishing the different functional relationships by means of different symbols besides the $f(\ )$. Thus one might replace the f by g or h or even a repetition of the symbol denoting the dependent variable and write $TR = f(Q)$, but $TC = g(Q)$ and $\Pi = h(Q)$ or even $\Pi = \Pi(Q)$. This means that total revenue is a function of quantity produced, that total cost is a *different* function of quantity produced, and that profit is a *still different* function of quantity produced.

Graphical representation. Functional relationships between 2 variables can easily be represented graphically. Thus Figure 1 represents a demand function, Figure 2 a supply function, Figure 3 a variety of cost functions (fixed, variable, and total), and Figure 4 a utility function.

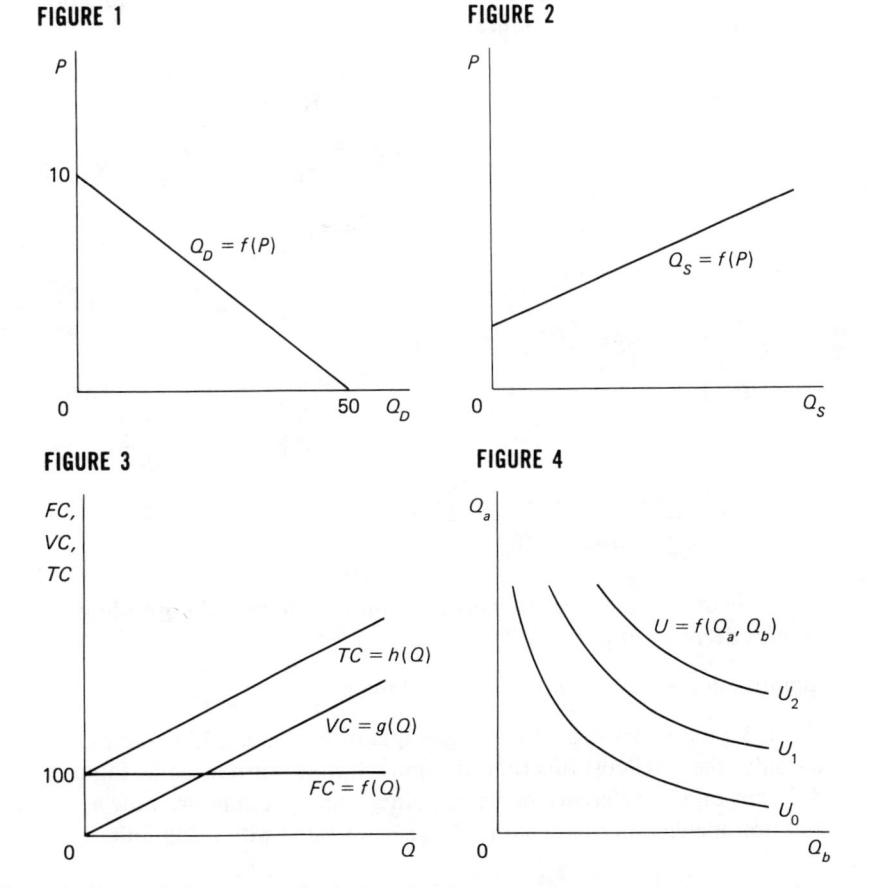

FIGURE 1

FIGURE 2

FIGURE 3

FIGURE 4

Precise equations. Any function can also be represented by a precise equation. Thus the demand function of Figure 1 can be written as $Q_D = 50 - 5P$, and the fixed-cost function of Figure 3 can be written as $FC = 100$. Similar and often more complicated equations can describe curvilinear functions. Thus, in Figure 5, the (quadratic) total-revenue function might be described by $TR = 300Q - 3Q^2$ and the (quadratic) total-cost function by $TC = 500 + 50Q + 2Q^2$.

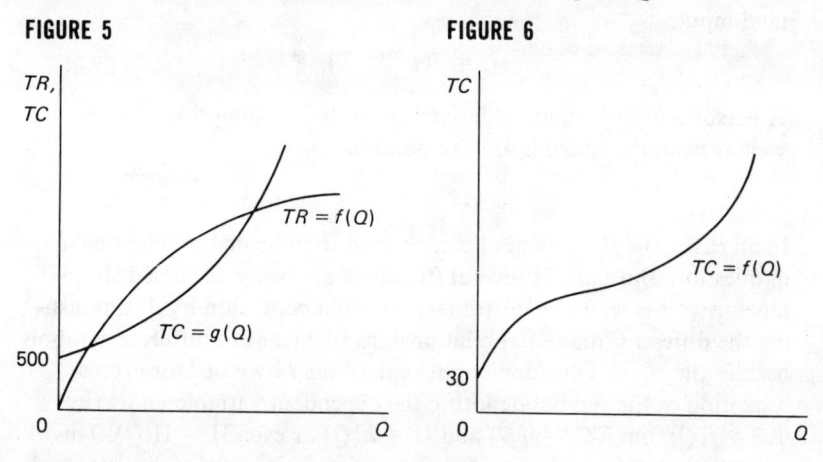

FIGURE 5

FIGURE 6

The (cubic) total-cost function in Figure 6, in turn, might be represented by an equation such as $TC = 30 + 10Q - 6Q^2 + 3Q^3$.

Marginalist thinking and slope. Microeconomists often engage in marginalist thinking. They want to know precisely by how much one variable changes, given a unit-change in another. In a two-dimensional graph, the answer is found by looking at the *slope* of a line. Slope measures the steepness of a line; it shows how much the line falls or rises while moving one unit to the right. Consider the demand function of Figure 1 which is reproduced as Figure 7 here.

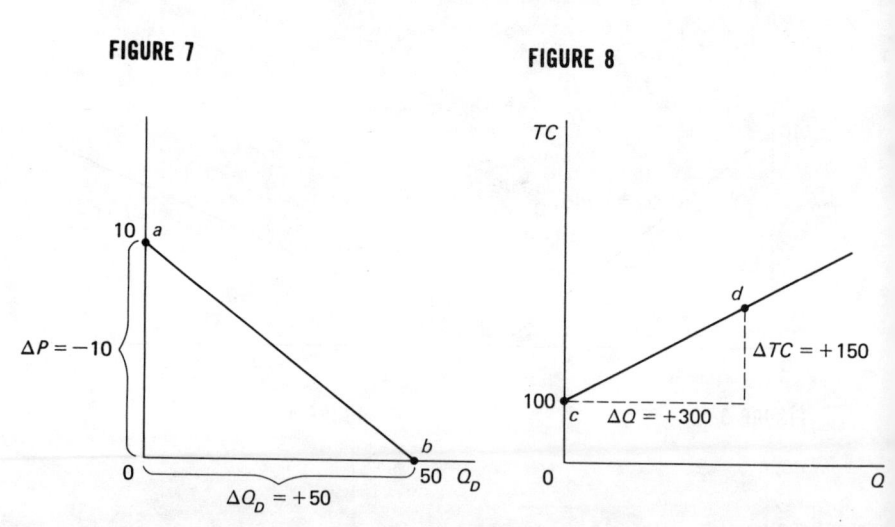

FIGURE 7

FIGURE 8

Focus on the difference between points a and b and notice how a 10-unit decrease in price (ΔP) is associated with a 50-unit increase in quantity demanded (ΔQ_D). Thus the slope is $\dfrac{\Delta P}{\Delta Q_D} = \dfrac{-10}{+50} = -.2$.

(Each \$.2 cut in price, perhaps, raises quantity demanded by 1 lb.) Or consider the total-cost function of Figure 3, now reproduced as Figure 8. Focus on the difference between points c and d and notice how a 150-unit increase in total coat (ΔTC) is associated with a 300-unit increase in output (ΔQ). Thus slope is $\dfrac{\Delta TC}{\Delta Q} = \dfrac{+150}{+300} = +.5$. (Each $.5 increase in total cost, perhaps, makes it possible to raise output by 1 lb.)

Consider, finally, Figure 9. While the slope of a straight line is the same everywhere, the slope of a curvilinear line such as this one varies from point to point. Yet the slope can still be measured at any given point (such as e or f) by the slope of a tangent to the curve. Thus the slope at e equals the slope of the dashed line passing through e, and the slope at f equals the slope of the dashed line passing through f.

Marginalist thinking and derivatives. The graphical measurement of slope quickly ceases to be possible when more than 2 variables are involved. Then a more general approach is needed to determine how much one variable changes, given a unit-change in another. This general approach (that applies to any number of variables) involves taking the *derivative* of the equation that represents the function of interest. Consider the 2-variable case first. The derivative of the 2-variable function $Y = f(X)$, for example, shows the small (positive or negative) change of the dependent variable (written as dY), given an infinitesimally small change of the independent variable (written as dX). In contrast to the difference quotient $\dfrac{\Delta P}{\Delta Q_D}$ (Figure 7) or $\dfrac{\Delta TC}{\Delta Q}$ (Figure 8) or, in general, $\dfrac{\Delta Y}{\Delta X}$, that gives the average rate of change of dependent variable Y with respect to independent variable X over some finite *range of X*, the derivative of $Y = f(X)$ is written as $f'(X) = \dfrac{dY}{dX}$ and gives the rate of change of Y with respect to X for a *given value of X* (at a specific *point*, such as d in Figure 8 or e in Figure 9). And note: The expression $f'(X) = \dfrac{dY}{dX}$ is called the *first derivative* of the function $Y = f(X)$. In Figure 9, it represents the positive slope at point e or the negative slope at point f. One can, however, also take the derivative of a derivative and note how the slope of a function (whatever its size) is itself changing. Note how, in Figure 9, as X is increased, the (positive) slope at e is declining (dashed tangents just to the right of e would be less steep), while the (negative) slope at f is rising (dashed tangents just to the right of f would take on a zero and then positive slopes). The *second derivative* of a function pinpoints this type of concavity (as at e seen from below) or this type of convexity (as at f seen from below). Given the function $Y = f(X)$, the second derivative is written as $f''(X) = \dfrac{d^2Y}{dX^2}$. All this is summarized in Box A.

FIGURE 9

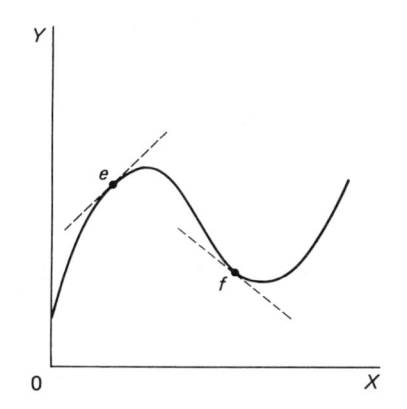

A.

original function	$Y = f(X)$
first derivative	$f'(X) = \dfrac{dY}{dX}$
second derivative	$f''(X) = \dfrac{d^2Y}{dX^2}$

Partial derivatives. When a function has multiple independent variables (each of which affects the value of the dependent variable Y), one can take *partial derivatives* of the function—one at a time for each independent variable. This procedure is equivalent to stating the *ceteris paribus* clause: the influence of an infinitesimally small change in one of the independent variables on the dependent variable Y is examined, while holding the other independent variables constant. To distinguish the result from the derivative of functions that only have one independent variable, the partial derivative is denoted by the lowercase Greek delta, δ, rather than the lowercase Roman d. For example, given the above function $Q = f(L, K, T)$, the expression $\delta Q/\delta L$ represents the rate of change of output with respect to the rate of change of labor input, while holding capital and land constant; it represents the marginal product of labor. Similarly, $\delta Q/\delta K$ represents the rate of change of output with respect to the rate of change of capital input, while holding labor and land constant; this partial derivative equals the marginal product of capital. Finally, $\delta Q/\delta T$ represents the rate of change of output, while holding labor and capital constant; it is the marginal product of land.

The total derivative. Finally, when a function has multiple independent variables and one wants to know the change in the independent variable given simultaneous small changes in *all* of the dependent variables, one can determine the *total derivative* or *differential*. For example, given the above function $U = f(Q_a, Q_b)$, the total derivative is

$$dU = \frac{\delta U}{\delta Q_a} \cdot dQ_a + \frac{\delta U}{\delta Q_b} \cdot dQ_b.$$

That is, the change in utility associated with a simultaneous small change (dQ_a) in the consumption of good a and a small change (dQ_b) in the consumption of good b equals the marginal utility of a times the change in the consumption of a plus the marginal utility of b times the change in the consumption of b.

The rules of differentiation. In this section, we review some of the major rules of differentiation that are applied in the Calculus Appendix of the text.

1. If the dependent variable, Y, is a constant, c (that, by definition, does not change), the derivative of this constant must be zero (Box B).

B.
Differentiating a constant.

Given $\qquad Y = f(X) = c$

then $\qquad f'(X) = \dfrac{dY}{dX} = 0$

Consider, for example, the fixed-cost function in Figure 3. $FC = f(Q) = 100$. Hence $f'(Q) = \dfrac{dFC}{dQ} = 0$. This result makes perfect sense; by definition, fixed cost does not vary with the quantity produced. The fixed-cost line has a zero slope.

2. If the dependent variable, Y, equals some value of the independent variable that is raised to the first degree or higher, the power rule applies (Box C).

C.
The power rule.

Given $$Y = f(X) = aX^b,$$
where a and b are constants,

$$f'(X) = \frac{dY}{dX} = baX^{b-1}$$

Given, for example, the exponential cost function $TC = f(Q) = 50Q^2$, marginal cost is $f'(Q) = \dfrac{dTC}{dQ} = 2(50)Q^{2-1} = 100Q^1 = 100Q$.

3. If the dependent variable, Y, equals the sum (or difference) of several terms, the sum/difference rule applies (Box D).

D.
The sum/difference rule.

Given $Y = f(X) = aX^b + cX^d$, where a, b, c, and d are constants,

$$f'(X) = \frac{dY}{dX} = baX^{b-1} + dcX^{d-1}$$

and similarly for a difference (replace + by − signs). The derivative of a sum (or difference) of several terms is the sum (or difference) of the derivatives of these terms.

Consider, for example, the total-cost and total-revenue functions of Figure 5.

Given $TC = g(Q) = 500 + 50Q + 2Q^2$, marginal cost is

$$g'(Q) = \frac{dTC}{dQ} = 0 + 1(50)Q^{1-1} + 2(2)Q^{2-1} = 50 + 4Q.$$

Given $TR = f(Q) = 300Q - 3Q^2$, marginal revenue is

$$f'(Q) = \frac{dTR}{dQ} = 1(300)Q^{1-1} - 2(3)Q^{2-1} = 300 - 6Q.$$

4. If the dependent variable, Y, equals the product of two terms, the product rule applies (Box E).

E.
The product rule.

Given $$Y = f(X) \cdot g(X),$$

$$\frac{dY}{dX} = f'(X) \cdot g(X) + g'(X) \cdot f(X)$$

The derivative of the product of 2 expressions equals the derivative of the first expression multiplied by the undifferentiated second expression plus the derivative of the second expression multiplied by the undifferentiated first expression.

For example, total revenue (TR) equals price (P) time quantity (Q), or $TR = P \cdot Q$. Marginal revenue, therefore, equals

$$MR = \frac{dTR}{dQ} = \left(\frac{dP}{dQ} \cdot Q\right) + \left(\frac{dQ}{dQ} \cdot P\right) = \left(\frac{dP}{dQ} \cdot Q\right) + P.$$

For the individual firm under perfect competition, when price does not vary with quantity, $(dP/dQ) = 0$; hence $MR = P$, which is a well-known proposition. Numerical exercise: Let $TR = P \cdot Q = (10 - .2Q) \cdot Q$, as implied by Figure 1 above. Then

$$MR = \frac{dTR}{dQ} = \left(\frac{dP}{dQ} \cdot Q\right) + P$$
$$= -.2Q + (10 - .2Q) = 10 - .4Q.$$

The same result would, of course, have been obtained by first simplifying the total-revenue function to $TR = 10Q - .2Q^2$ and then differentiating:

$$MR = \frac{dTR}{dQ} = 10 - [2(.2)Q^{2-1}] = 10 - .4Q^1 = 10 - .4Q.$$

5. If the dependent variable, Y, equals the ratio of two terms, the quotient rule applies (Box F).

F.
The quotient rule.

Given $$Y = \frac{f(X)}{g(X)},$$

$$\frac{dY}{dX} = \frac{f'(X) \cdot g(X) - g'(X) \cdot f(X)}{g(X)^2}$$

The derivative of the ratio of 2 expressions equals the derivative of the numerator multiplied by the denominator minus the derivative of the denominator multiplied by the numerator, all divided by the square of the denominator.

For example, average total cost (ATC) equals total cost (TC) divided by quantity (Q), or $ATC = \frac{TC}{Q}$. The change in average total cost with respect to changing quantity, therefore, equals

$$\frac{d\left(\frac{TC}{Q}\right)}{dQ} = \frac{\left(\frac{dTC}{dQ} \cdot Q\right) - \left(\frac{dQ}{dQ} \cdot TC\right)}{Q^2}$$

Consider just one implication: When ATC is declining with larger Q, the numerator of this expression must be less than zero (since the denominator cannot be). Thus

$$\left(\frac{dTC}{dQ}\cdot Q\right) - TC < 0. \quad \text{Dividing by } Q,$$

$$\frac{dTC}{dQ} - \frac{TC}{Q} < 0. \quad \text{Therefore,}$$

$$\frac{dTC}{dQ} < \frac{TC}{Q}$$

This result says that under the assumed condition (of declining ATC), marginal cost (dTC/dQ) is less than average total cost (TC/Q), which is a well-known proposition.

Optimizing. Much of microeconomics involves finding *optimum* values; usually these are associated with some maximum (as of utility or profit) or some minimum (as of physical input or cost). Such optimum points can be identified with the help of differential calculus. Consider the simple function of $Y = f(X)$. Whether it takes the form of Figure 10 or Figure 11, it is clear that the slope of the function (hence its derivative) equals zero at the optimum point if the optimum is a maximum (as at M) or a minimum (as at m). Thus the optimum, X_0, can be found, it seems, by taking the function's first derivative and setting it equal to zero. Clearly, at both M and m, $f'(X) = \frac{dY}{dX} = 0$, as the dashed horizontal tangents indicate. Yet the very fact that the function's derivative equals zero at *both* a maximum and a minimum implies a problem: Finding a first derivative and setting it equal to zero does not allow us to distinguish between the two possibilities. Consider the above total revenue function of $TR = 10Q - .2Q^2$ (that was, in turn, derived from the demand function of Figure 1). If we differentiate, find $(dTR/dQ) = 10 - .4Q$, and solve the equation $10 - .4Q = 0$, we get an apparently optimal $Q_0 = 25$, but how do we know whether this quantity corresponds to X_0 in Figure 10 (and, thus, *maximizes* total revenue) or to X_0 in Figure 11 (and, thus, *minimizes* total revenue)?

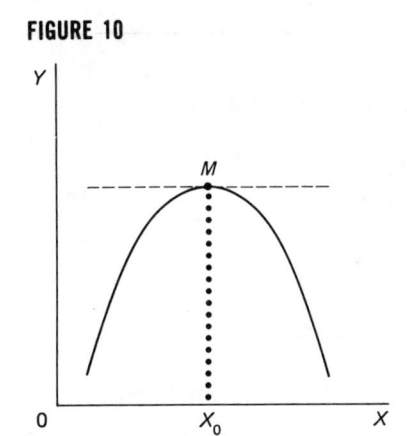

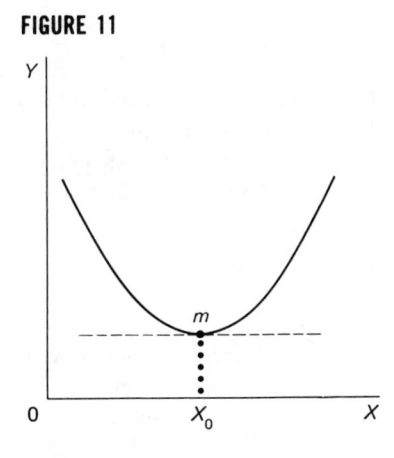

FIGURE 10

FIGURE 11

Indeed, matters are even worse: We cannot even be sure that the first derivative of a function, set equal to zero, pinpoints either a maximum or a minimum. As Figure 12 shows, this procedure might pinpoint neither, because at inflection point i it is also true that $(dY/dX) = 0$.

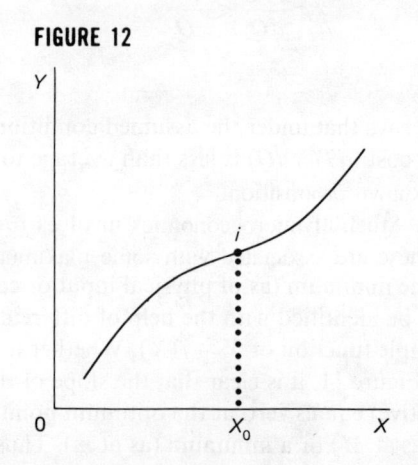

FIGURE 12

Fortunately, there is a way out of this dilemma: Once we have identified an apparent optimum point, X_0, by taking the first derivative of a function and setting it equal to zero, we can distinguish between such points as M, m, and i in Figures 10–12 by finding the second derivative $f''(X) = \dfrac{d^2Y}{dX^2}$ as well. Just as the first derivative indicates the rate of change of a function, so the second derivative indicates the rate of change of the rate of change. At a maximum, such as M, the rate of change (or slope) of the function is falling; note how the slope is changing from positive to negative; hence the second derivative is negative. At a minimum, such as m, the rate of change (or slope) is rising; note how the slope is changing from negative to positive; hence the second derivative is positive. At an inflection point, such as i, the slope does not reverse signs as it passes through; in Figure 12, for instance, the slope is positive both to the left and to the right of point i. Box G summarizes our most important conclusion:

G.

A maximum or minimum of a function, $Y = f(X)$, is identified by two conditions.

1st order (necessary) condition:
 The first derivative equals zero

$$f'(X) = \frac{dY}{dX} = 0$$

2nd order (sufficient) condition:
 The second derivative is negative for a maximum, positive for a minimum

$$f''(X) = \frac{d^2Y}{dX^2} < 0 \text{ at } M, > 0 \text{ at } m$$

Numerical example: Given our Figure 1 demand function of $P = 10 - .2Q$ and the implied total-revenue function of $TR = P \cdot Q = 10Q - .2Q^2$, the *1st order condition* is $\frac{dTR}{dQ} = 10 - .4Q = 0$. This implies a maximum *or* minimum at $Q_0 = 25$. The *2nd order condition* is $\frac{d^2TR}{dQ^2} = -.4$. This implies a total-revenue *maximum* at $Q_0 = 25$.

Two cautions are in order: First, some functions, such as that in Figure 13, have more than one maximum or minimum (relative to other points in the immediate vicinity). The conditions of Box G do not allow us to identify the highest of all these maxima or the lowest of all such minima. Strictly speaking, they only identify some *relative* maximum or minimum.

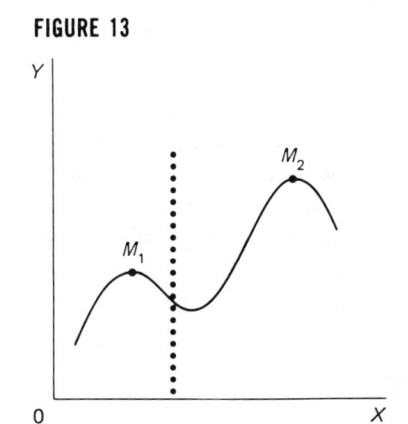

FIGURE 13

Second, when a function has more than one independent variable, the Box G conditions for finding relative maximum or minimum points are more complex, involving first and second and cross *partial* derivatives, but we can leave most of this material to more advanced courses. The next and last section of this Appendix considers it in part.

Constrained maximization. Oftentimes in microeconomics, the optimization of some variable must occur subject to some constraint. A household, for example, may wish to maximize utility from consuming quantities of goods a and b, $U = U(Q_a, Q_b)$, but is subject to the constraint of a given budget, B, and given prices, P_a and P_b, such that $P_a \cdot Q_a + P_b \cdot Q_b = B$. (For a physical analogy consider the task of finding the highest point on earth. Without any constraint on where to look, one would identify Mt. Everest, such as M_2 in Figure 13. Yet, when being constrained to looking only in North America, left of the dotted line, the answer is Mt. McKinley, such as M_1.) For constrained-maximization problems microeconomists follow a procedure named after Joseph Lagrange (1736–1813). They combine the constraint with the function to be maximized into a so-called Lagrangian expression, symbolized by £. Given the above example, they might seek to maximize not

$$U = U(Q_a, Q_b), \text{ but}$$
$$£ = U(Q_a, Q_b) + \lambda(B - P_aQ_a - P_bQ_b),$$

where the lowercase Greek lambda, λ, is a *Lagrangian multiplier* and the parenthetical expression following it represents the constraint that equals zero when the budget is fully spent on *a* and *b*. As numerous examples in the text's Calculus Appendix show, the maximum of this Lagrangian expression is found by taking the first partial derivatives with respect to the three unknowns (Q_a, Q_b, and the Lagrangian multiplier, λ) and setting them equal to zero. (For the taking of partial derivatives one follows the ordinary rules of differentiation, while treating all variables other than the two directly ivolved as constants.) Thus necessary conditions for a utility maximum are

1. $\dfrac{\delta \pounds}{\delta Q_a} = \dfrac{\delta U}{\delta Q_a} - \lambda P_a = 0$

2. $\dfrac{\delta \pounds}{\delta Q_b} = \dfrac{\delta U}{\delta Q_b} - \lambda P_b = 0$

3. $\dfrac{\delta \pounds}{\delta \lambda} = B - P_a Q_a - P_b Q_b = 0$

Equations (1) and (2) imply that the marginal utility of *a* divided by the price of *a* must equal the marginal utility of *b* divided by the price of *b*, which is a well-known utility-maximizing proposition. (Once more, we leave the derivation of *second* partial derivatives and *cross* partial derivatives and the associated proof of the existence of a maximum to more advanced courses. We do the same with respect to more complex problems that involve more than one constraint.)

Additional Readings

For further readings on calculus applications to microeconomics, students may wish to turn to any of the following:

Baumol, William J. *Economic Theory and Operations Analysis,* 4th ed. Englewood Cliffs, N.J.: Prentice-Hall, 1977, esp. chaps. 1–4.

Draper, Jean E. and Jane S. Klingman. *Mathematical Analysis: Business and Economic Applications.* New York: Harper and Row, 1967, esp. chaps. 1–4.

Yamane, Taro. *Mathematics for Economists: An Elementary Survey.* Englewood Cliffs, N.J.: Prentice-Hall, 1962, esp. chaps. 3–6.

CHAPTER 2

The Preferences of the Consumer

MULTIPLE-CHOICE QUESTIONS

Circle the letter of the *one* answer that you think is correct or closest to correct.

1. All the alternative combinations of two goods over which a consumer might conceivably exercise choice are referred to as
    a. the field of choice.
    b. the consumption-possibilities frontier.
    c. the budget line.
    d. either (b) or (c).

2. If a consumer's budget line between meat and sugar intercepts the vertical axis at 100 units of meat and the horizontal axis at 100 units of sugar,
    a. the consumer's budget equals $100.
    b. the price of a unit of meat equals that of a unit of sugar.
    c. meat as well as sugar costs $1 per unit.
    d. all of the above are true.

3. Imagine a graph depicting a consumer's possible allocations of a budget between meat and vegetables. If the price of meat doubled,
    a. the consumer could buy the same amount of vegetables as before, but only if any previous meat purchases were reduced.
    b. the consumer could still buy any previous amount of meat, but only if fewer vegetables were purchased.
    c. the budget line would shift to the right.
    d. both (a) and (b) would hold.

4. Imagine a graph depicting a consumer's possible allocations of a budget between apples and peaches. If the size of the budget rose along with the price of peaches, the budget line's intercept with the peach axis would
    a. move away from the origin.
    b. be unaffected.
    c. move toward the origin.
    d. do *one* of the above, but without additional information one cannot determine which.

5. As long as the principle of diminishing marginal utility is operating, any increased consumption of a good
    a. lowers total utility.
    b. produces negative total utility.
    c. lowers marginal utility and, therefore, total utility.
    d. lowers marginal utility, but may raise total utility.

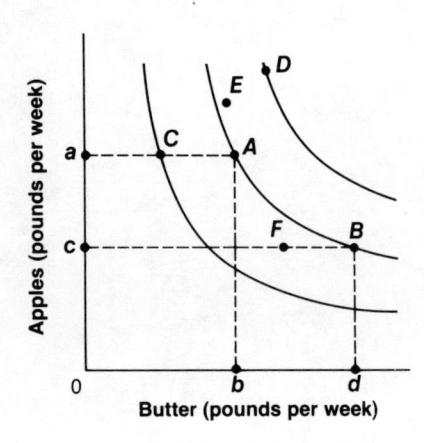

6. Among all the combinations of goods attainable by a consumer, the one that maximizes total utility is the one that
   a. maximizes the marginal utilities per dollar of each good.
   b. maximizes the marginal utilities per pound (or other physical unit) of each good.
   c. equates the marginal utilities per dollar of each good.
   d. equates the marginal utilities per pound (or other physical unit) of each good.

7. A utility contour (or consumption indifference curve) shows all the alternative combinations of two consumption goods that
   a. can be produced with a given set of resources and technology.
   b. yield the same total of utility.
   c. can be purchased with a given budget at given prices.
   d. equate the marginal utilities of these goods and, therefore, make the consumer indifferent between them.

When answering questions 8–10, consider the accompanying graph of a person's consumption-indifference curve:

8. This graph indicates that the consumer
   a. At A is indifferent between $0a$ of apples and $0b$ of butter.
   b. at A is consuming either $0a$ of apples or $0b$ of butter.
   c. is indifferent between $0a$ of apples plus $0b$ of butter on the one hand and $0c$ of apples plus $0d$ of butter on the other.
   d. is correctly described by all of the above.

9. This graph also indicates that the consumer prefers combination
   a. A to B.
   b. C to B.
   c. B to D.
   d. E to F.

10. This graph also shows the consumer's marginal rate of substitution in the AB range to be
    a. $0a$ of apples for $0d$ of butter.
    b. $0a$ of apples for $0b$ of butter.
    c. $0c$ of apples for $0d$ of butter.
    d. ac of apples for bd of butter.

11. As long as consumer choices are transitive, consumption-indifference curves
    a. cannot be vertical.
    b. cannot be horizontal.
    c. cannot intersect.
    d. cannot be positively sloped.

12. At any given point on a consumption-indifference curve, the absolute value of the slope equals
    a. unity—otherwise there would be no indifference.
    b. the marginal rate of substitution.
    c. the consumer's marginal utility.
    d. none of the above.

13. If a consumer's marginal rate of substitution equals 2 eggs for 1 hamburger,
    a. the consumer's indifference curve must be positively sloped.
    b. the consumer's indifference curve must be convex with respect to the origin of the graph.

c. the ratio of the consumer's marginal utility of 1 egg to that of 1 hamburger must equal $\frac{1}{2}$.
d. all of the above are true.

14. Bundles of goods on a consumption-indifference curve labeled I_4 yield
    a. twice the utility as bundles on a curve labeled I_2.
    b. greater utility than bundles on a curve labeled I_2.
    c. twice the marginal utility as bundles on a curve labeled I_2.
    d. the same utility as bundles on any other curve.

15. In the presence of declining marginal rates of substitution, consumers who again and again sacrifice a unit of one good cannot remain on their original consumption-indifference curves (that is, they cannot maintain their original levels of welfare) unless they receive as compensation
    a. again and again equal units of another good.
    b. ever smaller units of another good.
    c. ever larger units of another good.
    d. either (a), (b), or (c), depending on the tastes of the consumer involved.

16. If utility-maximizing consumers in Boston paid $1 per loaf of bread and $2 per pound of cheese, their marginal rate of substitution would be
    a. 1 loaf of bread for 2 pounds of cheese.
    b. 2 loaves of bread for 1 pound of cheese.
    c. $\frac{1}{2}$ loaf of bread for 1 pound of cheese.
    d. either (a) or (c).

17. Which one of the following is *not* a description of the consumer's optimum consumption bundle?
    a. The slope of the budget line equals the slope of the indifference curve.
    b. The marginal rate of substitution between goods *a* and *b* equals the ratio of the marginal utility of good *a* to that of *b*.
    c. The marginal rate of substitution between goods *a* and *b* equals the ratio of the price of *b* to that of *a*.
    d. The marginal utility per dollar of one good equals the marginal utility per dollar of any other good.

18. Consider a graph of a field of choice between bread and cigarettes. The imposition of a sales tax on cigarettes only would cause
    a. a parallel shift of a consumer's budget line toward the origin.
    b. a parallel shift of a consumer's indifference curves toward the origin.
    c. a movement of the budget line's intercept with the cigarette axis toward the origin.
    d. a movement of the budget line's intercept with the bread axis away from the origin.

19. When a subsidy in kind is replaced by a subsidy in cash of equal value
    a. the set of goods among which the recipient can choose increases.
    b. the set of goods among which the recipient can choose remains unchanged.
    c. an excess burden is imposed upon government.
    d. both (b) and (c) occur.

20. The imposition of rationing
    a. inevitably changes the amounts of affected goods people consume.
    b. lowers the utility of people who consume the affected goods.
    c. results in both (a) and (b).
    d. quite possibly results in neither (a) nor (b).

21. If a consumer's budget is $400 per month and the prices of apples and butter are $5 and $2, respectively, the consumer's budget-line equation is
    a. $Q_a = 80 - .4Q_b$.
    b. $Q_b = 200 - 2.5Q_a$.
    c. either (a) or (b).
    d. none of the above.

When answering questions 22 and 23, consider the accompanying table that gives points on three different consumption indifference curves.

22. This consumer's optimum clearly
    a. lies on indifference curve I_1.
    b. involves the consumption of $4a$ and $56b$.
    c. is correctly described by (a) and (b).
    d. cannot be determined without additional information.

23. This consumer's marginal rate of substitution of b for a equals, successively,
    a. $\frac{10}{19}, \frac{7}{22}, \frac{4}{29}, \frac{2}{48}$, etc., on I_0.
    b. $\frac{19}{10}, \frac{22}{7}, \frac{29}{4}, \frac{48}{2}$, etc., on I_0.
    c. 3.08, 13.18, 100, etc., on I_2.
    d. 4.56, 8.45, 14, 20.57, etc., on I_1.

Indifference Curves

I_0		I_1		I_2	
Q_a	Q_b	Q_a	Q_b	Q_a	Q_b
10	19	9	41	9.1	65
7	22	5.8	49	6.5	73
4	29	4	56	4.3	102
2	48	3.5	72	3.9	142
1.5	73	2.8	128		
1.3	102				
1	131				

24. The revealed-preference approach to the derivation of indifference curves is associated with the name of
    a. Edgeworth.
    b. Jevons.
    c. Samuelson.
    d. (a) and (c).

25. The revealed-preference approach to the derivation of indifference curves
    a. assumes stability over time of an observed person's tastes.
    b. assumes that all people have identical tastes.
    c. relies on repeated observations of the market behavior of a single person.
    d. is correctly described by both (a) and (c).

26. The *marginal rate of substitution* is
   a. the rate at which a consumer is willing to exchange, as a matter of indifference, a little bit of one variable (say, butter) for a little bit of another variable (say, meat).
   b. the rate at which a consumer is willing to exchange, in order to be better off, a little bit of one variable (say, butter) for a little bit of another variable (say, meat).
   c. the rate at which a producer is technically able to exchange, in the process of production, a little bit of one variable (say, jackets) for a little bit of another variable (say, pants).
   d. correctly described by both (a) and (c).

27. When economists assume that households seek to maximize utility, they are assuming that households are
   a. totally selfish.
   b. never subject to the will of other people.
   c. self-interested (which means selfish, altruistic, or both).
   d. (b) and (c).

TRUE-FALSE QUESTIONS

In each space below, write a *T* if the statement is true and an *F* if the statement is false.

_____ 1. All the alternative combinations of two goods lying on or above a consumer's budget line are unattainable.

_____ 2. All the alternative combinations of two goods lying above a consumer's budget line are not wanted.

_____ 3. With a budget of $800 per month, a consumer who faces prices of $10 per pound of meat and $2 per pound of bread can purchase 80 pounds of meat and 400 pounds of bread per month (as well as various other combinations of these goods).

_____ 4. By buying a car once every 5 years, a consumer can in fact consume $\frac{1}{5}$ car per year.

_____ 5. On or below a consumption-possibilities frontier, one can find all the alternative combinations of two goods that a consumer is able to buy in a given period at current market prices by fully using a given budget.

_____ 6. Jevons taught us that one must maximize marginal utility in order to maximize total utility.

_____ 7. If a consumer were maximizing utility while buying soap at $1 per bar and toothpaste at $2 per tube, the consumer's marginal utility per bar of soap would exceed that per tube of toothpaste.

_____ 8. According to the principle of diminishing marginal utility, successive additions of equal units of a good to the process of consumption eventually yield ever smaller additions to total utility—given a person's tastes and the quantities of all other goods consumed.

_____ 9. Mandated purchasing usually affects low-income people more than high-income people.

_____ 10. A budget of $500 and a budget-line equation of $Q_a = 100 - 5Q_b$ implies a price of b of $P_b = 5$.

PROBLEMS

1. Imagine a consumer with a monthly budget of $1,000. It is to be spent on food and housing. On the following graph, draw the consumer's budget line for prices of $10 per unit of food and $100 per unit of housing.

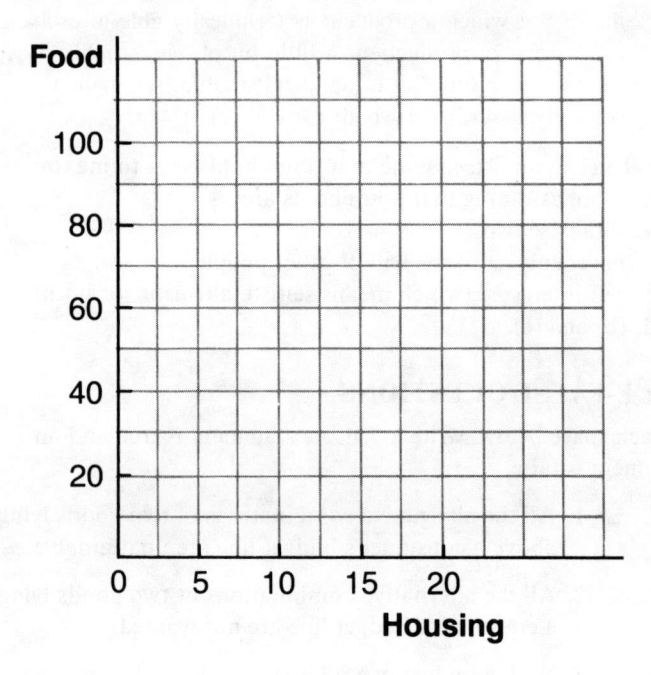

2. On the same graph draw a dashed line that indicates what would happen to the budget line if the budget halved, the price of food doubled, and the price of housing fell to $25 per unit.

3. Construct a numerical example that illustrates the principle of eventually diminishing marginal utility but that also allows for at first rising and then constant marginal utility. (*Hint:* Write down three column headings, labeled "units consumed," "total utility," "marginal utility"; then place numbers in each column).

4. Consider the accompanying data and assume that the marginal utility associated with the consumption of one good is independent of the quantity consumed of any other good. Let a unit of bread be priced at $2 and a unit of sausage at $1. If a consumer's budget were $13, how much of each good would a utility-maximizing consumer buy? Explain.

| Bread || Sausage ||
Units Consumed	Marginal Utility	Units Consumed	Marginal Utility
1	100	1	50
2	90	2	40
3	80	3	30
4	70	4	20
5	60	5	10
6	50	6	0

5. Imagine a consumer who was indifferent about the following combinations of food and housing (measured in respective units per month). Plot the data on the following graph and connect them with a smooth curve. What is this consumer's marginal rate of substitution between B and C? Does this consumer have a diminishing marginal rate of substitution? Explain.

	Food	Housing
A	100	2.5
B	50	5.0
C	20	10.0
D	6	15.2
E	2	20.0

6. On the following graph, redraw the solid budget line from question 1 and the indifference curve from question 5. Determine the consumer's optimum.

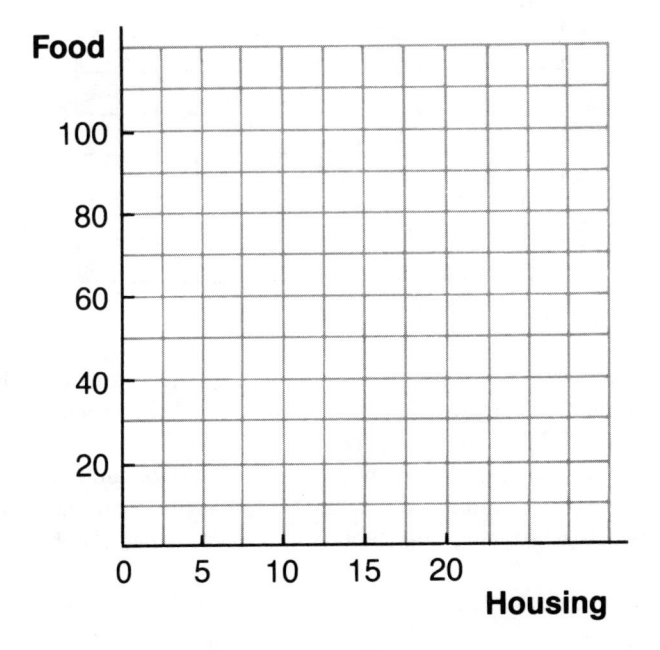

7. On the following graph, indicate why a lump-sum subsidy is better than a selective subsidy for a particular good only. (*Hint:* Use text Figure 2.7, "Lump-Sum vs. Selective Sales Tax," as a model.)

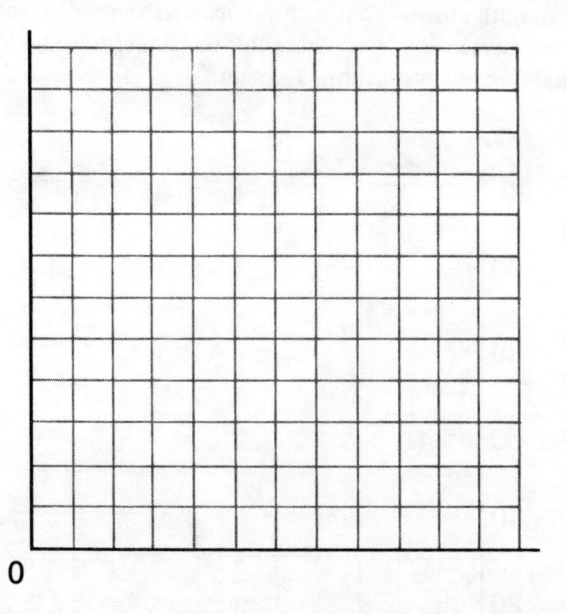

8. On the following graph, indicate why taxes in cash that reduce a consumer's budget by a fixed dollar amount may well be preferable to taxes in kind that reduce a consumer's consumption of a given good by a fixed quantity of equal value. (*Hint:* Use text Figure 2.8, "Cash or In-Kind Subsidy," as a model.)

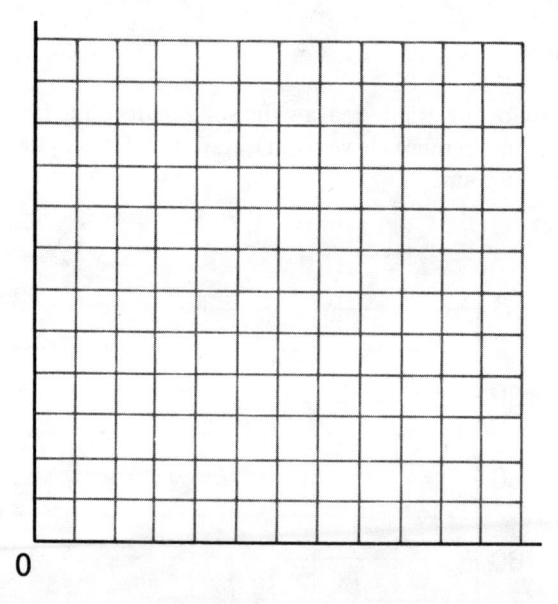

9. Redraw the graph from question 6 twice on the following grids.
   a. In the first grid, indicate how rationing of *both* goods could lower the consumer's utility.
   b. In the second grid show how rationing of *both* goods could leave the consumer's utility unaffected. (*Hint:* You may wish to review text Figure 2.9, "Rationing and Mandated Purchases.")

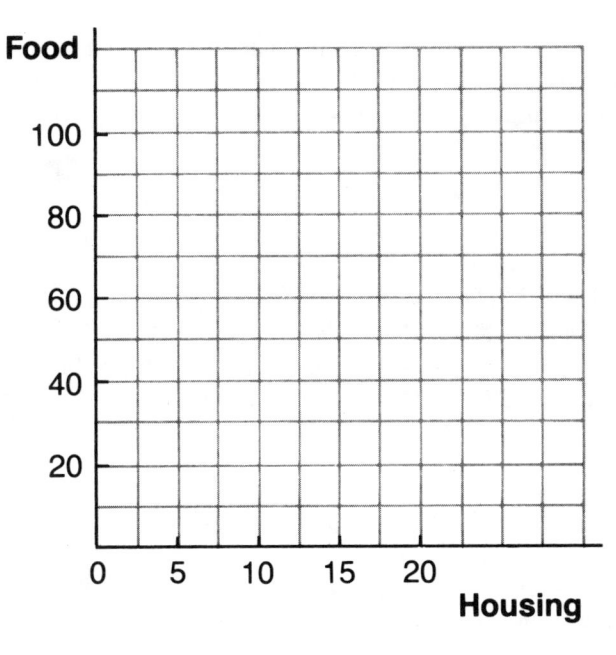

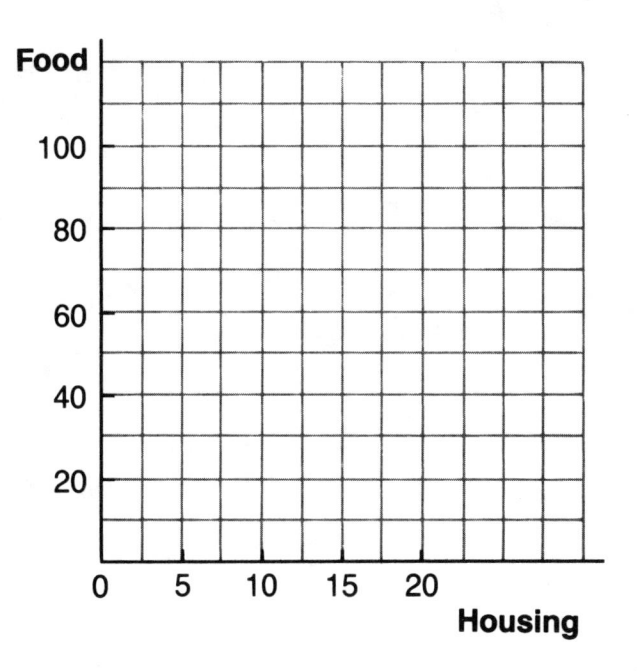

10. Consider the following graph that depicts a consumer's indifference curve between good x and all other goods, along with a budget line AB. Clearly, the consumer's optimum is at C and involves spending $0b$ on good x and $0a$ on all other goods. Assuming the person's work effort and pretax income stay the same, analyze
   a. the effect of a 50 percent income tax on the consumer's purchases of x, provided the tax law allows no deductions for purchases of x.
   b. the effect of the same tax on the consumer's purchases of x if spending on x is deductible from taxable income.
   c. the amount of tax collected by the government under (a) and (b).

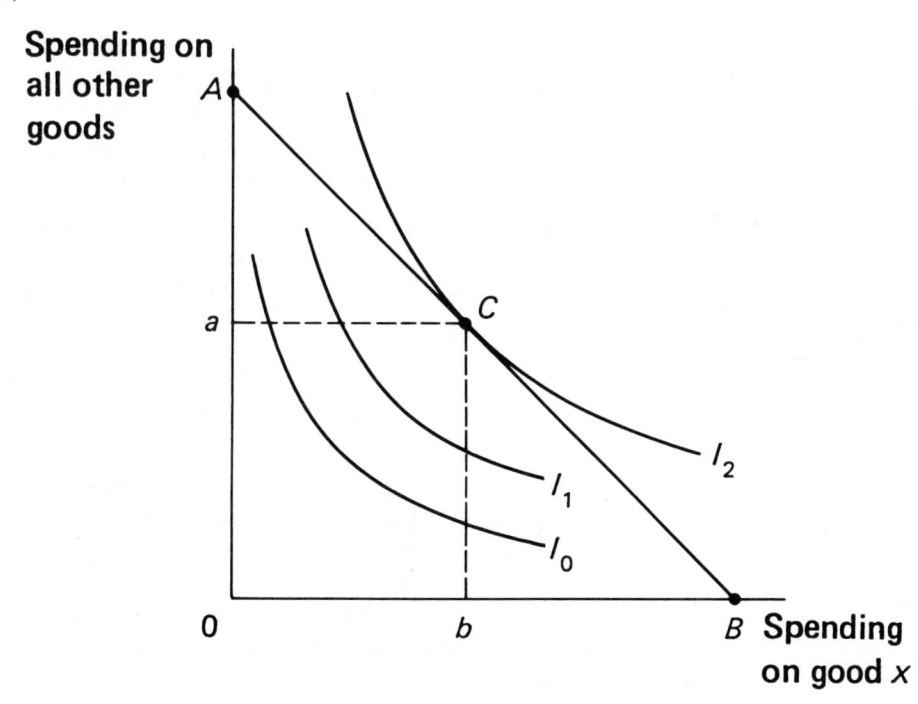

*11. Using calculus, determine the optimum quantities of a and b that a utility-maximizing consumer would choose, given a budget of 500, $P_a = 10$, $P_b = 15$, and a utility function of $U = Q_a \cdot Q_b$.

ANSWERS

Multiple-Choice Questions

1. a 2. b 3. a 4. d 5. d 6. c 7. b 8. c 9. d
10. d 11. c 12. b 13. c 14. b 15. c 16. b 17. b 18. c
19. a 20. d 21. c 22. d 23. c 24. c 25. d 26. a 27. c

True-False Questions

1. F (The ones lying on the line itself are attainable.)
2. F (They are not *attainable*.)
3. F (The consumer can purchase 80 pounds of meat *or* 400 pounds of bread per month *or* various other combinations.)
4. T
5. F (The consumer would not be using a given budget fully for combinations below the frontier.)
6. F (Jevons taught that one must *equate* the marginal utility *per dollar* of all goods in order to maximize total utility.)
7. F (The consumer's marginal utility per bar of soap would *be half* that per tube of toothpaste.)
8. T
9. T
10. F (Since the budget line is described by $Q_a = \dfrac{B}{P_a} - \dfrac{P_b}{P_a} Q_b$, the equation $Q_a = 100 - 5Q_b$ implies $(B/P_a) = 100$ and $(P_b/P_a) = 5$. Given $B = 500$, $P_a = 5$. Hence $P_b = 25$.)

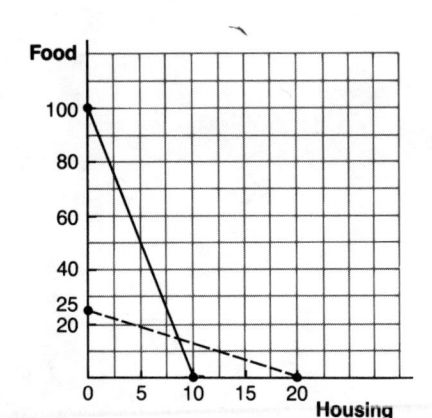

Problems

1. The solid line indicates that 100 units of food *or* 10 units of housing *or* any combination of the two on or below the line could be bought. Combinations *on* the line would exhaust the $1,000 budget, combinations *below* the line would not.

2. With $500 only, the consumer could buy 25 units of food at a price of now $20 per unit *or* 20 units of housing at a price of now $25 per unit *or* various combinations on or below the dashed line.

3. Note that diminishing marginal utility sets in after 3 units are consumed. See the table at the top of page 33.

4. Following Jevons' rule, the consumer would equate the marginal utility *per dollar* of each good and, therefore, would consume 5 units of bread ($10) and 3 units of sausage ($3). This combination would yield a marginal utility of 30 utils per dollar of both goods. Look at it this way: The consumer could spend each dollar on either bread or sausage. Each time, one-half unit of bread or 1 unit of sausage would be gained. Thus the 1st and 2nd dollar spent on bread would on average yield 100 utils ÷ 2, or 50

utils; the 3rd and 4th dollar spent on bread would on average yield 90 utils ÷ 2, or 45 utils; and so on down the marginal-utility-of-bread column. Now imagine the consumer spending 13 dollars in sequence, always looking for the highest possible increase in total utility. Such a consumer might spend the first and second dollar each on half a unit of bread (for extra utility of twice 50), the third dollar on one unit of sausage (for extra utility of 50), the fourth and fifth dollar each on half a unit of bread (for extra utility of twice 45), the sixth dollar on one unit of sausage (for extra utility of 40), the seventh and eighth dollar each on half a unit of bread (for extra utility of twice 40), the ninth and tenth dollar each on half a unit of bread (for extra utility of twice 35), the eleventh and twelfth dollar each on half a unit of bread (for extra utility of twice 30), and the thirteenth dollar on one unit of sausage (for extra utility of 30). No other allocation of money can yield a higher utility total. (Notice how a fourth unit of sausage would bring extra utility of 20 but would cost a dollar and require giving up half of the fifth unit of bread or 30 utils.)

Units Consumed	Total Utility	Marginal Utility
0	0	
		20
1	20	
		30
2	50	
		30
3	80	
		20
4	100	
		10
5	110	
		0
6	110	

5. The marginal rate of substitution between B and C is 30 units of food for 5 units of housing.
Yes, this consumer does have a diminishing marginal rate of substitution: Going from A toward E, the consumer is willing to sacrifice indifferently 50 F for 2.5 H; that is, 20 F for 1 H in range AB but only 6 F for 1 H in range BC, 2.69 F for 1 H in range CD, and .83 F for 1 H in range DE.
Similarly, going from E toward A, the consumer is willing to sacrifice indifferently, 4.8 H for 4 F; that is, 1.2 H for 1 F in range ED, but only .37 H for 1 F in range DC, .17 H for 1 F in range CB, and .05 H for 1 F in range BA.

6. The consumer's optimum corresponds to 50 units of food plus 5 units of housing. At the assumed prices of $10 per unit of food and $100 per unit of housing, this precisely exhausts the $1,000 budget.

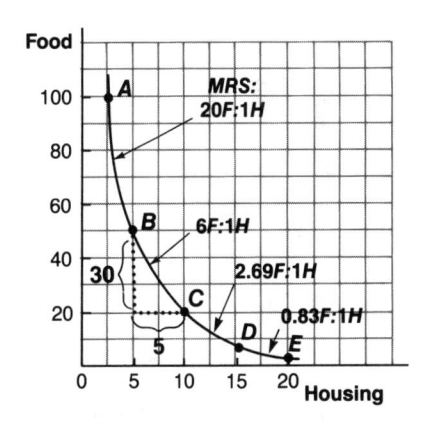

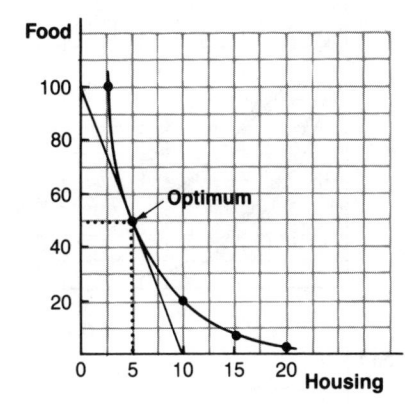

7. Let a consumer's original optimum be depicted by position c on budget line ab and indifference curve I_0. Let a subsidy for good A only reduce its price to the consumer, swinging the budget line to bd. The consumer then optimizes at e, with higher total utility corresponding to I_1. Dotted line ef represents the size of the subsidy, measured in terms of good A. (Without the subsidy, a consumer who wanted to buy the same amount of nonsubsidized good B as at e would have to move to f and give up ef of subsidized good A). Yet if the government provided the consumer with the cash equivalent of ef, the entire budget line would *shift* from ab to dashed parallel line gh. (The consumer would then consume combination i and gain even higher utility corresponding to I_2.) Thus a lump-sum subsidy of ga = ef is better.

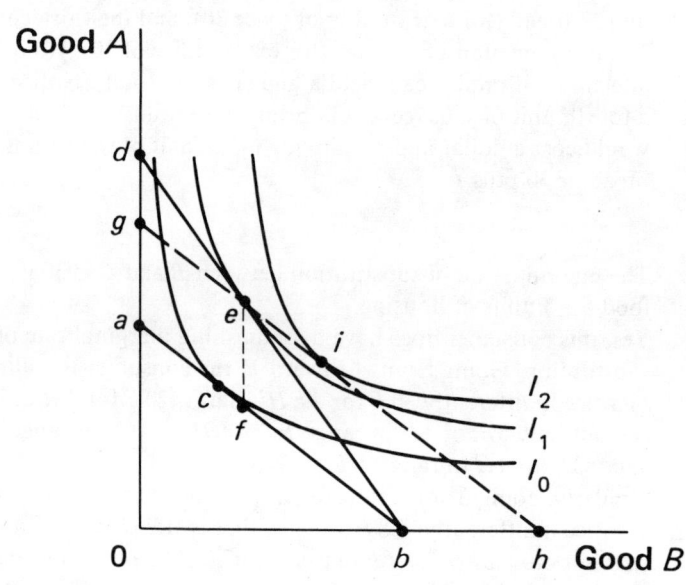

8. Let a consumer's original optimum be depicted by position c on budget line ab and indifference curve I_2. Let government decree that the consumption of good B be reduced by cd, an amount to be confiscated by government. The consumer would end up at d with the lower utility corresponding to I_0. (Higher quantities of good B would be illegal; even lower quantities of B and larger ones of A would yield even lower utility along line df). Yet if the consumer were taxed an equivalent amount dc = eb in cash, a new budget line fe would emerge. Then the consumer could choose combination h on indifference curve I_1 and receive a total utility in excess of I_0.

9. **a.** Let the maximum monthly ration of food equal 20 units and let that of housing equal 5 units. Then only combinations in the shaded area are legally available in spite of the consumer's ability to finance larger amounts. The consumer can at best achieve utility at a corresponding to I_0, far below that at b on original I_1.

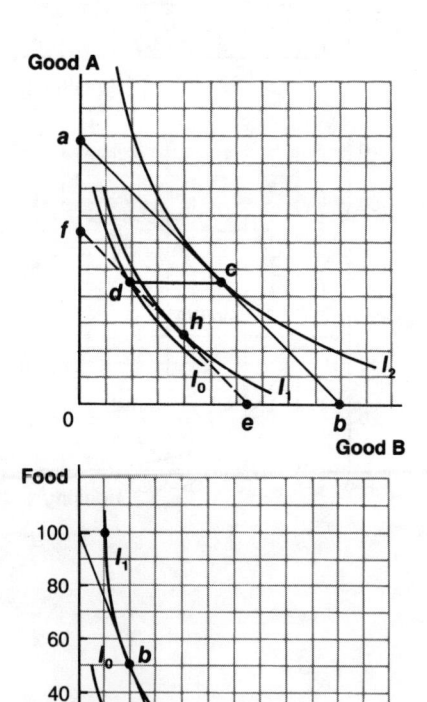

b. Let the maximum monthly ration of food equal 80 units and let that of housing equal 12.5 units. Once more all combinations in the shaded area are legally available, although the consumer is unable to finance combinations to the right of the budget line. As can be seen, the consumer can continue to buy the optimum combination b.

10. a. A 50 percent income tax shifts budget line AB to DE (a parallel shift with $OD = DA$). The consumer's optimum moves to F, and purchases of good x fall from $0b$ to $0f$.
   b. The budget line now becomes dashed line DB. (If the consumer chose to spend income only on goods other than x, no tax deduction could be made and 50 percent of income $0A$ would be taxed, placing the consumer at D. If the consumer chose to spend income on x only, no tax would have to be paid at all, placing the consumer at B.) The new optimum is at G; compared to the tax without deductions, purchases of x rise from $0f$ to $0g$.
   c. Under (a), because the consumer chooses F, the tax collected is $FH = DA$, or 50 percent of income. Under (b), as the consumer chooses G, the tax collected is lower amount GK.

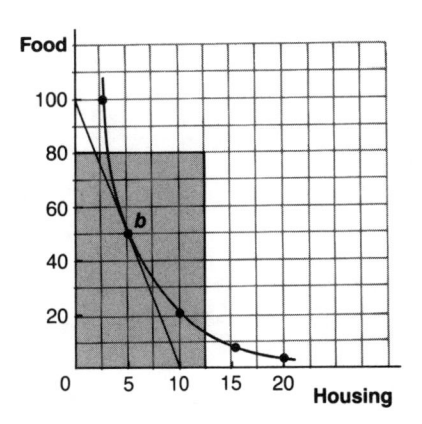

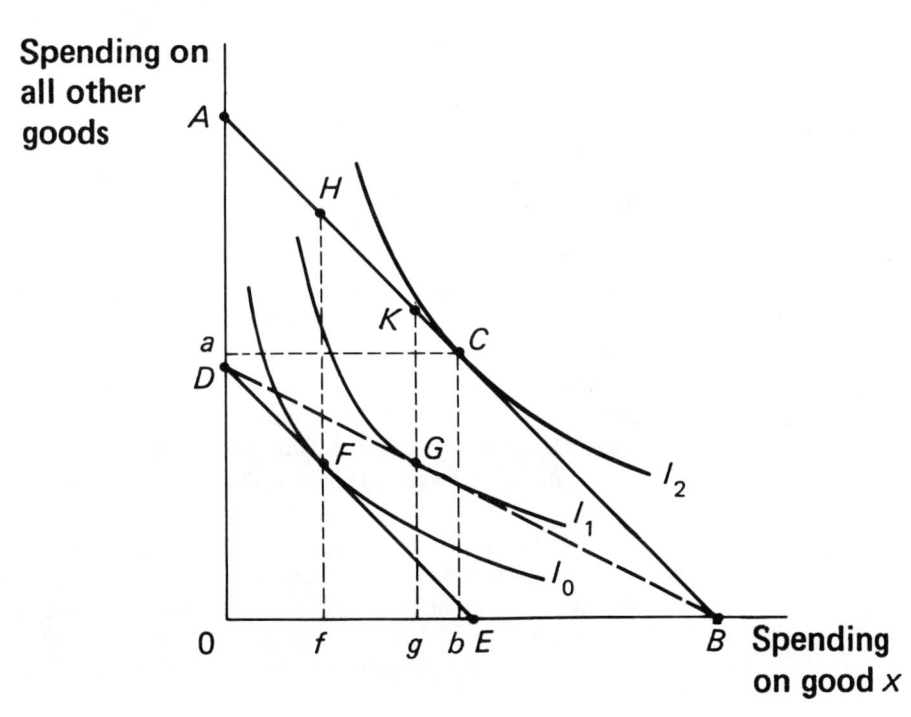

11. The procedure noted in Section 2B of the text's Calculus Appendix applies. The budget constraint is

$$500 - 10Q_a - 15Q_b = 0$$

The Lagrangian expression is

$$\pounds = Q_a \cdot Q_b + \lambda(500 - 10Q_a - 15Q_b).$$

Hence for a maximum

$$\frac{\delta £}{\delta Q_a} = Q_b - 10\lambda = 0 \tag{1}$$

$$\frac{\delta £}{\delta Q_b} = Q_a - 15\lambda = 0 \tag{2}$$

$$\frac{\delta £}{\delta \lambda} = 500 - 10Q_a - 15Q_b = 0 \tag{3}$$

It follows from (1) and (2) that

$$Q_b - 10\lambda = Q_a - 15\lambda \quad \text{and}$$
$$1.5Q_b - 15\lambda = Q_a - 15\lambda$$
$$1.5Q_b = Q_a$$

Given (3),
$$500 - 15Q_b - 15Q_b = 0$$
$$500 = 30Q_b$$
$$Q_b = 16.\overline{66}$$

Therefore, $Q_a = 25$.

BIOGRAPHY 2.1

Jeremy Bentham

Jeremy Bentham (1748–1832) was born in London, England. He was trained as a lawyer, but he retired early to devote his life to research and the ardent advocacy to his utilitarian philosophy. This philosophy was expounded in *An Introduction to the Principles of Morals and Legislation,* published in 1789, and in many other works. "Nature has placed man," said Bentham,[1] "under the empire of *pleasure* and of *pain*. . . . He who pretends to withdraw himself from this subjection knows not what he says. His only object is to seek pleasure and to shun pain."

As the review of the optimization principle in Chapter 1 indicated, modern economists follow in the footsteps of Bentham when they analyze human behavior based on comparisons of *benefits* and *costs* (Bentham's "pleasure" and "pain"). Yet, unlike Jevons and Edgeworth, modern economists do not share Bentham's faith in the cardinal measurability of utility. Nor do they believe that a social utility maximum can be found by making interpersonal comparisons of utility. Nevertheless, the Benthamite search for some kind of social optimum continues (but will be more fully addressed in Parts Six and Seven).

BIOGRAPHY 2.2

William S. Jevons

William Stanley Jevons (1835–1882) was born in Liverpool, England, the son of an iron merchant. He studied mathematics, the natural sciences, and metallurgy at University College, London, and then

---

[1]Jeremy Bentham, *The Theory of Legislation* (New York: Harcourt, Brace and Co., 1931), p. 2.

became an assayer at the Royal Mint in Sidney, Australia. After he returned to London, his interest turned to logic, philosophy, and political economy. In later years, he taught political economy at Owens College, Manchester, and at University College, London.

His writings covered a wide range of subjects, reflecting his training and life history. Uniformly, these works were exact, lucid, original. He wrote on gold mining in Australia (where he lived from 1853–59) and on Britain's dwindling coal reserves (a book that is useful even today). He wrote on money and finance (in particular, the effect of gold discoveries on the general price level). He developed the sunspot theory of business cycles, tracing periodic sunspot activity to the weather and from there to agricultural production and economic activity in general. (Later, he incorporated other causes of the cycle as well.) And he wrote a famous treatise on logic and the scientific method.

Yet his immortality was achieved by his pioneering application of mathematics to economics. In a paper read to the British Association for the Advancement of Science in 1862, he introduced the concept of the "final degree of utility," now known as *marginal utility*. This single decisive achievement made scientific history. His subsequent *Theory of Political Economy* (1871) contains a systematic exposition of the theory of consumer optimization based on the marginal utility concept.

The achievement of Jevons was genuinely original, although he had three forerunners of whom he was unaware. W. F. Lloyd of England in 1834, J. Dupuit of France in 1844, and H. H. Gossen of Germany in 1854 had each developed the notion of marginal utility, but no one had paid attention at the time. And even after Jevons, the concept was independently derived by Carl Menger of Austria in 1871 and by Léon Walras of France in 1874 (see Biography 13.2). At that point, the "marginal revolution" in economics was irreversible. In Britain and the Commonwealth, for a period of half a century, practically all elementary students both of logic and of political economy were brought up on Jevons.

BIOGRAPHY 2.3

Francis Y. Edgeworth

Francis Ysidro Edgeworth (1845–1926), son of a British father and Spanish mother, was born in Edgeworthstown, Ireland, on an estate where his ancestors established themselves at the time of Queen Elizabeth. He studied classics and mathematics at Trinity College, Dublin, and later at Oxford. He was steeped in Milton, Pope, Virgil, and Homer and would quote them on numerous occasions throughout his life.

He spent early years in London, first as a barrister, then as a teacher of logic and political economy at King's College. In 1891, he was appointed professor of political economy at Oxford, and there he stayed for the rest of his life, teaching, writing, and editing the prestigious *Economic Journal*.

For decades, he was a prolific exponent of the application of mathematics to the social sciences. *Mathematical Psychics: An Essay on the Application of Mathematics to the Moral Sciences* (1881) was his first contribution to economics. This book contains two of his most enduring achievements, the discoveries of the indifference curve (discussed in this chapter) and of the contract curve (to be discussed in

Chapter 13). Practically all of his other work on economic theory has been collected in *Papers Relating to Political Economy,* 3 vols. (1925).

As a mathematician, Edgeworth devoted much effort to measurement—of ethical value (or utility), of belief (or probability), of evidence (or statistics), and of economic value (or index numbers). Yet at times he also entertained the possibility of mathematical reasoning without numerical data.[2] "We cannot *count* the golden sands of life," he said, "we cannot *number* the 'innumerable' smiles of seas of love; but we seem to be capable of observing that there is here a *greater,* there a *less,* multitude of pleasure units, mass of happiness; and that is enough."

Thus Edgeworth, speaking like a true prophet, clearly pointed the way beyond the cardinal and toward the ordinal labeling of his indifference curves. His words, although dressed in poetic garb, say precisely what this chapter says: the theory of consumer choice does not depend on our ability to label each indifference curve with cardinal numbers (17 utils, 33 utils, and the like); ordinal numbers (such as 1st, 2nd, and 3rd), which tell us whether the utility of one set of goods is equal to, greater than, or less than that of another, will serve the purpose just as well.

---

[2] Francis Y. Edgeworth, *Mathematical Psychics* (London: C. Kegan Paul, 1881), Part I.

CHAPTER 3

The Demand for Goods

MULTIPLE-CHOICE QUESTIONS

Circle the letter of the *one* answer that you think is correct or closest to correct.

1. The price-consumption line is derived while assuming
   a. that a consumer's tastes are changing, but that prices are not.
   b. that a consumer's budget is changing, but that prices are not.
   c. that many variables affecting a consumer are changing, except price and quantity consumed.
   d. none of the above.

2. The "law" of downward-sloping demand refers to
   a. the downward-sloping shape of most consumption-indifference curves.
   b. the tendency of people normally to buy larger quantities of something when its price is lower, all else being equal.
   c. the alternative amounts of something a person (or group of persons) would buy during a given period at all conceivable prices of the item, all else being equal.
   d. all of the above.

3. The *demand for good* x
   a. never refers to a single quantity number.
   b. can easily be a function of the price of *y*.
   c. is a function of the price of *x*.
   d. is correctly described by all of the above.

4. A *change in quantity demanded* of good *x* can be the result of a change in
   a. the price of *x*.
   b. the price of *y*.
   c. the consumer's taste.
   d. any of the above.

5. The *income-consumption line*
   a. indicates how the optimum quantities of two consumption goods change in response to a change in income, all else being equal.
   b. indicates the alternative amounts of a good a person (or group of persons) would buy during a given period at all conceivable incomes, all else being equal.
   c. is another name for the Engel curve.
   d. depicts a rare exception to the "law" of downward-sloping demand (Giffen's paradox).

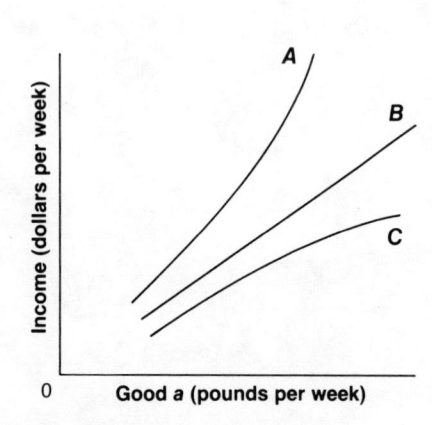

6. The tendency for food expenditures to take a smaller percentage of income the larger is the income is generally referred to as
   a. the "law" of downward sloping demand.
   b. Engel's law.
   c. Giffen's paradox.
   d. the income-consumption line.

7. An upward-sloping Engel curve is
   a. inevitable.
   b. typical, but not inevitable.
   c. impossible.
   d. implied by a downward-sloping demand curve.

Consider the accompanying graph when answering questions 8–9.

8. Which of the following is true about line A?
   a. Line A pictures an income-consumption line.
   b. Line A pictures a price-consumption line.
   c. Line A tells us that good a is a normal good as well as a necessity.
   d. Both (a) and (c).

9. Which of the following is false?
   a. All three lines tell us that good a is a normal good.
   b. Line C tells that good a is a luxury.
   c. Line C tells us that good a is an inferior good.
   d. Line A tells us that good a is a necessity.

10. The *income effect*
    a. always makes a consumer buy more of a good with a lowered price, all else being equal (because lowered price implies higher real income).
    b. always makes a consumer buy less of a good with an increased price, all else being equal (because increased price implies lower real income).
    c. is correctly described by (a) and (b).
    d. is correctly described by neither (a) nor (b).

When answering questions 11–13, consider the graph on page 41, in which a consumer's optimum moves from a to b as the price of bread rises:

11. The (Hicksian) substitution effect of the higher price is measured by
    a. the horizontal difference between c and b.
    b. the horizontal difference between a and d.
    c. the vertical difference between d and b.
    d. the vertical difference between a and d.

12. The (Hicksian) income effect of higher price is measured by
    a. the vertical difference between d and b.
    b. the vertical difference between a and c.
    c. the horizontal difference between d and b.
    d. the horizontal difference between c and d.

13. Bread is clearly shown to be
    a. a luxury good.
    b. a normal good.
    c. a necessity.
    d. a good subject to Giffen's paradox.

Bread (pounds per week)

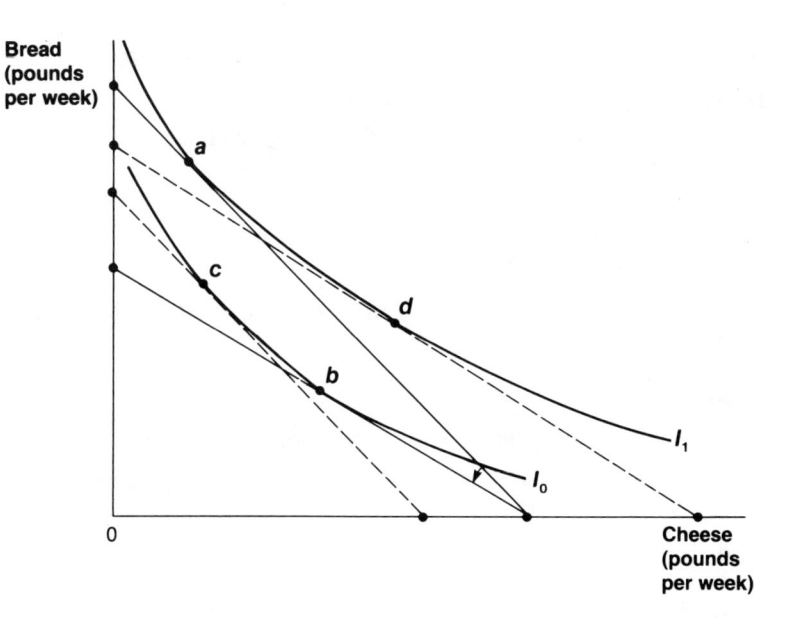

[Figure: Indifference curves I_0 and I_1 with points a, b, c, d and budget lines, showing Cheese (pounds per week) on horizontal axis]

14. When the substitution effect of a lowered price is counteracted by the income effect, the good in question is
    a. an inferior good.
    b. a substitute good.
    c. an independent good.
    d. a normal good.

15. The substitution and income effects of a price cut are
    a. counteracted by the bandwagon effect.
    b. reinforced by the snob effect.
    c. counteracted by the snob effect.
    d. affected as noted in (a) and (b).

16. In the presence of the Veblen effect, market demand is likely to be
    a. downward-sloping at very high prices.
    b. downward-sloping at very low prices.
    c. upward-sloping at some prices.
    d. all of the above.

17. When market demand is estimated with the help of price-quantity data pertaining to different populations during the same past period of time,
    a. the identification problem is avoided.
    b. the procedure is a time-series study.
    c. the procedure is a cross-section study.
    d. both (a) and (c) are true.

When answering questions 18–20, refer to the following graph.

18. The arc elasticity in section *bc* equals
  a. the ratio of *gc* to 0*e* + 0*f*, all divided by the ratio of −*bg* to *eg* + *eb*.
  b. *gc* divided by *bg*.
  c. *bg* divided by *gc*.
  d. none of the above.

19. If *b* lies one third of the way from *a* to *d*, the point elasticity at *b*
  a. equals $|3|$.
  b. equals $|2|$.
  c. equals $|\frac{1}{3}|$.
  d. cannot be determined without additional information.

20. If *c* lies halfway between *a* and *d*, and distance 0*a* is half distance 0*d*, the point elasticity at *c* equals
  a. $|\frac{1}{2}|$
  b. $|2|$.
  c. $|1|$.
  d. none of the above.

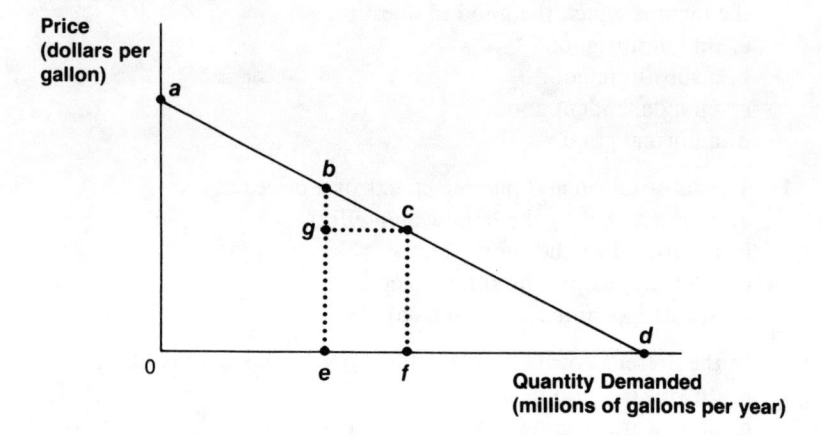

21. The own-price elasticity of demand equals zero
  a. at the horizontal intercept of every straight and downward sloping demand line, regardless of its slope.
  b. at every point of a horizontal demand line.
  c. at every point of a demand line that is a rectangular hyperbola.
  d. in all of the above cases.

22. When the own-price elasticity of demand exceeds unity,
  a. a decrease in price lowers the total expenditures of consumers.
  b. a decrease in price raises the total revenues of firms.
  c. a 21 percent change in price may well cause a 15 percent change in quantity demanded.
  d. both (a) and (c) can be true.

23. The income elasticity of demand
  a. is negative for normal goods.
  b. is positive for inferior goods.
  c. equals the relative change in a good's quantity demanded divided by the relative change in the income of consumers, all else being equal.
  d. is correctly described by all of the above.

24. Empirical studies usually yield own-price elasticities of demand that are
   a. low for goods that have excellent substitutes.
   b. high for goods that have excellent substitutes.
   c. lower the more narrowly a good is defined.
   d. correctly described by (b) and (c).

25. If a good's income-elasticity-of-demand estimate equaled
   a. 2.46, an economist would call the good a necessity.
   b. .37, an economist would call the good a luxury.
   c. −.50, an economist would call the good an inferior one.
   d. −.50, an economist would call the good a complementary one.

TRUE-FALSE QUESTIONS

In each space below, write a *T* if the statement is true and an *F* if the statement is false.

_____ **1.** A price-consumption line indicates how the optimum quantities of two consumption goods change in response to a consumer's changing tastes, given the prices of the two goods.

_____ **2.** As economists use the term, a consumer's *change in demand* for good *x* cannot result from a change in the price of *x*.

_____ **3.** All else being equal, goods of which a consumer consumes larger physical quantities at higher incomes are called *normal goods*.

_____ **4.** The Engel curve, being upward-sloping, exhibits Giffen's paradox.

_____ **5.** Just as Giffen's paradox is always associated with inferior goods, inferior goods always produce Giffen's paradox.

_____ **6.** The income effect reinforces the substitution effect in the case of normal goods.

_____ **7.** The snob effect can never overpower the combined substitution and income effects.

_____ **8.** A good's own-price elasticity of demand is measured by the change in the good's quantity demanded divided by the change in the good's price, all else being equal.

_____ **9.** The steeper a demand line, the lower is its elasticity.

_____ **10.** Whenever the own-price elasticity of demand exceeds unity, a decrease in price increases the total expenditures of consumers.

PROBLEMS

1. Consider the following graph. Panel (a) pictures a consumer's optimum at point A. Panel (b) shows this consumer's demand curve for cheese.

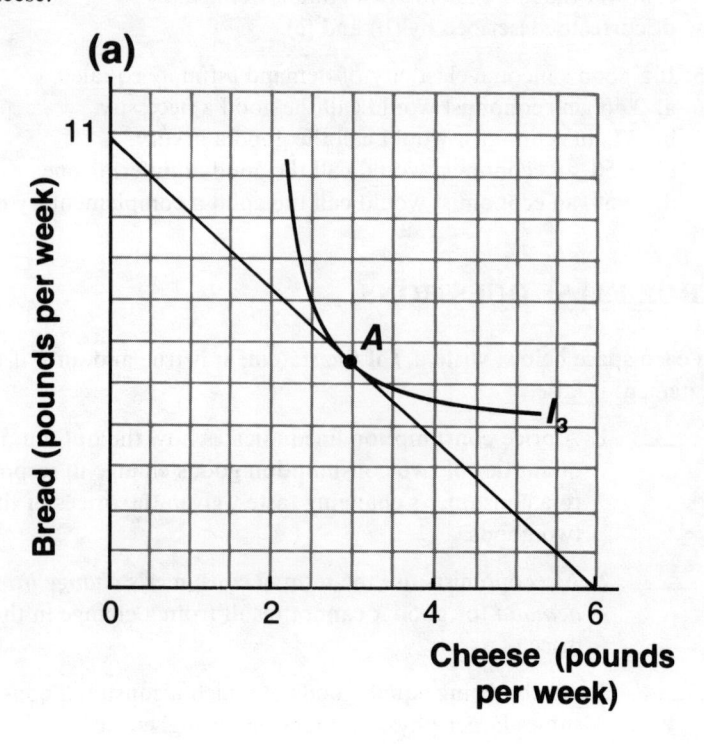

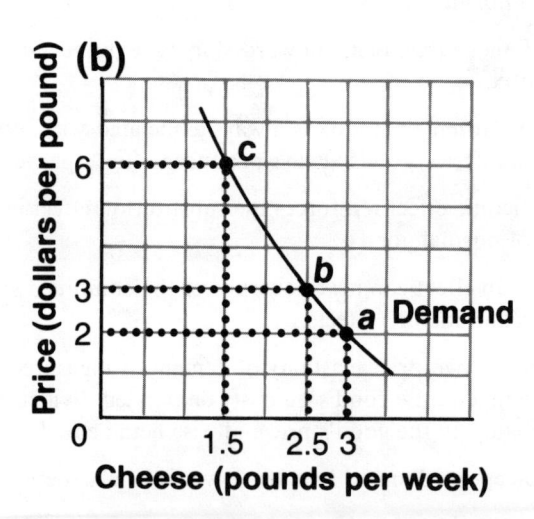

   a. What is the size of this consumer's budget?
   b. In panel (a), draw budget lines corresponding to points b and c on the demand curve.
   c. In panel (a), draw two indifference curves (I_2 and I_1) that are consistent with the consumer's demand curve.
   d. Draw this consumer's price-consumption line in panel (a).
   e. How much bread is this consumer buying when cheese costs $3 per pound?

2. In the following graph, the original budget line and indifference curve from question 1 are reproduced.

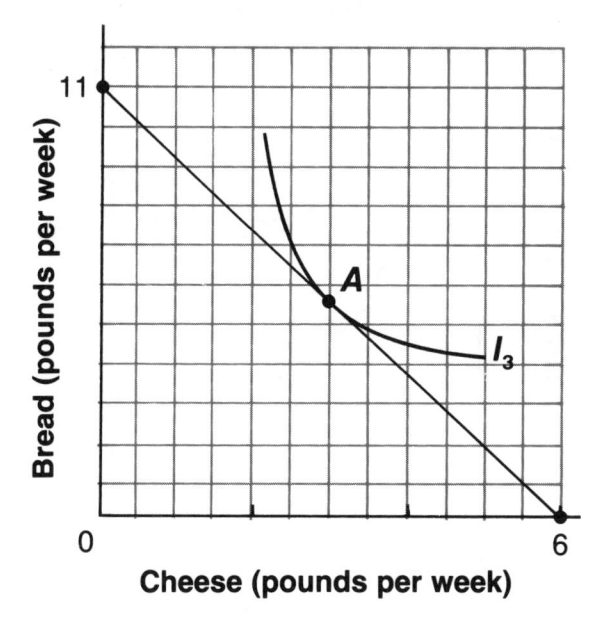

   a. Redraw the budget line for a cheese price of $6 per pound, as well as indifference curve I_1, which is tangent to it. Then indicate the (Hicksian) substitution and income effects associated with the move from the original optimum on I_3 to the new optimum on I_1.
   b. Is cheese a normal good for this consumer?

3. Have another look at panel (a) in the graph for the answer to question 1.
   a. Compared to cheese, is bread an independent, complementary, or substitute good? Explain.
   b. How would one have to redraw indifference curves I_2 and I_1 to get a different answer?

4. Consider graph (a) below. Assuming that the consumer's income equals the budget and that goods *a* and *b* are each priced at $2 per pound, draw Engel curves for each good in graph (b).

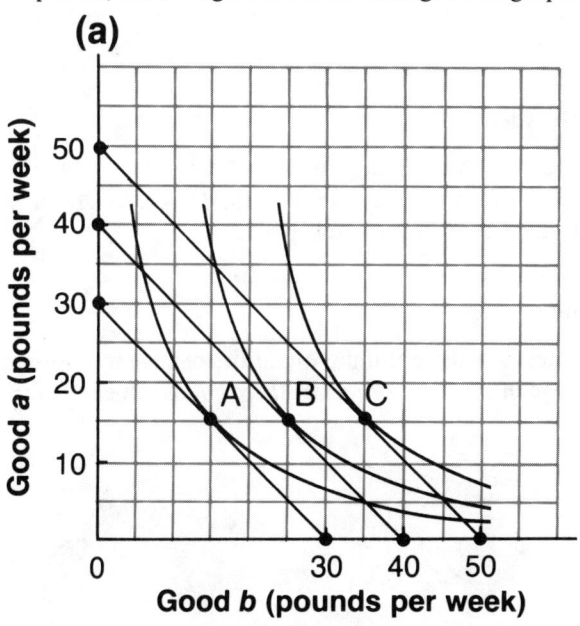

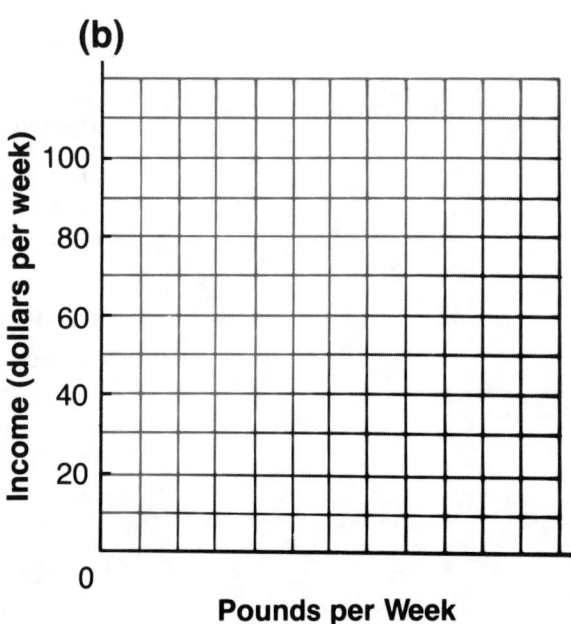

	Price per Pound	Millions of Pounds Traded
1979	50¢	10
1980	10¢	40
1981	30¢	25

5. Consider the data in this table about the U.S. market for peaches when answering questions (a) and (b).
   a. Plot the data in the following graph.

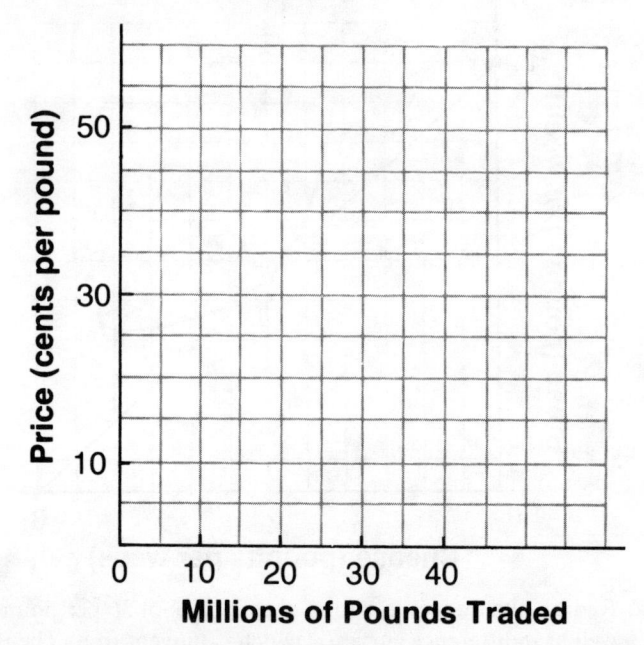

   b. Draw a demand curve for the U.S peach market from 1979–81 in the graph for 5(a).

6. Consider the answer graph in problem 1:
   a. What is the arc elasticity on the demand curve for cheese for a move from *b* to *c*?
   b. How would you determine the point elasticity at *b*?
   c. What is the cross-price elasticity for bread when the price of cheese rises from $3 to $6 per pound?

7. Consider the answer to question 3:
   a. What is the income elasticity for good *a* as income rises from $60 to $80 per week?
   b. What is the corresponding income elasticity for good *b*?

ANSWERS

Multiple-Choice Questions

1. d   2. b   3. d   4. a   5. a   6. b   7. b   8. c   9. c
10. d   11. d   12. a   13. b   14. a   15. c   16. d   17. c   18. a
19. b   20. c   21. a   22. b   23. c   24. b   25. c

True-False Questions

1. F   (It indicates how the optimum quantities change in response to a change *in the price of one of these goods,* all else being equal.)
2. T
3. T

4. F (The Engel curve relates quantity purchased to *income;* Giffen's paradox relates it to *price.*)
5. F (The first part of the sentence is correct, but the second one is false.)
6. T
7. T
8. F (The words "percentage change" or "relative change" must twice be substituted for "change" in this sentence.)
9. F (Elasticity is not slope; furthermore, elasticity often differs at every point on a demand line. It is, therefore, advisable never to talk of the elasticity of a demand line, but only of the elasticity at a point or small section of it: point vs. arc elasticity.)
10. T

Problems

1. a. $12 per week. Panel (a) shows the consumer in fact consuming 3 pounds of cheese per week (point *A*); panel (b) indicates that this quantity is associated with a price of $2 per pound (point *a*). The budget line in panel (a) shows that the consumer could at most buy 6 pounds of cheese per week. Hence the budget must equal $2 times 6, or $12 per week.
   b. The budget lines are shown in the following graph. They indicate that the consumer could at most buy 4 (or 2) pounds of cheese per week if price rose to $3 (or $6) per pound.

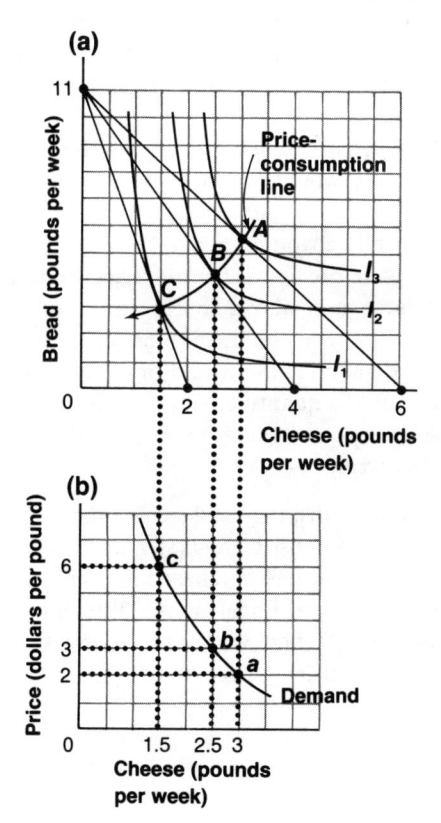

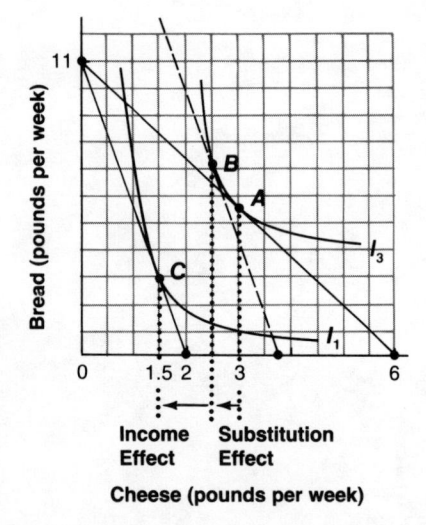

c. As shown in the graph for 1(b), I_2 must be tangent to the budget line for $3 per pound of cheese at B because 2.5 pounds of cheese are demanded at that price. I_1 must be tangent to the budget line for $6 per pound of cheese at C because 1.5 pounds of cheese are demanded at that price.
d. See the graph for 1(b).
e. As point B shows, slightly over 4 pounds per week.

2. a. The steeper solid budget line in this graph reflects the higher price of cheese. So does the parallel dashed line. Without the fall in the consumer's real income implied by the higher price of cheese, this dashed line tells us, the consumer could maintain the original utility associated with I_3 by buying combination B (less of the more expensive cheese and more of the *relatively* less expensive bread). The horizontal distance between A and B, therefore, measures the substitution effect of the higher price of cheese. The consumer's actual budget line, however, is now the solid line going through point C, the consumer's actual new choice. The parallel shift of the hypothetical dashed budget line to the new actual one reflects the consumer's loss of real income because of the higher price of cheese. The horizontal distance between hypothetical choice B and actual choice C measures the income effect of this higher price.
b. Yes, because the income effect reinforces the substitution effect. Lower real income because of the higher price lowers the quantity of cheese bought.

3. a. Bread is a complementary good in this example: As the price of cheese rises, not only does the consumer buy less cheese, but also less bread. (Compare the vertical heights of points A, B, and C.)
b. To make bread an *independent* good, the I_2 and I_1 tangency points B and C would have to be on the respective budget lines to the left of A *at the same height as* A. (The price-consumption line would have to be parallel to the horizontal axis.) To make bread a *substitute* good, tangency points B and C would have to be above and to the left of A. (The price-consumption line would have to be upward-sloping to the left.) Note: The demand curve would of course, be different in such cases because the quantities of cheese associated with prices of $3 and $6 per pound would then be smaller than 2.5 and 1.5 pounds per week.

4. Given the assumptions, the total weekly budgets associated with optima A, B, and C must be $60, $80, and $100, respectively. The same amount of good *a* is purchased in each case: 15 pounds per week; hence the vertical Engel curve. Different amounts of good *b* are purchased: 15, 25, and 35 pounds per week, respectively; hence the upward-sloping Engel curve.

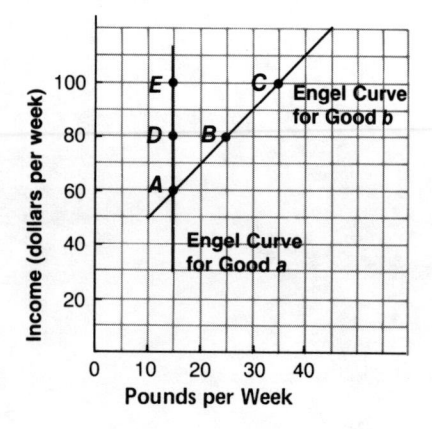

5. **a.** The data plots are shown in the accompanying graph as points *a, b,* and *c.*
    b. If you followed the request and drew anything like the dashed line, you are almost certainly wrong! You do not have sufficient information to draw a demand curve. (Note the text discussion surrounding Figure 3.9, "The Identification Problem.") Quite conceivably, the true demand lines in the three years looked like d_1 through d_3, but we do not know.

6. **a.** Using the text's Box 3.B formula,

$$\epsilon_D^{o-p} = \frac{\frac{Q_2 - Q_1}{Q_2 + Q_1}}{\frac{P_2 - P_1}{P_2 + P_1}} = \frac{\frac{1.5 - 2.5}{1.5 + 2.5}}{\frac{6 - 3}{6 + 3}} = \frac{\frac{-1}{4}}{\frac{3}{9}} = -\frac{1}{4}\left(\frac{9}{3}\right) = -.75$$

b. By placing a tangent on the demand curve at *b* and taking the ratio *bx* over *by* along the tangent, where *x* is its intercept with the horizontal axis and *y* is its intercept with the vertical one.

c. The quantity of bread then falls from 4.125 pounds per week (point *B*) to 2.75 pounds per week (point *C*). Hence, according to the text's Box 3.C,

$$\epsilon_D^{c-p} = \frac{\frac{\Delta Q_B}{\overline{Q}_B}}{\frac{\Delta P_c}{\overline{P}_c}} = \frac{\frac{-1.375}{3.4375}}{\frac{3}{4.5}} = \frac{-.4}{.66} = -.6$$

Note: With a weekly budget of $12 and maximum possible bread purchases of 11 pounds per week, a bread price of $1.09 per pound is implied. When cheese costs $3 per pound (and 2.5 pounds are bought), $7.50 are spent on cheese; hence $4.50 are left for bread—enough for 4.125 pounds. When cheese costs $6 per pound (and 1.5 pounds are bought), $9 are spent on cheese; hence $3 are left for bread—enough for only 2.75 pounds.

7. **a.** Between points *A* and *D* on the graph, according to the text's Box 3.D,

$$\epsilon_D^Y = \frac{\frac{\Delta Q_{D_a}}{\overline{Q}_{D_a}}}{\frac{\Delta Y}{\overline{Y}}} = \frac{\frac{0}{15}}{\frac{+20}{70}} = 0$$

b. Between points *A* and *B* on the graph,

$$\epsilon_D^Y = \frac{\frac{\Delta Q_{D_b}}{\overline{Q}_{D_b}}}{\frac{\Delta Y}{\overline{Y}}} = \frac{\frac{+10}{20}}{\frac{+20}{70}} = \frac{10}{20} \cdot \frac{70}{20} = \frac{700}{400} = +1.75$$

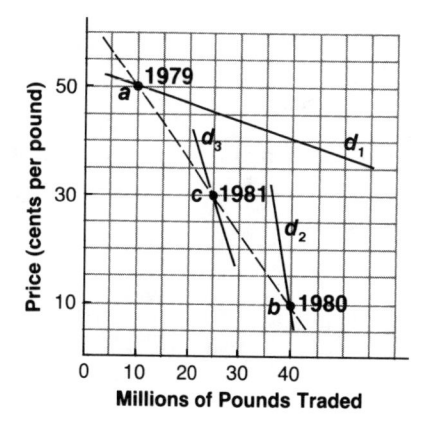

BIOGRAPHY 3.1

Evgeny E. Slutsky

Evgeny Evgenievich Slutsky (1880–1948) was born in Novoe, Russia. He studied physics and mathematics at the University of Kiev but was expelled from the university in 1901 because of participation in student revolts. After spending three years at the Institute of Technology in Munich, Germany, he was allowed to return to Kiev, where he graduated with a gold medal in 1911. Near the time of the Soviet Revolution, he received a degree in political economy from the University of Moscow. By 1920, he was a full professor at the Kiev Institute of Commerce; later he worked at the Central Statistical Board in Moscow and the University of Moscow. He was a member of the Mathematical Institute of the Academy of Sciences of the U.S.S.R.

One of Slutsky's many articles has had the most lasting influence, but the article remained unknown to most economists and mathematicians for many years. Slutsky's great contribution to the theory of consumer behavior appeared under the title "Sulla teoria del bilancio del consumatore" in the Italian journal *Giornale degli Economisti*, July 1915, pp. 1–26. Publication in Italian, in wartime, and in highly mathematical form helped keep the article out of the limelight. Slutsky's achievement was to show that any change in price has two effects. The *substitution effect* refers to the change in quantity as a result of changed relative prices with real (not money) income fixed. This substitution effect is measured while the consumer maintains a given level of welfare. The *income effect* refers to the change in quantity as a result of changed real income. This income effect shifts the consumer from one welfare level to another. The two effects are independent and additive. Slutsky defined them, however, somewhat differently than did Hicks, as Figure 3.5 illustrates. While Hicks viewed a given level of welfare or real income as a given total of utility (represented by an indifference curve), Slutsky viewed it as a given set of goods (represented by a point in the field of choice).

BIOGRAPHY 3.2

John R. Hicks

John Richard Hicks (1904–1989) was born in Leamington Spa, England. Educated at Oxford, where he studied mathematics, philosophy, politics, and economics, he later became a lecturer at the London School of Economics, a fellow at Cambridge, and a professor, first at Manchester and then at Oxford. While at Oxford in 1964 he was knighted, and in 1972 he was awarded the Nobel Memorial Prize in Economic Science (jointly with Kenneth Arrow).

His major works include *The Theory of Wages* (1932, revised 1963), *Value and Capital* (1939, revised 1946), *A Contribution to the Theory of the Trade Cycle* (1950), *A Revision of Demand Theory* (1956), *Capital and Growth* (1965), and *Capital and Time* (1973). His major achievements are the skillful refinement and application of three theories: the economic theory of the consumer (indifference curve analysis, the concept of the marginal rate of substitution, classification of the effect of price changes into substitution and income effects); the theory of general equilibrium among a multitude of markets; and dynamic theory (about combinations of accelerator, multiplier, and lagged linear functions producing oscillations overlaid on patterns of growth).

CHAPTER 4

The Technology of the Firm

MULTIPLE-CHOICE QUESTIONS

Circle the letter of the *one* answer that you think is correct or closest to correct.

1. According to the law of (eventually) diminishing returns (to a variable input), all else being equal,
   a. successive increases in this input cannot yield successively equal increases in output.
   b. successive increases in this input reduce total output.
   c. successive equal increases in this input will eventually reduce the input's marginal product.
   d. all of the above are true.

2. If 1 orchard, 7 workers, and 3 tons of fertilizer yield 1,000 bushels of peaches, while 1 orchard, 7 workers, and 4 tons of fertilizer yield 1,300 bushels,
   a. the average product of labor equals 1,150 bushels.
   b. the marginal product of labor cannot be calculated.
   c. the average product of fertilizer equals 1,150 bushels.
   d. the marginal product of fertilizer cannot be calculated.

When answering questions 3–7, refer to the graph on page 52.

3. The marginal product of labor is rising with increased use of labor until
   a. 10 workers are employed.
   b. 20 workers are employed.
   c. 30 workers are employed.
   d. 40 workers are employed.

4. The average product of labor is falling with increased use of labor once
   a. 10 workers are employed.
   b. 20 workers are employed.
   c. 30 workers are employed.
   d. 40 workers are employed.

5. As long as fewer than 30 workers are employed,
   a. the average product of labor exceeds the marginal product of labor.
   b. the marginal product of labor exceeds the average product of labor.
   c. the marginal product of labor is rising.
   d. both (a) and (c) are true.

6. Between points *d* and *e*, increased use of labor means
   a. negative marginal product of labor.
   b. falling average product and falling marginal product of labor.
   c. marginal product of labor below average product of labor.
   d. all of the above.

7. Maximum average product of labor corresponds to
   a. point *a*.
   b. point *b*.
   c. point *c*.
   d. point *d*.

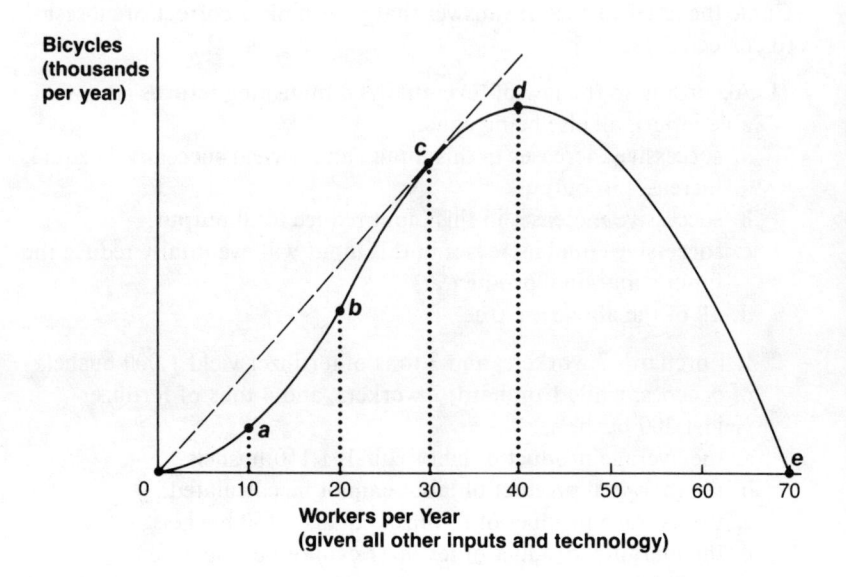

8. *Technical efficiency* refers to a situation in which a firm
   a. finds it possible to produce a given output with less of one or more inputs without increasing the amounts of other inputs.
   b. cannot possibly produce a larger output from the inputs it is in fact using.
   c. is equating the marginal and average products of its inputs.
   d. is simultaneously maximizing the average products of all inputs.

9. A production-indifference curve shows
   a. all the alternative combinations of two inputs that yield the same maximum total product.
   b. all the alternative combinations of two products that can be produced by using a given set of inputs fully and in the best possible way.
   c. all the alternative combinations of two products among which a producer is indifferent because they yield the same profit.
   d. both (b) and (c).

10. A negatively sloped isoquant implies
    a. products with negative marginal utilities.
    b. products with positive marginal utilities.
    c. inputs with negative marginal products.
    d. inputs with positive marginal products.

11. The *marginal rate of technical substitution* is
    a. the rate at which a producer is able to exchange, without affecting the quantity of output produced, a little bit of one input for a little bit of another input.
    b. the rate at which a producer is able to exchange, without affecting the total cost of inputs, a little bit of one input for a little bit of another input.
    c. the rate at which a producer is able to exchange, without affecting the total of inputs used, a little bit of one output for a little bit of another output.
    d. a measure of the ease or difficulty with which a producer can substitute one technique of production for another.

12. In the presence of a diminishing marginal rate of technical substitution between labor and capital, output can be kept unchanged only if
    a. equal successive sacrifices of capital go hand in hand with ever smaller increases of labor.
    b. equal successive sacrifices of capital go hand in hand with ever smaller sacrifices of labor.
    c. equal successive increases in labor go hand in hand with ever smaller increases in capital.
    d. equal successive increases in labor go hand in hand with ever smaller sacrifices of capital.

13. If the capital-labor ratio changes from 100 to 150, while the marginal rate of technical substitution between capital and labor changes from 50 to 100, the elasticity of input substitution
    a. cannot be calculated.
    b. remains unchanged.
    c. equals 2.
    d. equals $\frac{3}{5}$.

14. The elasticity of input substitution
    a. is smaller the easier input substitution is.
    b. is larger the easier input substitution is.
    c. is constant along any given isoquant.
    d. equals the reciprocal of an isoquant's slope.

15. If a simultaneous and equal percentage decrease in the use of all physical inputs leads to a larger percentage decrease in physical output, a firm's production function is said to exhibit
    a. decreasing returns to scale.
    b. constant returns to scale.
    c. increasing returns to scale.
    d. diseconomies of scale.

16. If a firm triples all inputs, and output triples as well, the firm is subject to
    a. constant returns to scale.
    b. increasing returns to scale.
    c. economies of scale.
    d. both (b) and (c).

17. Which of the following may well account for the existence of increasing returns to scale?
    a. Increasing returns to a variable input.
    b. A high elasticity of input substitution.
    c. Specialization of inputs.
    d. All of the above.

When answering questions 18–19, refer to the graph below.

18. The production function pictured by the above set of isoquants exhibits
    a. negative marginal products (because of the negative slope of the isoquants).
    b. constant returns to scale (because of the even numbering of isoquants).
    c. decreasing returns to scale (because $BC < AB < 0A$, etc).
    d. increasing returns to scale (in spite of $BC < AB < 0A$, etc).

19. The numbers attached to the isoquants (10, 20, 30) might refer to
    a. bicycles produced per year.
    b. total annual revenue.
    c. total annual profit.
    d. marginal rates of technical substitution.

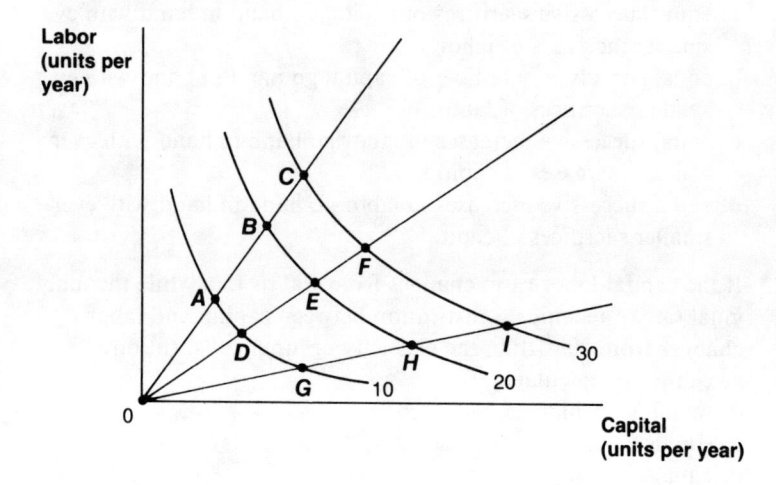

20. The major function of a firm's administrative hierarchy is
    a. to prevent detailed information from reaching the upper ranks of management.
    b. to channel detailed information to the upper ranks of management.
    c. to eliminate diseconomies of scale.
    d. both (b) and (c).

21. Empirical estimates of production functions were pioneered by
    a. Gossen.
    b. Dupuit.
    c. von Thünen.
    d. Leibenstein.

22. According to the "survivor principle,"
    a. only firms with constant returns to scale survive in a competitive industry in the long run.
    b. if, over time, firms of large size are supplanted by smaller ones, we can suspect the presence of increasing returns to scale in small firms.
    c. only firms with a constant elasticity of input substitution survive in a competitive industry in the long run.
    d. competition among different sized firms in an industry will, in the long run, allow only the technically most efficient firms to survive.

23. Production functions have been estimated empirically with the help of
   a. time-series studies.
   b. consumer interviews.
   c. market experiments.
   d. all of the above.

24. The Cobb-Douglas production function
   a. always has an elasticity of input substitution below unity.
   b. is a special type of constant-elasticity-of-substitution production function.
   c. always has an elasticity of input substitution above unity.
   d. is correctly described by (a) and (b).

25. Given a Cobb-Douglas production function of $Q = AL^{.75} \cdot K^{.25}$,
   a. a .75 percent increase in labor input will increase output by 1 percent, all else being equal.
   b. a 1 percent decrease in capital input will decrease output by .25 percent, all else being equal.
   c. constant returns to scale prevail.
   d. both (b) and (c) are true.

TRUE-FALSE QUESTIONS

In each space below, write a *T* if the statement is true and an *F* if the statement is false.

_____ 1. The set of all the production methods known to people is the *production function*.

_____ 2. A period of one year or less is generally referred to by economists as *the short run*.

_____ 3. While labor is a variable input, capital and natural resources are fixed inputs.

_____ 4. When marginal anything is rising, average anything cannot be falling.

_____ 5. The total-product curve depicts alternative output levels each of which is technically efficient.

_____ 6. For isoquants that allow input substitution, the elasticity of input substitution is always positive.

_____ 7. Statements about returns to scale are necessarily statements about the long run.

_____ 8. When a firm is subject to increasing returns to scale, the average products of all inputs fall when scale is reduced.

_____ 9. Given a Cobb-Douglas production function of $Q = A \cdot L^a K^b$, one knows that $(a \cdot b) > 1$ implies increasing returns to scale (given that a and b are positive).

_____ 10. When the elasticity of capital-labor substitution equals .72, the underlying production function is a Cobb-Douglas function.

PROBLEMS

1. Given fixed quantities of other inputs (not shown), current technology allows a firm to use the different amounts of a variable input shown in column (1) and produce the different amounts of output shown in column (2):

	Units of Variable Input Used (1)	Maximum Units of Associated Output (2)	Marginal Product (3)	Average Product (4)
(A)	10	100		
(B)	20	200		
(C)	30	250		
(D)	40	275		
(E)	50	285		

a. Calculate the marginal products of the variable input.
b. Calculate the average products of the variable input.
c. In graph (a), plot the total-product curve as a smooth line through the available data.
d. In graph (b), plot the marginal and average product curves in analogous fashion.

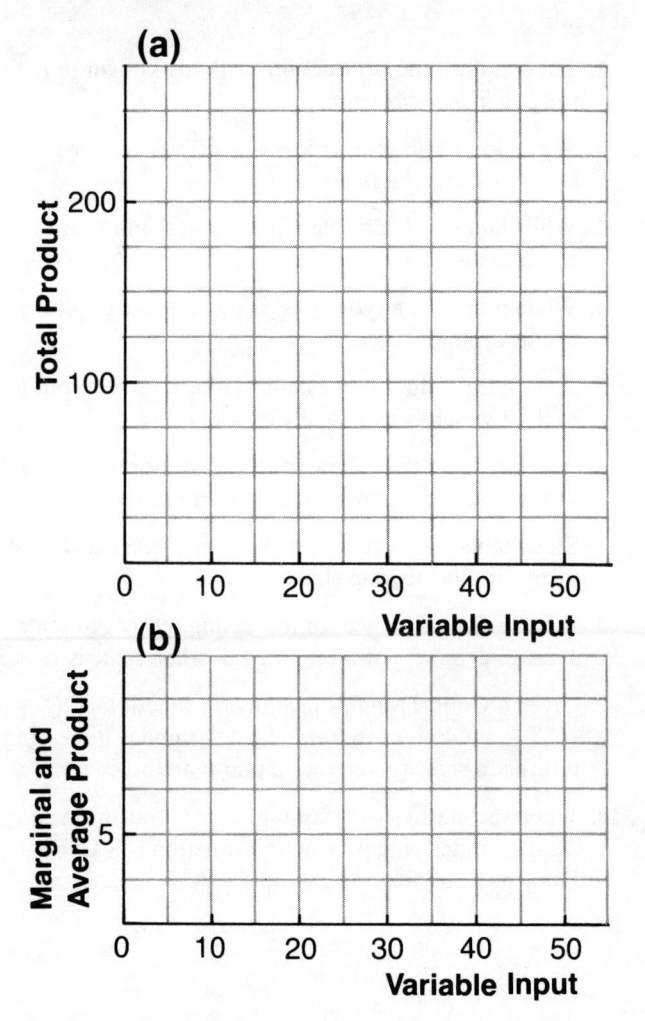

2. Consider text Table 4.2, "A Complex Production Function."
   a. Calculate the marginal products of labor as labor is increased from 1 to 7 units, while capital is held constant at 1 and 6 units.
   b. Calculate the marginal products of capital as capital is increased from 1 to 7 units, while labor is held constant at 3 and 7 units.

3. Imagine a producer who could turn each of the indicated input combinations (measured in respective units per year) into the identical maximum quantity of meat (100 pounds per year).
   a. Plot the data from the table in the following graph and connect them with a smooth curve.

	Hay	Grain
A	100	2.5
B	50	5.0
C	20	10.0
D	6	15.2
E	2	20.0

   b. What is this producer's marginal rate of technical substitution between C and D?
   c. Is the producer subject to a diminishing marginal rate of technical substitution? Explain.
   d. Calculate the elasticity of input substitution between A and B, given a marginal rate of technical substitution between hay and grain of $85H:1G$ at A and $8.5H:1G$ at B.

4. Refer to the data in columns (1) and (2) of the table in question 1.
   a. How would these data have to change if the producer were subject to constant returns to scale and all (fixed and variable) inputs tripled?
   b. How would these data have to change if the producer were subject to increasing returns to scale and all (fixed and variable) inputs halved?

5. Consider the following graph. If you were told that the isoquants in it depicted decreasing returns to scale, what would be a possible output quantity label for I_0? For I_2? Explain.

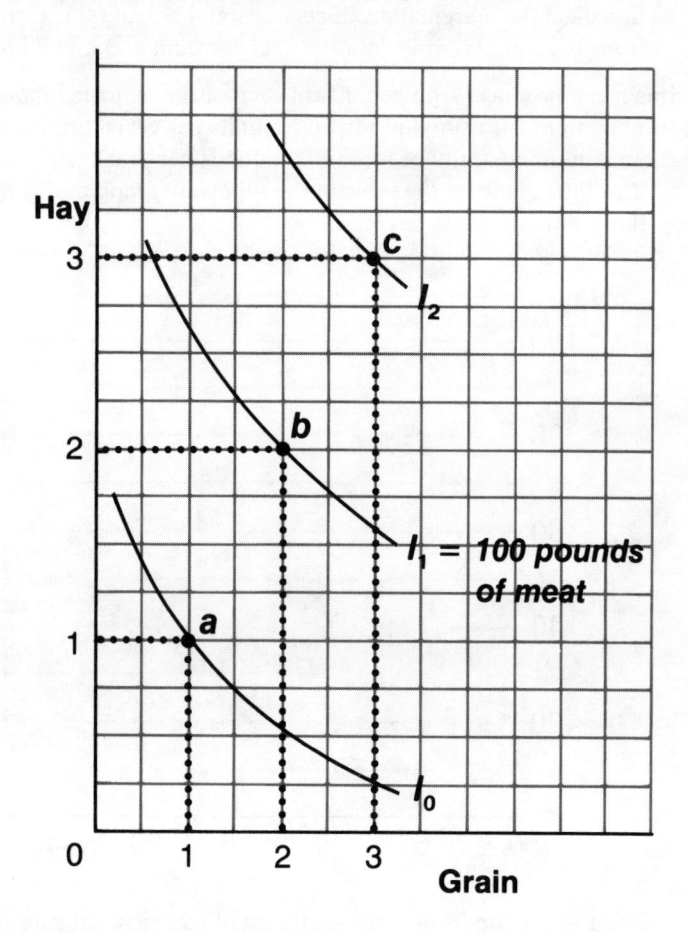

6. Prove that economic inefficiency prevails when one producer's marginal rate of technical substitution equals 5 units of capital for 1 unit of labor, while another producer's rate equals 1 unit of identical capital for 1 unit of identical labor. Does it make a difference whether both producers belong to the same industry (and both produce apples, say) or whether they produce totally different products (such as apples and turpentine)? Explain.

7. For each of the following production functions construct a table like text Table 4.2 "A Complex Production Function" and determine whether the law of diminishing returns holds and what kinds of returns to scale exist. Use K and L values from 1 through 4.
   a. $Q = \sqrt{K \cdot L}$
   b. $Q = K^2 \cdot L$
   c. $Q = K + L$

ANSWERS

Multiple-Choice Questions

1. c 2. b 3. b 4. c 5. b 6. d 7. c 8. b 9. a
10. d 11. a 12. d 13. d 14. b 15. c 16. a 17. c 18. d
19. a 20. a 21. c 22. d 23. a 24. b 25. d

True-False Questions

1. F (This is the definition of *technology*.)
2. F (The definition has nothing to do with chronological time. The short run refers to a time period so short—whatever its length—that a firm cannot vary the quantity of at least one of its inputs.)
3. F (Any type of input can be fixed or variable.)
4. F (Not "is rising," but "is above average.")
5. T
6. T
7. T
8. T
9. T (Because in that case $a + b$ also exceeds 1 and *that* means increasing returns to scale.)
10. F (For a Cobb-Douglas production function the elasticity always equals 1.)

Problems

1. a.–b.

	Units of Variable Input Used (1)	Maximum Units of Associated Output (2)	Marginal Product (3)	Average Product (4)
(A)	10	100		100 : 10 = 10
			$\frac{+100}{+10} = 10$	
(B)	20	200		200 : 20 = 10
			$\frac{+50}{+10} = 5$	
(C)	30	250		250 : 30 = 8.333
			$\frac{+25}{+10} = 2.5$	
(D)	40	275		275 : 40 = 6.875
			$\frac{+10}{+10} = 1$	
(E)	50	285		285 : 50 = 5.7

c.–d. Note: Strictly speaking, the marginal products calculated here are averages of marginal products within the input ranges shown. This produces the stair-step line once marginal diverges from average product.

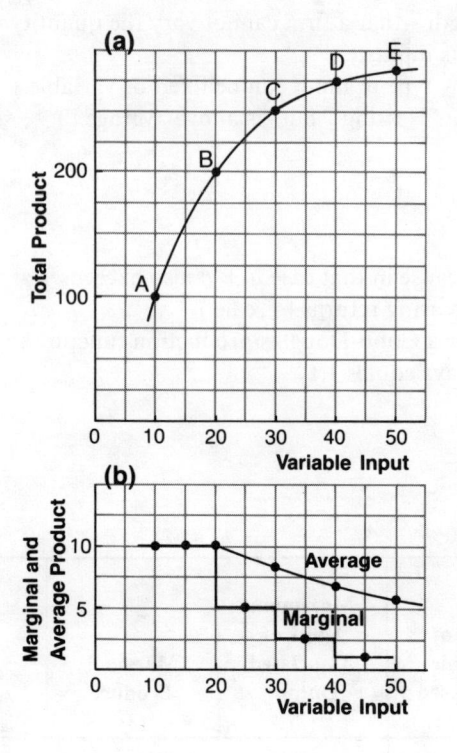

2. **a.** Marginal products of labor (in bushels of wheat per year) with capital at 1 are 117-89-76-67-60-55 and with capital at 6 are 286-219-186-163-147-136.
   b. Marginal products of capital (in bushels of wheat per year) with labor at 3 are 203-155-131-115-104-96 and with labor at 7 are 309-237-200-176-160-146.

3. **a.**

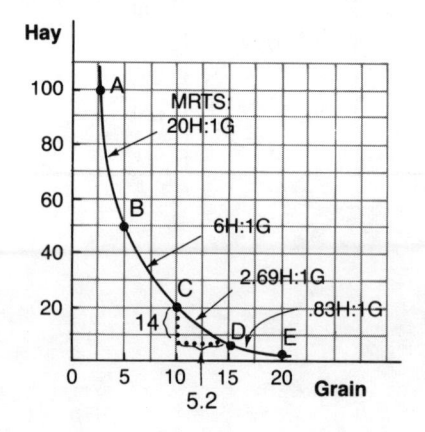

b. 14 units of hay for 5.2 units of grain or $2.69H$ for $1G$.
   c. Yes, this producer is subject to a diminishing marginal rate of technical substitution: Going from A toward E, the producer can keep output unchanged while sacrificing $50H$ and using an extra $2.5G$; that is, by sacrificing $20H$ for $1G$ in range AB, but only $6H$ for $1G$ in range BC, $2.69H$ for $1G$ in range CD, and $.83H$ for $1G$ in range DE. Similarly, going from E toward A, the marginal rate of technical substitution is $4.8G$ for $4H$, or $1.2G$ for $1H$ in range ED, but only $.37G$ for $1H$ in range DC, $.17G$ for $1H$ in range CB, and $.05G$ for $1H$ in range BA. (Note the similarity of this problem to problem 5 of Chapter 3.)
   d. The elasticity of input substitution σ equals the percentage change in the hay-grain ratio used, divided by the associated percentage change in the marginal rate of technical substitution between these inputs. At A, the hay-to-grain ratio equals $40:1$ at B, it equals $10:1$. Thus the numerator of the text's Box 5.A formula is $\frac{-30}{25} = -1.2$. The $MRTS$ at A and at B must be estimated from the above graph as the isoquant's slope at A and B, respectively. If your estimates were 50 and 10 (admittedly rough guesses), the formula's denominator would be $\frac{-40}{30} = -1.33$, and σ would be estimated as $\frac{-1.2}{-1.33} = .9$.

4. **a.** Each entry in columns (1) and (2) would triple.
   b. Each entry in column (1) would halve, but each entry in column (2) would be cut by more than 50 percent.

5. For I_0 any number above 50 pounds of meat (and below 100 pounds) will do: Note that input combination b uses 2 units of hay plus 2 units of grain, while yielding 100 pounds of meat. Combination a, on the other hand, uses exactly half of these inputs and would yield 50 pounds of meat under constant returns to scale. Thus it must yield more than 50 pounds under decreasing returns to scale. Then a doubling of all inputs (at a) does *not* double output (at b). For I_2 a number below 150 pounds of meat (but above 100 pounds) is called for: note that input combination c uses 50 percent more of all inputs than does b. Thus b must yield less than 50 percent extra output.

6. The first producer could release $5K$ to the second one in exchange for $1L$. The output of the first producer (whose $MRTS$ equals $5K$ for $1L$) would remain unchanged. The output of the second producer (whose $MRTS$ equals $1K$ for $1L$) would rise: this producer could keep output unchanged if the sacrifice of $1L$ were compensated by the addition of $1K$. In fact, $5K$ were received. Given a positive marginal product of capital, output would rise. It makes no difference whether both belong to the same industry or not. In one case, the apple output of the first producer does not fall, but the apple output of the second producer rises. So the world has received more apples from given resources (differently allocated). In the other case, the apple output of the first producer does not fall, but the turpentine output of the second producer rises. So the world has received more turpentine from given resources (differently allocated). In either case, someone can be made better off (with the extra output) *without anyone else having to be made worse off* (because no other output has to be reduced).

7. a.

Capital, K	Output, $Q = \sqrt{K \cdot L}$			
4	2	2.83	3.46	4
3	1.73	2.45	3	3.46
2	1.41	2	2.45	2.83
1	1	1.41	1.73	2
Labor, L	1	2	3	4

The law of diminishing returns holds. For example, holding capital at $K = 3$, successive increases of L by 1 unit raise output by .72, .55, .46. Holding labor at $L = 4$, successive increases of K by 1 unit raise output by .83, .63, .54.

There are constant returns to scale. For example, doubling, tripling, quadrupling all inputs doubles, triples, quadruples output.

b.

Capital, K	Output, $Q = K^2 \cdot L$			
4	16	32	48	64
3	9	18	27	36
2	4	8	12	16
1	1	2	3	4
Labor, L	1	2	3	4

The law of diminishing returns does not hold. For example, holding capital at $K = 4$, successive increases of L by 1 unit raise output by 16, 16, 16.

There are increasing returns to scale. For example, starting at $K = L = Q = 1$, doubling all inputs raises Q 8-fold; tripling all inputs raises Q 27-fold; quadrupling all inputs raises Q 64-fold.

c.

Capital, K	Output, $Q = K + L$			
4	5	6	7	8
3	4	5	6	7
2	3	4	5	6
1	2	3	4	5
Labor, L	1	2	3	4

The law of diminishing returns does not hold. For example, holding labor at $L = 3$, successive increases of K by 1 unit raises output by 1, 1, 1.

There are constant returns to scale.

BIOGRAPHY 4.1

Johann H. von Thünen

Johann Heinrich von Thünen (1783–1850) was born into an old feudal family in the Grand Duchy of Oldenburg, Germany. On his father's estate, Kanarienhausen, he developed an early interest in agriculture and mathematics. He attended an agricultural college near Hamburg, and later the University of Göttingen but never graduated; he preferred the career of a practical farmer. Nevertheless, he had many insights of genius. Above all else, these insights concerned the production function, the location of economic activities, and the distribution of income. His thoughts are preserved in his single major book, *The Isolated State in Relation to Agriculture and Political Economy,* which appeared in four installments between 1826 and 1863.

Unlike any of his contemporaries except Cournot (see Biography 7.1) and Gossen, von Thünen applied mathematics to economic analysis. On his estate in Mecklenburg, he kept meticulous farm accounts from 1810–20, costing every plot of land, every bushel of rye, every cow and goose. By doing so, von Thünen became the first investigator who put his occupational life into the service of scientific economic research. He thus became the patron saint of *econometrics* (the application of statistical methods to the study of economics). The data he collected served as the empirical basis for his discovery of the law of eventually diminishing marginal products.

His agricultural activities also led von Thünen to a brilliant and original vision, which many consider his peak achievement. He envisioned an extended domain, of circular form and uniform fertility, isolated from the rest of the world, free from all obstacles to or special facilities for transport (such as mountains or rivers), with a single source of demand for agricultural products—a town in its center. He demonstrated how the uniform variation of transport costs with distance, which follows from his assumptions, would bring about a regional specialization among different products and different techniques of producing the same product. This specialization would reveal itself in the establishment of a series of concentric rings around the town, each being the optimal location for a different type of activity. Products that were perishable, imposed heavy transport costs, or had to be cultivated intensively would be produced near the population center; others would be produced farther away. (He pictured seven rings of activities outwards from the town: horticulture, forestry for building and fuel, cereal production by crop rotation, cereal production by alternating crops and pasture, cereal production via the three-field system, stock farming, and hunting.) Von Thünen showed that differential rent would arise to reflect differential advantage of location. He later introduced differential fertility and additional towns as well. He thus anticipated the location theories of Alfred Weber and August Lösch and proved himself superior to his British contemporary, David Ricardo, whose rent theory was based on fertility differences alone, a single product (corn), and zero transport costs.

In a third major accomplishment, von Thünen developed a theory of income distribution based on the concept of marginal productivity. He stated clearly and explicitly that the (real) wage of all workers, in a large firm employing many workers, would tend to equal the marginal product of the last worker employed. Consider Table 4.1 of the text, row (G): If a 6th worker raises annual output by 700 bushels of apples per year, no employer would pay more than 700 bushels as a wage. If all workers are alike and 6 are hired, all get this wage regardless of the chronological order in which they are hired. Von Thünen applied this thinking to capital as well, suggesting that the profit of capital would equal the marginal product of the last small portion of capital employed. Von Thünen was, however, baffled by these results: If his theory of actual wage (or profit) determination was correct, all units, except for the labor (or capital) unit hired last, would be receiving less than the marginal products associated with the chronological order of their hiring. See Table 4.1, column (4). This seemed unfair. Such thoughts led von Thünen into the realm of normative statements. Living in the stormy days when social revolution, incited by the misery of workers, seemed imminent, he looked for an *ethical principle* that would reconcile the claims of workers (demanding "the whole produce of labor") with those of the owners of capital and land (offering "bare subsistence" to the workers). Von Thünen suggested a compromise, a "natural wage," w, equal to the geometric mean between a worker's subsistence requirements, a, and total product, p. In his own eyes, the formula $w = \sqrt{ap}$ was his highest achievement. He had it engraved on his tombstone to indicate not how wages actually were determined, but how they ought to be determined in a just society.

Indeed, von Thünen exhibited great concern for his own workers. At a time when most employers treated farm hands like cattle, he supported a doctor, nurse, and cottage-hospital on his estate for the free treatment of all workers and their families. He provided sick pay and retirement pensions. In return, however, he required punctilious performance of duty, paying workers by piece rates whenever possible. In his book, von Thünen also introduced the concept of human capital:

> The reluctance to view a man as capital is especially ruinous of mankind in wartime; here capital is protected, but not man, and in time of war we have no hesitation in sacrificing one hundred men in the bloom of their years to save one cannon. In a hundred men at least twenty times as much capital is lost as is lost in one cannon. But the production of the cannon is the cause of an expenditure of the state treasury, while human beings are again available for nothing by means of a simple conscription order. . . . When the statement was made to Napoleon, the founder of the conscription system, that a planned operation would cost too many men, he replied: "That is nothing. The women produce more of them than I can use."

Unfortunately for economic science, von Thünen's many original ideas never had the influence they deserved. He was a prophet with little honor in any country. In Germany, economic theorists and political liberals were equally despised; von Thünen was both, and his lack of academic status didn't help. In Britain, Ricardo's brilliant advocacy of policies eclipsed the German thinker's superior theoretical ability. It took nearly a century after Ricardo before a British economist of like stature, Alfred Marshall (see Biography 6.1), would say: "I loved von Thünen above all my other masters."[1]

---

[1] A. C. Pigou, ed. *Memorials of Alfred Marshall* (London: Macmillan, 1925), p. 360.

Readings on Thünen

Dickinson, H. D. "Von Thünen's Economics." *Economic Journal,* December 1969, pp. 894–902.

Hall, Peter, ed. *Von Thünen's Isolated State.* Oxford: Pergamon Press, 1966.

> A translation of volume I of Thünen's work on location theory and rent, originally published in 1826, plus excerpts from volume II, section 1 (1850) and section 2 (1863).

Samuelson, Paul A. "Thünen at Two Hundred," *Journal of Economic Literature,* December 1983, pp. 1468–88.

Schneider, Erich. "Johann Heinrich von Thünen." *Econometrica,* January 1934, pp. 1–12.

BIOGRAPHY 4.2

Harvey Leibenstein

Harvey Leibenstein (1922–   ) was born in Russia but came to Canada at an early age. He studied at Northwestern University and Princeton and became professor of economics first at the University of California (Berkeley) and in 1967 at Harvard.

The titles of his works reflect his major interests in population economics and X-efficiency theory: *A Theory of Economic Demographic Development,* (1954), *Economic Backwardness and Economic Growth,* (1957), *Economic Theory and Organizational Analysis,* (1960), *Beyond Economic Man: A New Foundation for Microeconomics,* (1976), *General X-Efficiency Theory and Economic Development,* (1978), *Inflation, Income Distribution, and X-Efficiency Theory* (1980), *Inside the Firm: The Inefficiencies of Hierarchy* (1987).

Leibenstein's critics, such as George Stigler (see Biography 15.1 and the "Selected Readings" section of Chapter 4) are apt to dismiss the very concept of X-inefficiency as nonsensical. They tend to view Leibenstein's "missed opportunities within firms" for utilizing existing resources as effectively as they might be used as uncontrollable. In most cases, they contend, the alleged gap between a firm's actual and potential output can in fact not be closed, except at an inordinate cost (such as placing an overseer next to every worker). Therefore, it is not worthwhile to get the extra output. Why spend an extra $5 in better resource administration only to get an extra $1 of output? Leibenstein, of course, disagrees. He thinks firms miss plenty of opportunities to raise output in a cost-effective way.

CHAPTER 5

Costs and the Supply of Goods

MULTIPLE-CHOICE QUESTIONS

Circle the letter of the *one* answer that you think is correct or closest to correct.

1. The opportunity cost of a good equals
   a. the minimum payment necessary to keep all the resources making the good in their present employment
   b. the minimum payment that the resources making the good could get elsewhere.
   c. both (a) and (b).
   d. neither (a) nor (b).

2. In the short run, a firm's fixed cost
   a. is zero.
   b. cannot be escaped.
   c. can be escaped only by cutting production to zero.
   d. is not correctly described by any of the above.

3. When average total cost rises from $10 to $30 as total production rises from 100 to 300 units, average variable cost
   a. cannot be calculated.
   b. equals $10.
   c. equals $20.
   d. equals $30.

4. If a woman invests $10,000 of her money in a retail store and then quits the best job available to her (in which she was earning $20,000 a year), the retail store has annual fixed costs of at least
   a. $30,000.
   b. $21,000, if the interest rate is 10 percent per year.
   c. $22,000, if the interest rate is 10 percent per year.
   d. $33,000, if the interest rate is 10 percent per year.

When answering questions 5–8, refer to the graph on the top of page 67.

5. When total product equals $0A$,
   a. variable cost equals BC.
   b. average variable cost equals BC divided by $0A$.
   c. fixed cost equals CD.
   d. all of the above are true.

6. When total product equals $0A$, the associated marginal cost
   a. cannot be determined from this graph.
   b. exceeds average total cost.
   c. equals DA divided by $0A$.
   d. equals the slope of the variable-cost curve at C.

7. According to this graph,
   a. marginal cost is positive at all levels of output.
   b. marginal cost is falling whenever total product rises.
   c. marginal cost exceeds average total cost at all levels of output.
   d. all of the above are true.

8. According to this graph,
   a. average total cost exceeds marginal cost at all levels of output.
   b. average total cost exceeds average variable cost at all levels of output.
   c. average fixed cost is the same at all levels of output.
   d. average fixed cost exceeds average variable cost when total product equals 0A.

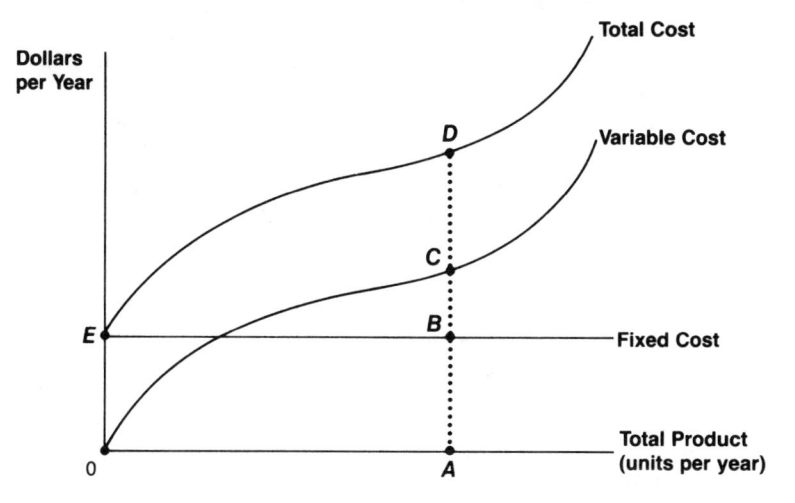

When answering questions 9–11, refer to the graph below:

9. Line B represents
   a. marginal cost.
   b. average variable cost.
   c. average fixed cost.
   d. average total cost.

10. The vertical difference, at any level of output, between lines B and C represents
    a. marginal cost.
    b. average variable cost.
    c. average total cost.
    d. average fixed cost.

11. When output equals 0f,
    a. total cost equals 0f times fe.
    b. fixed cost equals 0f times fe.
    c. variable cost equals 0f times fe.
    d. marginal cost equals ed.

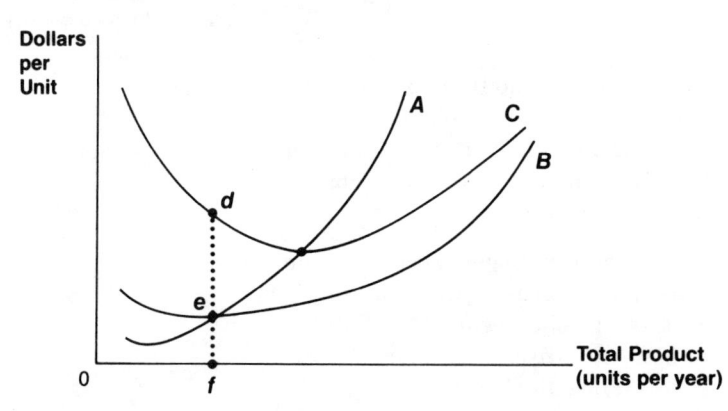

12. At the point where a straight line from the origin is tangent to the variable-cost curve
    a. marginal cost equals average total cost.
    b. marginal cost equals average fixed cost.
    c. marginal cost equals average variable cost.
    d. average total cost is minimized.

13. A firm operating in a perfectly competitive market maximizes its profit by adjusting
    a. its output price until it exceeds average total cost as much as possible.
    b. its output price until it exceeds marginal cost as much as possible.
    c. its output until its marginal cost equals output price.
    d. its output until its average total cost is minimized.

When answering questions 14–16, refer to the graph below.

14. Assuming this firm maximizes profit, it will
    a. produce $0A$ at a price of P_4.
    b. produce $0D$ at a price of P_1.
    c. incur a total cost of GC times $0C$.
    d. do all of the above.

15. Assuming this firm maximizes profit, it will
    a. produce $0A$ regardless of price.
    b. produce $0A$ only at price P_4.
    c. make a zero profit at price P_2.
    d. make a profit of KL times $0D$ at price P_1.

16. Assuming this firm maximizes profit, it will
    a. incur fixed cost of KL at a price of P_1.
    b. incur variable cost of HC at a price of P_2.
    c. supply varying quantities in the short run, depending on the price—quantities that one can read off on line EI.
    d. make a loss of FM times $0B$ at price P_3.

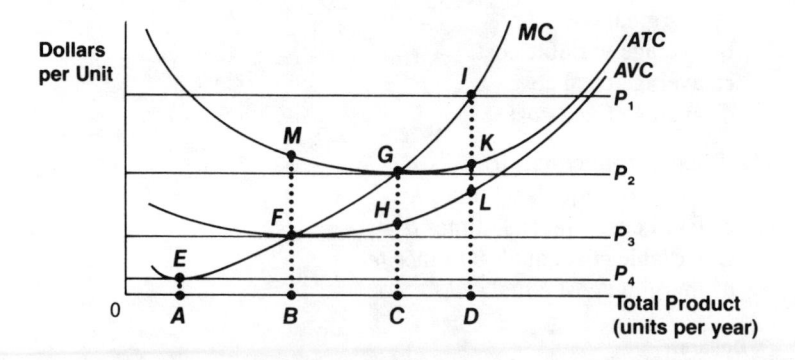

17. In the short run, no firm operates with a loss, unless
    a. variable cost equals fixed cost.
    b. variable cost falls short of fixed cost.
    c. total revenue covers variable cost.
    d. total revenue covers fixed cost.

18. For a firm operating in a perfectly competitive market, its short-run supply is identical with the rising arm of
    a. its marginal-cost curve.
    b. its average-fixed-cost curve.
    c. its average-total-cost curve.
    d. none of the above.

19. A good's short-run supply curve is shifted to the right by
   a. a fall in the good's price.
   b. a rise in the prices of inputs used to make the good.
   c. an improvement in the technology of making the good.
   d. none of the above.

20. Imagine a graph depicting a firm's isocost line between inputs of labor and capital. If the price of labor doubled,
   a. the firm could buy the same amount of capital as before, but only if any previous nonzero labor purchases were reduced.
   b. the firm could still buy any previous amount of labor, but only if less capital were purchased.
   c. the isocost line would shift to the right.
   d. both (a) and (b) would hold.

21. If a profit-maximizing firm's marginal product of labor equals 1 ton of output, while the marginal product of capital equals 7 tons of output and the use of capital is priced at $14 per unit, then
   a. the price of labor must be $2.
   b. the price of labor must be $7.
   c. the price of labor must be $14 as well.
   d. none of the above are true.

22. A firm's long-run average-total-cost line is
   a. identical to its long-run marginal-cost line.
   b. also its long-run supply curve.
   c. in fact the average-total-cost curve of the optimal plant.
   d. tangent to all the curves of short-run average total cost.

23. When decreasing returns to scale exist,
   a. the long-run average-total-cost curve is downward-sloping as output is raised.
   b. the envelope curve is above the curve of long-run marginal cost.
   c. it is cheaper to produce a given output by running a larger plant below capacity than a smaller plant at capacity.
   d. it is cheaper to produce a given output by running a smaller plant above capacity than a larger plant at capacity.

24. All else being equal,
   a. favorable technological externalities make a long-run market-supply line flatter than a simple horizontal addition of given upward-sloping long-run marginal-cost curves would suggest.
   b. unfavorable technological externalities make a long-run market-supply line flatter than a simple horizontal addition of given upward-sloping long-run marginal-cost curves would suggest.
   c. favorable pecuniary externalities make a long-run market-supply line steeper than a simple horizontal addition of given upward-sloping long-run marginal-cost curves would suggest.
   d. both (a) and (c) are true.

25. At every point on a straight line of supply, the price elasticity of supply exceeds unity, as long as the supply line in quadrant I of the typical graph
   a. intercepts the vertical axis.
   b. is vertical.
   c. goes through the origin.
   d. intercepts the horizontal axis.

TRUE-FALSE QUESTIONS

In each space below, write a *T* if the statement is true and an *F* if the statement is false.

_____ 1. *Implicit costs* are hidden costs that the owners of firms incur when using the services of their own resources in their own firms instead of hiring them out to collect at least the minimum income available elsewhere.

_____ 2. *Variable cost* is associated with the use of inputs the quantity of which can be varied and the variation of which changes the level of production during a given period.

_____ 3. When variable cost rises from $30,000 to $40,000 per year as production rises from 100 to 200 units, marginal cost equals $100.

_____ 4. For a zero-profit business, total revenue just equals average total cost.

_____ 5. At the shutdown point, a firm's product price equals minimum average variable cost.

_____ 6. A firm's capacity output corresponds to that output level which minimizes the marginal cost of production.

_____ 7. The optimum plant is that plant, among all conceivable ones, which is operated at capacity.

_____ 8. When increasing returns to scale exist, it is cheaper to produce a given output by running a larger plant below capacity than a smaller plant at capacity.

_____ 9. When the production function of one firm is affected, favorably or unfavorably, by the operation of other firms, a pecuniary externality is said to exist.

_____ 10. A good's price elasticity of supply is the percentage change in quantity supplied divided by the percentage change in price, all else being equal.

PROBLEMS

1. Consider the following data for a firm. Calculate, for each potential output volume, the total cost, average fixed cost, average variable cost, average total cost, and marginal cost, filling in columns (4)–(8).

Total Product (tons of tobacco per year) (1)	Fixed Cost (dollars per year) (2)	Variable Cost (dollars per year) (3)	Total Cost (4)	Average Fixed Cost (5)	Average Variable Cost (6)	Average Total Cost (7)	Marginal Cost (8)
0	10,000	0					
500	10,000	10,000					
1,000	10,000	20,000					
1,300	10,000	30,000					
1,400	10,000	40,000					

2. Plot the data of column (1) against those of columns (2)–(8) in the appropriate parts of the following graph. (Plot marginal-cost data at midpoints of applicable output ranges.) Connect the data plots with straight lines and label these lines.

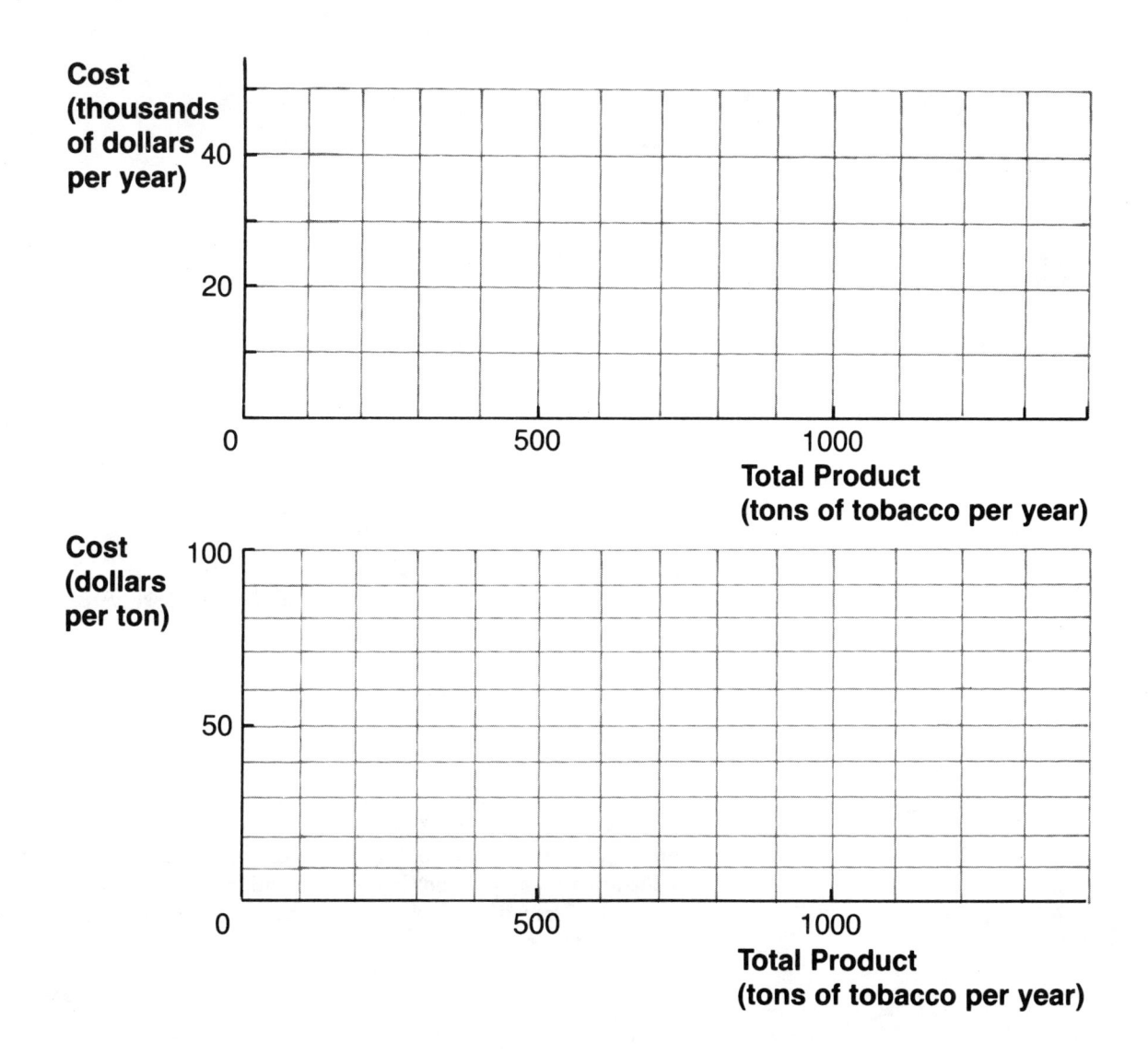

Cost (thousands of dollars per year) vs **Total Product (tons of tobacco per year)**

Cost (dollars per ton) vs **Total Product (tons of tobacco per year)**

3. Now assume that tobacco is sold in a perfectly competitive market at $15 per ton. Enter this information in each of the two graphs above in such a way that the firm's most profitable output volume becomes immediately obvious.

4. Now identify, in each of the two above graphs, the profit-maximizing firm's total revenue, total cost, variable cost, fixed cost, and profit (or loss).

5. Identify the firm's break-even point, shutdown point, and short-run supply.

Capital (units per year)

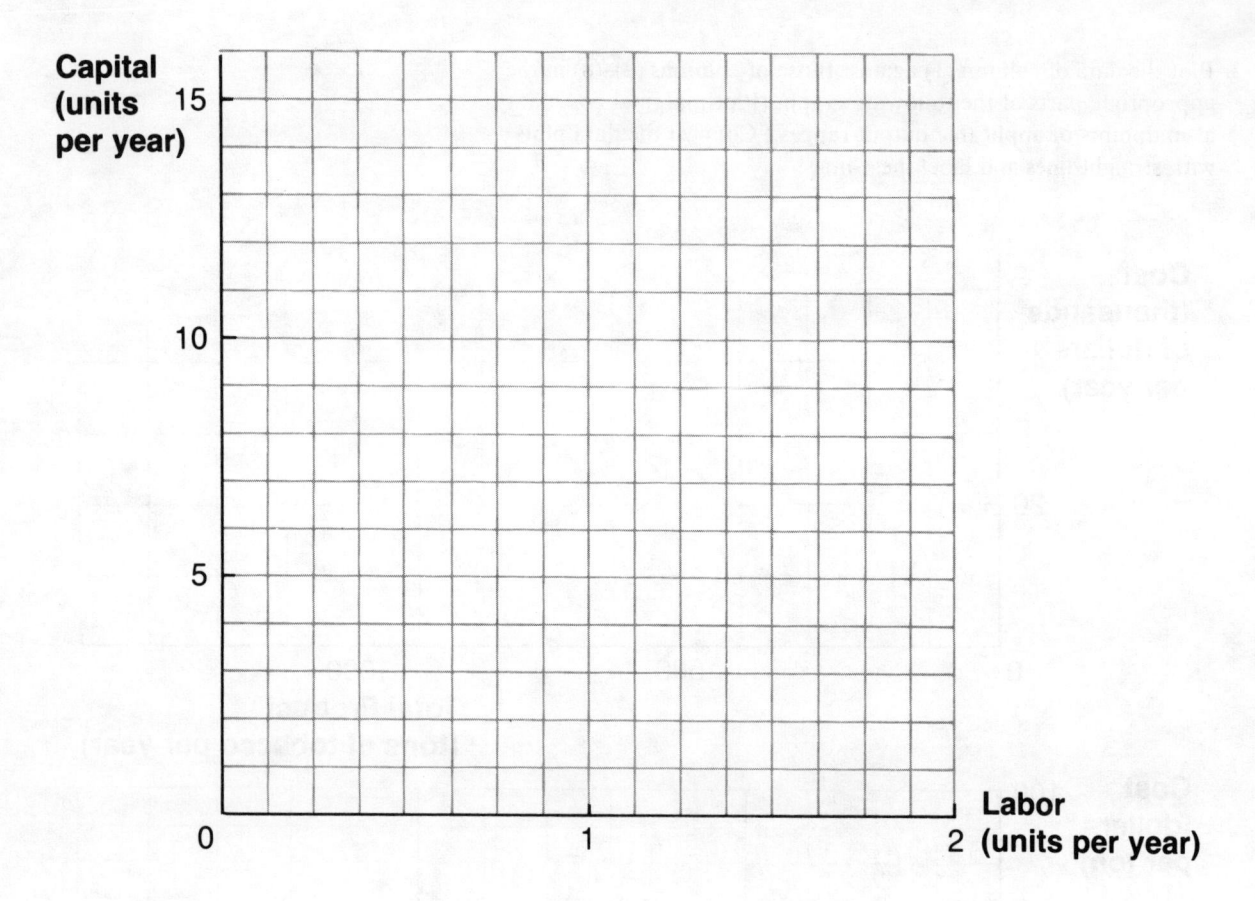

[Blank graph with Capital on y-axis (0 to 15) and Labor on x-axis (0 to 2, units per year)]

6. In the above graph, draw isocost lines for $1,000, $2,000 and $3,000 per month on the assumption that using a unit of capital costs $200 per month and using a unit of labor costs $2,000 per month. Then draw in a set of three isoquants in such a way that we can identify the capital-labor input combination that a profit-maximizing firm would use when incurring any one of the total costs given above.

7. Consider the following graph of a firm's long-run average-total-cost curve. Calculate and plot the long-run marginal cost associated with total product ranges of 100–200, 300–400, 600–700.

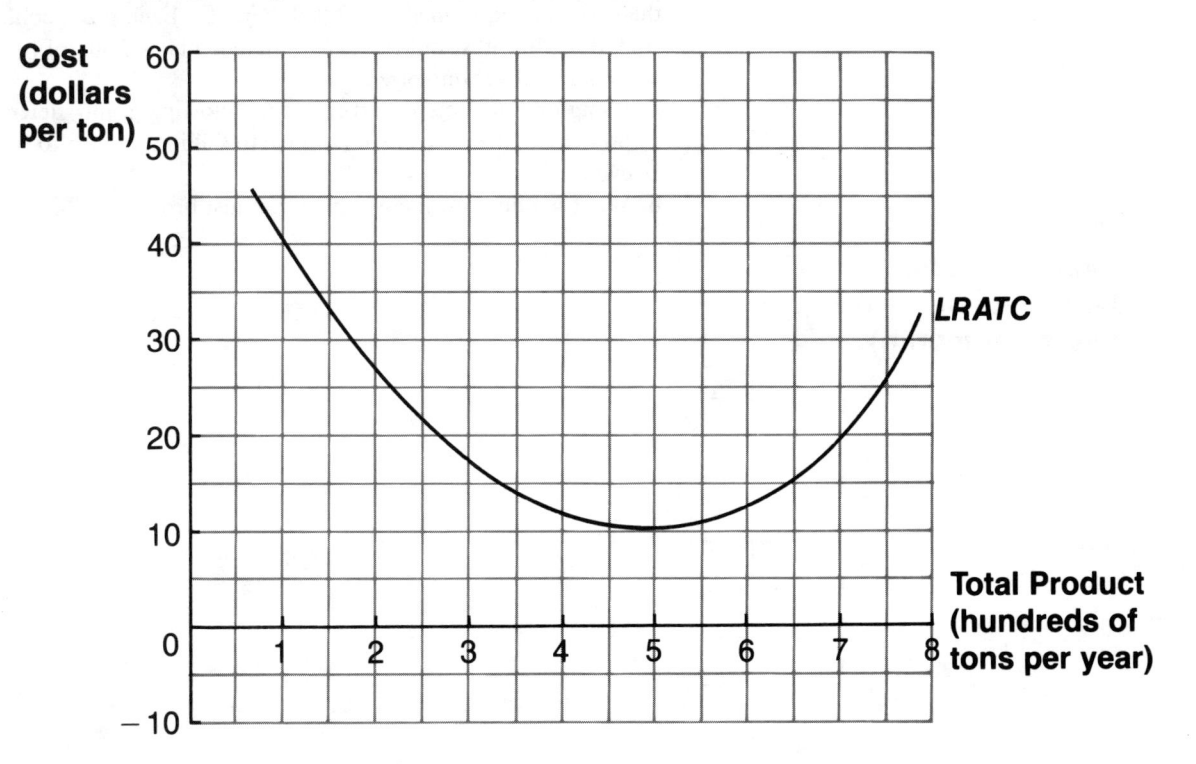

8. Consider the supply lines in the following graph and determine the price elasticity of supply at points a, b, c, d, and e.

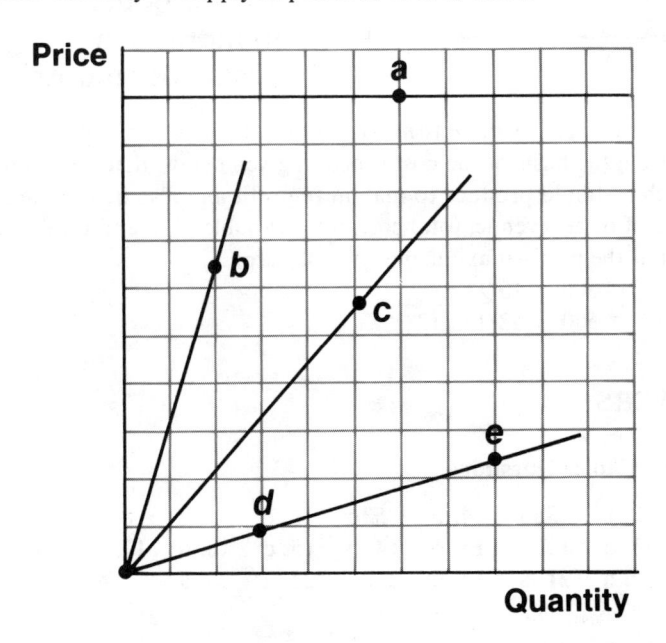

9. Suppose that the best job you could get working for someone else would bring you $700 a month. Suppose also that you were thinking of opening a pizza parlor. You could rent a building with all the necessary equipment at $2,000 a month. Each pizza could be sold at $5, but it would take $2 of ingredients and 30¢ of fuel to make and deliver one pizza.
   a. Using first arithmetic and then the following graph, determine how many pizzas you would have to sell in a month to break even.
   b. What would your income be if you just broke even?

Revenue or Cost (thousands of dollars per month) vs **Total product (hundreds of pizzas per month)**

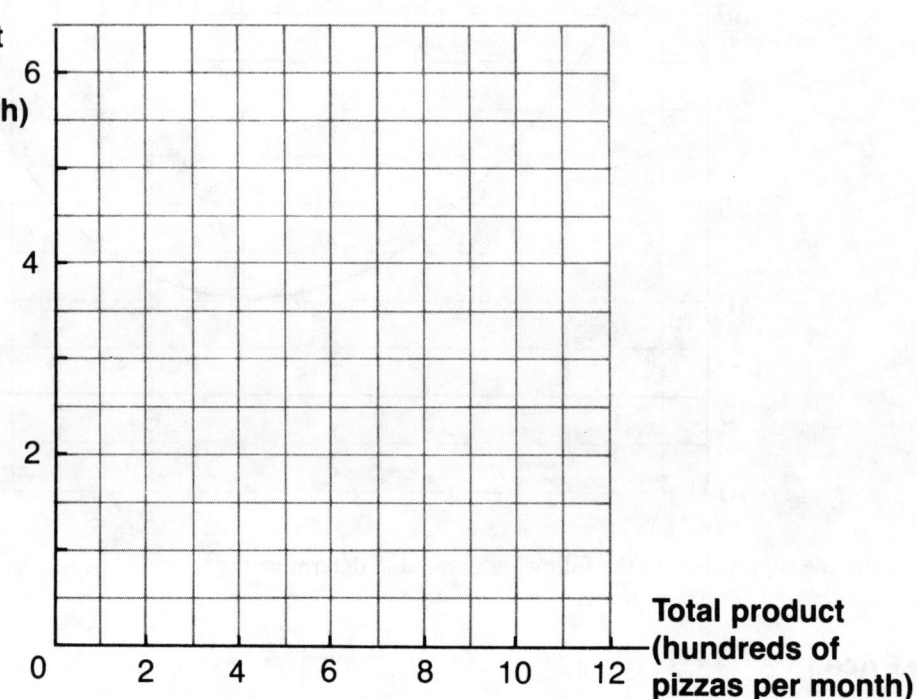

*10. A perfectly competitive firm faces a market price for its product of 100. On the basis of the cost function given below, determine how much it should produce to maximize profit and also determine the sizes of total revenue, total cost, total variable cost, and total fixed cost at the profit-maximizing level of output.
   a. $TC = 300 + 10Q^2$
   b. $TC = 500 - 52Q + 12Q^2$

ANSWERS

Multiple-Choice Questions

1. a	2. b	3. a	4. b	5. c	6. d	7. a	8. b	9. b
10. d	11. c	12. c	13. c	14. b	15. c	16. d	17. c	18. d
19. c	20. a	21. a	22. d	23. d	24. a	25. a		

True-False Questions

1. F (These costs equal the *maximum* income available elsewhere.)
2. T
3. T
4. F (Total revenue just equals *total cost*.)
5. T

6. F (It corresponds to that level which minimizes the *average total cost* of production of any *given* plant.)
7. F (It's the plant, among all conceivable ones, *with the lowest possible minimum average total cost*.)
8. T
9. F (This describes a *technological* externality.)
10. T

Problems

1.

Total Cost (dollars per year) (4)	Average Fixed Cost (5)	Average Variable Cost (6)	Average Total Cost (7)	Marginal Cost (8)
10,000	?	?	?	
				20
20,000	20	20	40	
				20
30,000	10	20	30	
				33.33
40,000	7.69	23.08	30.77	
				100
50,000	7.14	28.57	35.71	

2. Note: In the following graphs, because so few data plots are available, all points on the straight lines for output volumes other than those given in the table will only be rough approximations. This explains the curious facts that MC can lie above AVC for some quantities without AVC rising and that MC does not precisely go through the ATC minimum—both logical impossibilities. More detailed data would eliminate such effects.

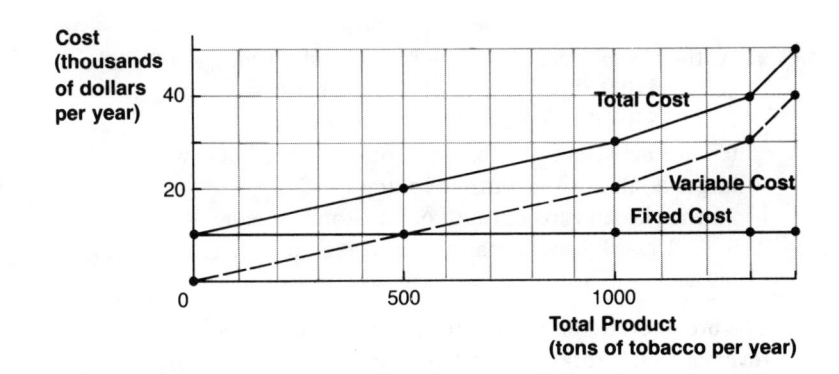

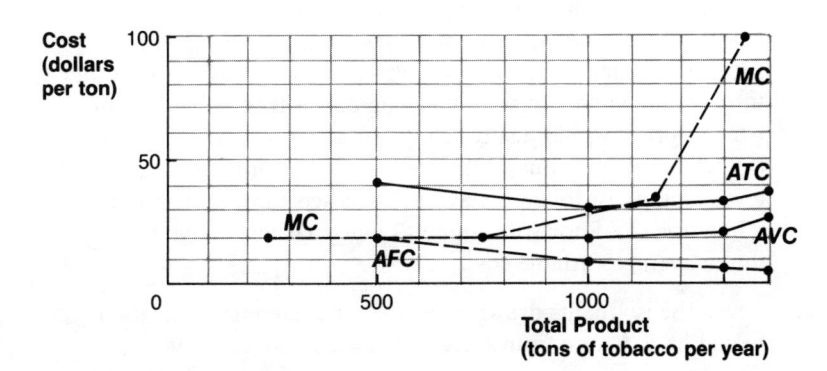

3. **a.** Zero product brings total revenue of 0; 1,000 tons of product, total revenue of $15,000 per year (point *A*); etc. We note that the firm is unable even to cover variable cost at any output volume. The firm, therefore, shuts down at once and produces nothing.

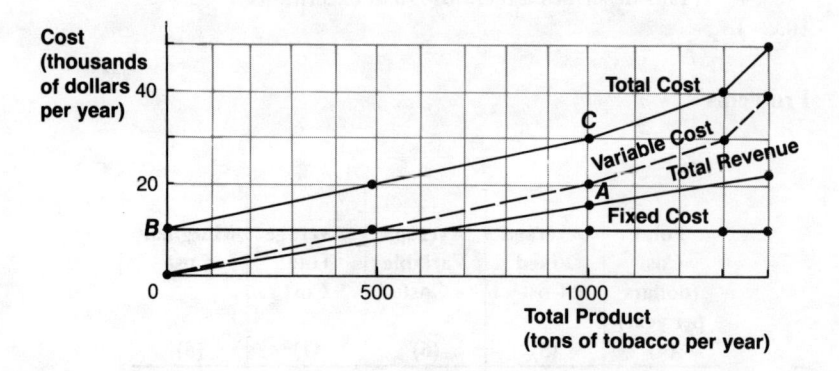

b. The $15 price lies below marginal cost at all output levels. Thus the same conclusion emerges as above.

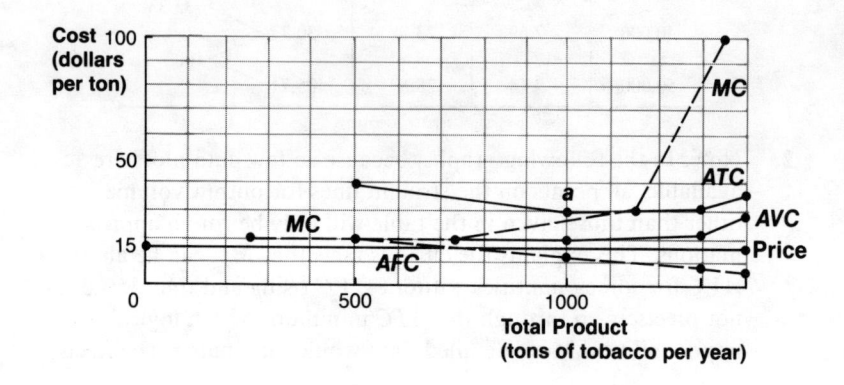

4. **a.** At the chosen zero output volume, total revenue = 0, total cost = $10,000 per year (0*B*), variable cost = 0, fixed cost = $10,000 per year (0*B*). Note how the loss (vertical difference between total cost and total revenue lines) would be larger at all positive output volumes.

b. At the chosen zero output volume, total revenue = $0 \times \$15 = 0$; other data cannot be read off the graph as it is drawn.

5. The break-even point refers to the output level at which total revenue equals total cost and at which price equals average total cost. As the graphs are drawn (all points on the straight lines between the fat dots merely being guesses), this seems to be a $30-per-ton and 1,000-tons-per-year combination. Note that the total-revenue line, if it were to swing upwards to reflect a $30-per-ton price, would just touch the total-cost line at *C* in graph (a); while the price line, if it were to shift upwards to the $30 level, would just touch the average-total-cost line at its minimum *a* in graph (b). Short-run supply is marginal cost above minimum average variable cost.

6. Given the isoquants drawn here, the input combinations used would be *a, b,* or *c,* respectively. These combinations would

assure the highest possible output for any given cost, hence the highest revenue (given product price) and the highest profit.

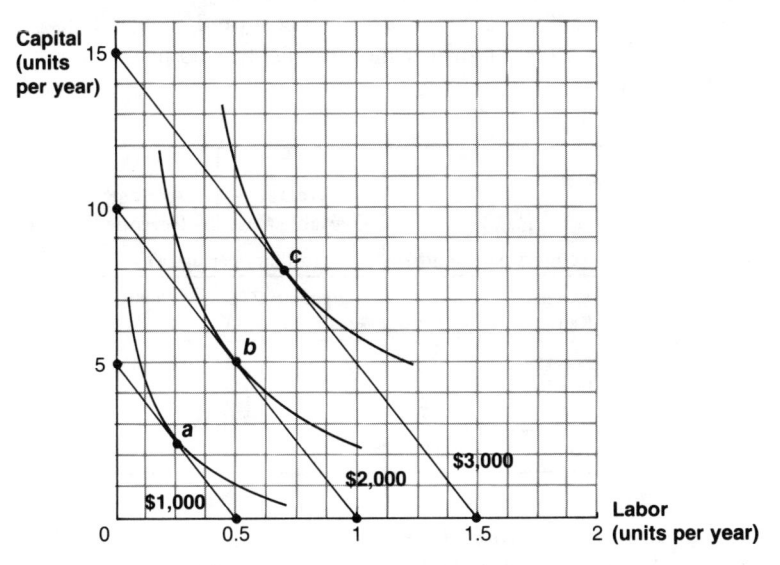

7. **a.** Long-run average total cost per ton changes from $40 (point *a*) to $27 (point *b*) in the 100- to 200-ton annual output range. Thus total cost changes from $4,000 to $5,400 per year, or by +$1,400 per 100 units of extra output. Thus marginal cost is, on the average, $14 per ton in this range (point *A*). Note: Marginal below average pulls average down.

b. Long-run average total cost per ton changes from $17 (point *c*) to $11 (point *d*) in the 300- to 400-ton annual output range. Thus total cost change from $5,100 to $4,400 per year, or by −$700 per 100 units of extra output. Thus marginal cost is on the average −$7 per ton in this range (point *B*). Note: we are here not concerned with the realism of negative marginal costs, but with the logical implications of the *LRATC* curve as drawn.

c. Long-run average total cost per ton changes from $12 (point *e*) to $18.50 (point *f*) in the 600- to 700-ton annual output range. Thus total cost changes from $7,200 to $12,950 per year, or by +$5,750 per 100 units of extra output. Thus marginal cost is on the average $57.50 per ton in this range (point *C*). Notes: Marginal cost must, of course, equal average cost at the latter's minimum. We also see that marginal above average pulls average up.

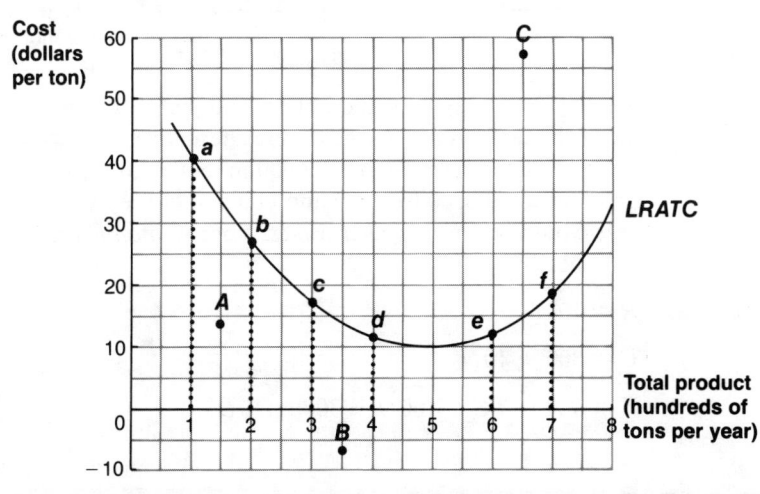

8. At *a,* it is infinite; at all other designated points, it is unity.

9. **a.** *Arithmetic Solution:* Each pizza sold yields $5 − $2.30 = $2.70 toward fixed cost (and profit). Thus $\frac{\$2,700}{\$2.70} = 1,000$ pizzas per month must be sold just to break even.

Fixed Cost (dollars per month)		Average Variable Cost (dollars per pizza)	Average Revenue
Explicit Rent	2,000	Ingredients 2.00 Fuel .30	5.00
		2.30	
Implicit Forgone wages	700		
Total	2,700		

9. **a.** *Graphical solution:* Fixed Cost can be graphed directly from the above table. Variable cost is zero at zero output, $2,300 for 1,000 pizzas, etc. Thus total cost is $2,700 at zero output (point *a*) and $2,700 + $2,300 = $5,000 for 1,000 pizzas per month (point *b*). Connecting *a* and *b* yields the total-cost line. The total-revenue line reflects the $5-per-pizza price: Zero revenues go with zero pizzas (point 0); $5,000 of revenue goes with 1,000 pizzas (point *b*); etc. Thus the break-even point is seen to involve the sale of 1,000 pizzas per month.

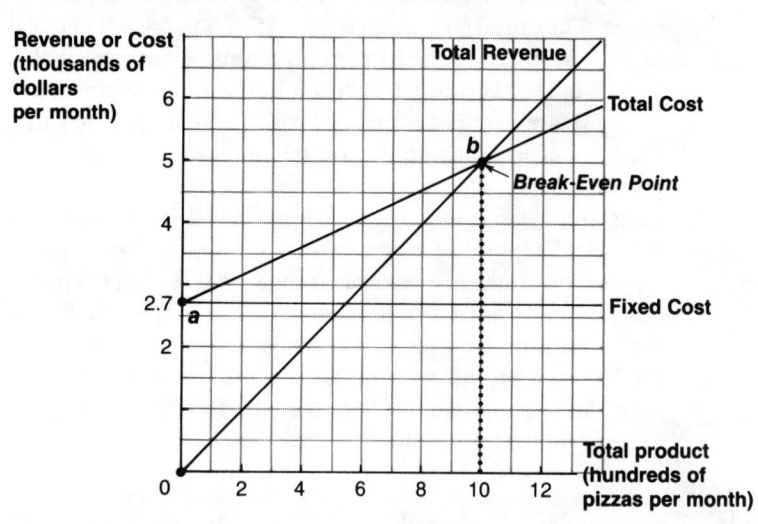

b. Caution: Your income would not be zero, but $700 a month, exactly the same as in your next-best alternative. You would take in $5,000 in revenue, pay $2,000 in rent, $2,000 for ingredients, and $300 for fuel, and you would keep $700. These would be your *implicit wages.* Your *economic profit* would be zero.

10. **a.** $TR = 100Q$

$TC = 300 + 10Q^2$

$\Pi = TR - TC = 100Q - 300 - 10Q^2$

For maximum profit, $\dfrac{d\Pi}{dQ} = 100 - 20Q = 0$

$$Q = 5$$

Maximum confirmed by $\dfrac{d^2\Pi}{dQ^2} = -20.$

Profit-maximizing $Q = 5$. Then $TR = 500$, $TC = 550$, $TVC = 250$, $TFC = 300$. Note that the profit of -50 is preferable, in the short run, to the profit of -300 when shutting down.

b. $TR = 100Q$

$TC = 500 - 52Q + 12Q^2$

$\Pi = TR - TC = 100Q - 500 + 52Q - 12Q^2 = -500 + 152Q - 12Q^2$

For maximum profit, $\dfrac{d\Pi}{dQ} = 152 - 24Q = 0$

$$Q = 6.\overline{33}$$

Maximum confirmed by $\dfrac{d^2\Pi}{dQ} = -24.$

Profit-maximizing $Q = 6.\overline{33}$. Then $TR = 633.\overline{33}$, $TC = 652$, $TVC = 152$, $TFC = 500$. Thus total profit is $-18.\overline{66}$, which is preferable, in the short run, to the profit of -500 when shutting down.

BIOGRAPHY 5.1

Jacob Viner

Jacob Viner (1892–1970) was born in Montreal, Canada and studied at McGill and Harvard. For many decades, he taught economics, first at the University of Chicago and later at Princeton. He was also a frequent consultant to the U.S. government. His presidency of the American Economic Association in 1939 and the receipt of the Association's Walker Medal were only two of a large number of honors bestowed upon him (among them honorary degrees from thirteen institutions of higher learning).

Viner's major interest was international trade. Consider this partial listing of his works: *Dumping: A Problem in International Trade* (1923), *Canada's Balance of International Indebtedness, 1900–1913* (1924), *Studies in the Theory of International Trade* (1937), *The Customs Union Issue* (1950), *International Economics* (1951), *International Trade and Economic Development* (1952). Viner also wrote on other subjects: *The Long View and the Short* (1958), *The Role of Providence in the Social Order* (1972).

Among microeconomists, Viner is best known for his brilliant article on "Cost Curves and Supply Curves," which appeared in the *Zeitschrift für Nationalökonomie* (September 1931, pp. 23–46). In this article, Viner introduced much of the material contained in this chapter, but he also made a single, though famous, mistake (which should not detract from his achievement). He instructed his draftsman to draw a smooth curve of long-run average total cost, as in panel (b) of text Figure 5.9.

The curve was to pass through all the minimum points of short-run average total cost (such as *f, g, h, i,* and *k*) without ever rising above a short-run curve. His draftsman objected that this couldn't be done. Viner insisted; the result was a rather impossible graph.

Viner had confused the minimum (short-run) average total cost achievable in a given plant (such as *g* for plant #5 or *i* for plant #31) with the minimum (long-run) average total cost achievable for a given rate of production (such as *t* for 150,000 and *u* for 350,000 bushels of apples per year). The latter may well require the underutilization or overutilization of some plant. Two decades later, when his justly famous article was readied for reprinting, Viner declined the opportunity to revise it.[1]

> I do not take advantage of the opportunity [to revise]. . . . The error of Chart IV is left uncorrected so that future teachers and students may share the pleasure of many of their predecessors of pointing out that if I had known what an "envelope" was I would not have given my excellent draftsman the technically impossible and economically inappropriate assignment. . . .

APPENDIX 5A LINEAR PROGRAMMING

The conventional theory of the firm, as presented in Chapters 4 and 5, is rather general. It covers the short run and the long run. It considers relationships between output and other variables (such as inputs, revenue, cost, or profit) that are linear or curvilinear. It extends over any range of output. Many of the day-to-day problems of business managers, however, are concerned with the short run only and with linear relations over fairly narrow ranges of output. The mathematical technique of **linear programming** is ideally suited for solving managerial problems under such circumstances. The technique was developed since World War II: in the Soviet Union by Leonid V. Kantorovich (a Biography of whom appears later in this Appendix) and in the United States by George B. Dantzig and Tjalling C. Koopmans. The technique involves the maximization (or minimization) of a linear function of variables, subject to constraints that limit what can be done, as will be explained below.

While the term *programming* simply refers to a systematic type of decision making, the adjective *linear* reminds us that all relevant relationships are those that can be represented by straight lines (as, for example, in Analytical Example 5.1, "Break-Even Analysis"). The fact that relationships are linear implies constant returns to scale and input and output prices that do not vary with a firm's output level. All of this is likely to be true over limited output ranges, even for firms that do not operate in perfectly competitive markets. Let us consider a number of typical applications of the technique.

A Minimization Problem

One frequent problem faced by management is finding that input mix among a variety of possible input mixes which *minimizes* the total cost

---

[1] "Supplementary Note (1950)" in American Economic Association; *Readings in Price Theory* (Chicago: Irwin, 1952), p. 227. *See also* Machlup, Fritz; Samuelson, Paul; and Baumol, William J. "In Memoriam: Jacob Viner (1892–1970)." *Journal of Political Economy,* January/February 1972, pp. i–15.

of producing a specified output.[2] Consider the case of a firm that wishes to produce 3 tons of paint per day. Assume all relevant relationships are linear: Up to the desired output quantity, the prices of the firm's inputs (labor and capital) are constant at $600 per unit of labor and $1,800 per unit of capital, and constant returns to scale prevail in the firm's production function. The four different production processes that are available are shown in columns (A) to (D) of Table 5A.1.

TABLE 5A.1

Alternative Ways to Produce a Product

Type of Input	Units of Input Needed for the Production of 1 Ton of Paint per Day, Using Process			
	(A)	(B)	(C)	(D)
Labor	20	10	6	3
Capital	1	2	3	5

The information in Table 5A.1 can be graphed, as in Figure 5A.1. If paint is produced by process (A), labor, L, must be combined with capital, K, in a fixed ratio of 20 to 1, as shown in the graph by the ray labeled "Production Process A." Point A_1 indicates that $20\,L$ plus $1\,K$ can produce 1 ton of paint by process (A). Point A_2 indicates that doubling the inputs doubles output, and A_3 tells us that tripling input triples output. Output is similarly measured along the other production rays. Using process (B), which combines labor and capital in a fixed ratio of 10 to 2, the input combinations needed to produce one, two, and three tons of paint are shown by points B_1, B_2, and B_3, respectively. Again, constant returns to scale prevail: Distances $0B_1$, B_1B_2, and B_2B_3 are equal to each other, just as $0A_1$ equals A_1A_2 and A_2A_3, in turn.

FIGURE 5A.1
Production-Process Rays and Isoquants

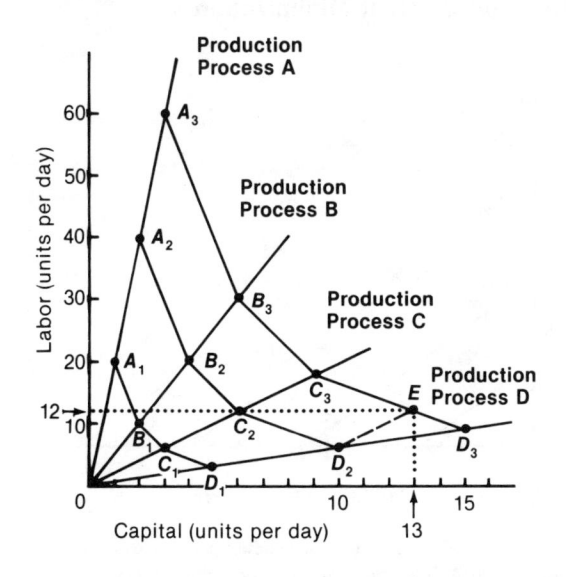

The four straight rays from the origin represent the limited number of production processes available to a firm. Under constant returns to scale, equal distances (such as $0A_1$, A_1A_2, and A_2A_3) along any one ray represent equal output quantities. Isoquants can be constructed by connecting input combinations producing the same output (such as A_1, B_1, C_1, and D_1).

---

[2] See Section 5D of the text's Calculus Appendix for a mathematical presentation.

Note: While equal distances along any *one* ray represent equal output quantities, equal distances along *different* rays do not because A_1, B_1, C_1, and D_1 each represent input combinations capable of producing 1 ton of paint per day, and distances of $0A_1$, $0B_1$, $0C_1$, and $0D_1$ are clearly different.

We can, however, join points of equal output on the four production rays to give us *isoquants* for various output levels. Line $A_2B_2C_2D_2$, for instance, shows all the combinations of inputs capable of producing at most 2 tons of paint per day. Any point on a segment between two production process rays, such as E, indicates an output quantity that can be produced by a combination of the two processes represented by the two rays the segment connects. Note how point E indicates the production of 3 tons of paint by use of $12L$ and $13K$, a combination that does not appear in any of the four processes in Table 6A.1. Yet, the feat can be accomplished by producing 1 ton by means of process C (using $6L$ plus $3K$) and 2 more tons by means of process D (using $6L$ and $10K$). This can be determined graphically by constructing a line from E parallel to ray C (dashed line) or parallel to ray D (not shown). The dashed line happens to intersect ray D at D_2, indicating that the input combination corresponding to E can produce 3 tons if 2 of them are produced by process D (a line through E parallel to ray D would intersect ray C at C_1, indicating E would also require 1 ton produced by process C).

To solve this minimization problem, we need information not only about possible input combinations but also about *input prices*. The slope of each *isocost line* (Figure 5A.2) reflects the assumed input prices of $600 per unit of labor and $1,800 per unit of capital. When the 3-ton isoquant of Figure 5A.1 (A_3 B_3 C_3 D_3) is superimposed upon the isocost lines, the solution to our problem emerges at once: 3 tons of paint per day are produced most cheaply by using production process C. Any other input combination capable of producing 3 tons per day (such as F) would cost more.

FIGURE 5A.2 Cost Minimization

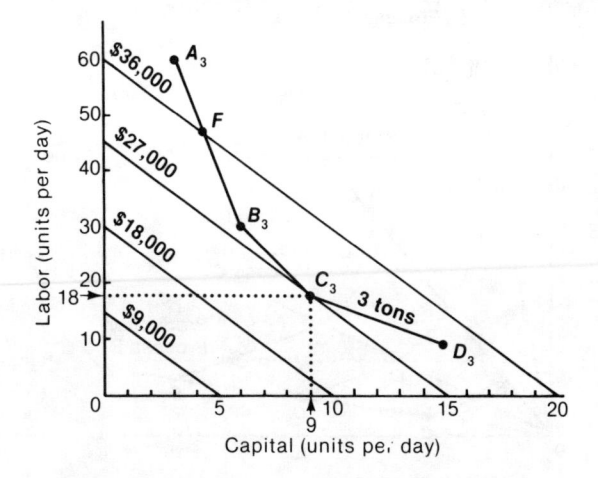

The cost of producing 3 tons of product per day can be minimized (at $27,000 per day), if production process C is utilized. The optimum combination of inputs at C_3 uses 18 units of labor and 9 units of capital.

Note: The solution to every linear programming problem can always be read at a "corner," such as C_3. (What if the isocost line had had a slightly different slope and coincided with an isoquant *segment*, such as $B_3 C_3$? Then corners B_3 and C_3 would have been equally good at giving us the answer, and all points in-between would have been equivalent.)

While the above example is an extremely simple one (that could have been solved without the graphs), many real-world problems are considerably more complex (and cannot be solved with the help of graphs at all). A more complex example follows.

A Maximization Problem

Another typical problem faced by management is finding that product mix which maximizes the value of output produced with the available inputs.[3] Table 5A.2 clearly points to alternative ways to use given inputs. The firm in question could produce either refrigerators or washers (but each with a single production process only.)

TABLE 5A.2

Alternative Ways to Use Given Inputs

Type of Input	Units of Inputs Needed for the Production of		Quantity of Inputs Available (units per day)
	1 Refrigerator	1 Washer	
	(1)	(2)	(3)
Labor #1 (hours)	10	0	350
Labor #2 (hours)	0	5	125
Raw materials (pounds)	50	100	3,000
Machine time (hours)	5	4	200

The technology embodied in columns (1) and (2) of the table and the fixed quantities of inputs available to the firm in column (3) restrict the production possibilities to the product combinations in the unshaded area of Figure 5A.3. Because there are only 350 units of type #1 labor in a day and 10 units of such labor are needed to make one refrigerator, a maximum of 35 refrigerators can be produced in a day. The vertical line limits refrigerator production but not the production of washers, which does not require this type of labor at all. The availability of only 125 units of labor #2, similarly, only limits the production of washers (to a maximum of 25 per day); hence the horizontal line. Because raw materials and machine time must be used for either product, these inputs limit the production of both. Some 3,000 units of raw materials can make 60 refrigerators *or* 30 washers; some 200 hours of machine time can make 40 refrigerators *or* 50 washers; hence the sloped lines. Note how the combination of these constraint lines produces a production possibilities frontier for our firm. Given its present technology, the firm's inputs are insufficient to produce any of the product mixes lying in the shaded area of the graph.

---

[3]See Section 5C of the text's Calculus Appendix for a mathematical treatment concerning the maximization of *physical* output. Given output price, this is equivalent to maximizing the *value* of output.

FIGURE 5A.3 Input Constraints and Production Possibilities

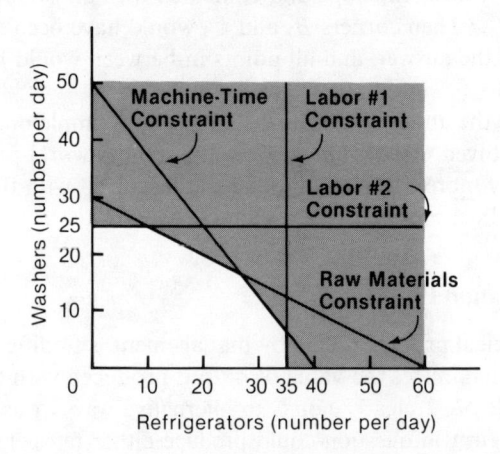

Given technology and fixed input quantities, a firm's production possibilities (indicated by the unshaded area) can be determined.

As in the minimization example, a graphical solution can be found easily. Figure 5A.4 shows a family of **isorevenue lines** that are akin to isocost lines. Instead of showing input combinations that cost the same amount, each of these lines shows all the alternative combinations of two outputs that the firm is able to sell in a given period at current market prices, *while receiving the same total revenue.* The lines in Figure 5A.4 have been drawn on the assumption that the firm can sell each washer at $280 and each refrigerator at $200. When the production-possibilities frontier of Figure 5A.3 is superimposed upon the isorevenue lines, the solution to our problem is immediately apparent: Revenue is maximized when $16\frac{2}{3}$ washers and $26\frac{2}{3}$ refrigerators are produced per day. Any other feasible output combination (such as *P, Q* or *N*) would bring in less revenue. As in the minimization problem, the solution emerges at a "corner" (point *M*).

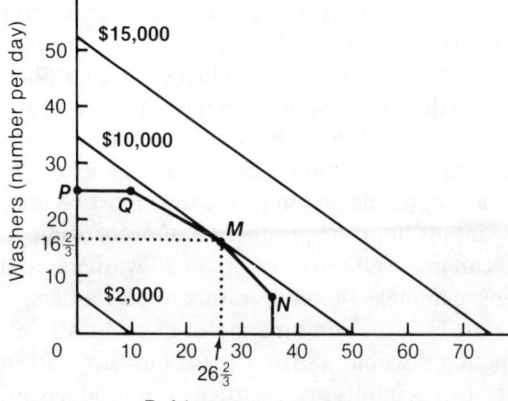

The revenue from using the given inputs underlying the production possibilities frontier shown here can be maximized (at $10,000 per day) by producing output combination *M*. This optimum combines the production of $16\frac{2}{3}$ washers and $26\frac{2}{3}$ refrigerators per day.

Naturally, in more complicated problems (involving, perhaps, hundreds of products, each one producible by a variety of processes), graphical solutions are out of the question. George B. Dantzig was the first to develop a mathematical routine for solving such complicated problems called the **simplex method.** The simplex method does not get its name, however, because it is simple, but because *simplex* refers to the *n*-dimensional analogue of a triangle, and a comupter can be programmed to search the "corners" of such triangles for the optimum solution to linear programming problems.

The Simplex Method and Shadow Prices

Our product-mix problem is simple enough to lend itself to a demonstration of the algebraic solution that would normally be performed by a computer. For this purpose, it is useful to expand the data of Table 5A.2 in Table 5A.3, "Activity Analysis." Our firm is engaged in six "activities" defined in such a way that as a group they must utilize completely the inputs available. These activities include the production of refrigerators, the production of washers, and the "production" of unemployment of each one of the four inputs. Columns (1)–(6) show how much of each type of input must be utilized to perform each of the six activities at the level of 1 unit: The production of 1 refrigerator (*R*), column (1) tells us, uses up 10 hours of labor #1, zero hours of labor #2, 50 pounds of raw materials, and 5 hours of machine time. Similarly, the production of 1 washer (*W*) uses up zero hours of labor #1, 5 hours of labor #2, 100 pounds of raw materials, and 4 hours of machine time. The "production" of 1 unit of unemployment of labor #1 (U_{L1}) uses up 1 hour of labor #1 and none of the other inputs. Finally, the "production" of 1 unit of unemployment of labor #2 (U_{L2}), of raw materials (U_{RM}), or of machine time (U_M) each uses only one unit of the respective inputs. Altogether, of course, only the input quantities shown in column (7), listing the constraints (*C*), can be used.

TABLE 5A.3

Activity Analysis

Type of Input	1 Refrigerator (1)	1 Washer (2)	Labor #1 (3)	Labor #2 (4)	Raw Materials (5)	Machine Time (6)	Constraints (Units of Inputs available per day) (7)
(A) Labor #1 (hours)	10	0	1	0	0	0	350
(B) Labor #2 (hours)	0	5	0	1	0	0	125
(C) Raw materials (pounds)	50	100	0	0	1	0	3,000
(D) Machine time (hours)	5	4	0	0	0	1	200

Columns (1)–(6) header: Units of Inputs Needed for the Production of; Columns (3)–(6) sub-header: 1 Unit of Unemployment of

The information contained in each of the table columns can be written as a column vector, such as

$$R = \begin{bmatrix} 10 \\ 0 \\ 50 \\ 5 \end{bmatrix} \text{ or } U_M = \begin{bmatrix} 0 \\ 0 \\ 0 \\ 1 \end{bmatrix} \text{ or } C = \begin{bmatrix} 350 \\ 125 \\ 3{,}000 \\ 200 \end{bmatrix}$$

Correspondingly, the performance of any one of these activities at a higher rate raises the input utilization in proportion:

$$3R = 3 \begin{bmatrix} 10 \\ 0 \\ 50 \\ 5 \end{bmatrix} = \begin{bmatrix} 3 \times 10 \\ 3 \times 0 \\ 3 \times 50 \\ 3 \times 5 \end{bmatrix} = \begin{bmatrix} 30 \\ 0 \\ 150 \\ 15 \end{bmatrix}$$

This indicates that the production of 3 refrigerators uses 30 hours of labor #1, zero hours of labor #2, 150 pounds of raw materials, and 15 hours of machine time.

To determine the optimal levels of each of the six activities, we need, first, the firm's **objective function,** a statement of the goal (or objective) that is to be achieved. Since we assumed that the firm will seek a maximum value (V) of output produced from its given resources in column (7), and since output prices were assumed to be $200 per refrigerator and $280 per washer, we can state the firm's objective as the *maximization* of

$$V = 200a + 280b, \tag{5A.1}$$

where a and b, respectively, denote the numbers of refrigerators and washers produced.

We can, second, designate the firm's production possibilities by a single equation, in which the letters a through f refer to the unknown levels of the six possible activities:

$$aR + bW + cU_{L1} + dU_{L2} + eU_{RM} + fU_M = C. \tag{5A.2}$$

According to this equation, producing a units of refrigerators plus b units of washers, while keeping c, d, e, and f units, respectively, of labor #1, labor #2, raw materials, and machine time unemployed, must exactly meet the constraint of available inputs. The capital letters are, of course, shorthand ways of writing down columns (1)–(7) of Table 5A.3. Equation (5A.2), therefore, should be envisioned to imply the following:

$$a \begin{bmatrix} 10 \\ 0 \\ 50 \\ 5 \end{bmatrix} + b \begin{bmatrix} 0 \\ 5 \\ 100 \\ 4 \end{bmatrix} + c \begin{bmatrix} 1 \\ 0 \\ 0 \\ 0 \end{bmatrix} + d \begin{bmatrix} 0 \\ 1 \\ 0 \\ 0 \end{bmatrix}$$

$$+ e \begin{bmatrix} 0 \\ 0 \\ 1 \\ 0 \end{bmatrix} + f \begin{bmatrix} 0 \\ 0 \\ 0 \\ 1 \end{bmatrix} = \begin{bmatrix} 350 \\ 125 \\ 3{,}000 \\ 200 \end{bmatrix}$$

This can be expanded into four equations, one for each of the table rows:

$$10a + 0b + 1c + 0d + 0e + 0f = 350 \quad (5A.3)$$

$$0a + 5b + 0c + 1d + 0e + 0f = 125 \quad (5A.4)$$

$$50a + 100b + 0c + 0d + 1e + 0f = 3{,}000 \quad (5A.5)$$

$$5a + 4b + 0c + 0d + 0e + 1f = 200 \quad (5A.6)$$

These equations are easy to interpret. For example, equation (5A.3) refers to hours of labor #1: 10 hours times the unknown quantity a of refrigerators, plus 1 hour times the unknown quantity c of unemployed labor #1, must equal the 350 hours of labor #1 available. Equations (5A.4) through (5A.6) make similar statements about the other 3 types of input.

Our firm's optimum production program can be found by solving equations (5A.3)–(5A.6), subject to the goal stated in equation (5A.1). But note that the *four* equations (5A.3)–(5A.6) contain *six* unknowns; hence a large number of answers are possible. Some of these answers (namely all those providing negative values for a to f) can be ruled out as economic nonsense: One cannot produce negative refrigerators or negative washers, and the "production" of negative unemployemnt for inputs would, in effect, amount to using greater input quantities than are available. The *simplex method* provides a systematic procedure for finding the best solution among all the remaining combinations of zero or positive unknowns.

Finding the Basic Feasible Solution. The simplex method begins by establishing a "basic feasible solution"; that is, *any* solution of n equations and n unknowns that does not violate the constraints. If, for example, we arbitrarily set a and b equal to zero, we are left with four equations and four unknowns, summarized by

Basic feasible solution.
$$cU_{L1} + dU_{L2} + eU_{RM} + fU_M = C \quad (5A.7)$$

In this case, as an inspection of equations (5A.3)–5A.6) immediately reveals, $c = 350$, $d = 125$, $e = 3{,}000$, and $f = 200$. The firm "produces" nothing but unemployment; no refrigerators or washers are produced; the value of output equals zero.

Improving Upon the Basic Solution. Step 2 of the simplex method (see Table 5A.4) tests for possible changes in this basic solution. The simplex method, in effect, applies the optimization principle. *Any activity not included in a given solution,* the simplex method suggests, *should be added, and other activities requiring equivalent resources should be deleted, if the net effect is to improve upon the achievement of the goal.* That goal, in our case, is to maximize V.

TABLE 5A.4

Simplex Method: Step 2

Activity That Might be Added to Program (1)	Equivalent Activities in Present Program (2)	Marginal Benefit of Addition (value of 1) (3)	Marginal Cost of Addition (value of 2) (4)	Marginal Net Benefit of Addition (5) = (3) − (4)
R	$10U_{L1} + 50 U_{RM} + 5U_M$	$200	0	+$200
W	$5U_{L2} + 100 U_{RM} + 4U_M$	$280	0	+$280
U_{L1}	U_{L1}	0	0	0
U_{L2}	U_{L2}	0	0	0
U_{RM}	U_{RM}	0	0	0
U_M	U_M	0	0	0

Consider, for example, adding the production of refrigerators to the four activities contained in equation (5A.7). To find the amounts of other activities that are equivalent to the production of 1 refrigerator, we write

$$R = wU_{L1} + xU_{L2} + yU_{RM} + zU_M. \quad (5A.8)$$

This, of course, is shorthand for four equations:

$$\begin{bmatrix} 10 \\ 0 \\ 50 \\ 5 \end{bmatrix} = w \begin{bmatrix} 1 \\ 0 \\ 0 \\ 0 \end{bmatrix} + x \begin{bmatrix} 0 \\ 1 \\ 0 \\ 0 \end{bmatrix} + y \begin{bmatrix} 0 \\ 0 \\ 1 \\ 0 \end{bmatrix} + z \begin{bmatrix} 0 \\ 0 \\ 0 \\ 1 \end{bmatrix}$$

It follows that $w = 10$, $x = 0$, $y = 50$, and $z = 5$. Therefore,

$$R = 10U_{L1} + 0U_{L2} + 50U_{RM} + 5U_M. \quad (5A.9)$$

If the firm wants to add 1 R to its operations, it must reduce labor #1 unemployment by 10 units, raw material unemployment by 50 units and machine-time unemployment by 5 units. As the first line of Table 5A.4 shows, the addition would raise total revenue by $200 (the price of a refrigerator); the deletion would have no effect on revenue; the net benefit of this move would equal $200. The remaining lines show analogous equivalencies for the other five activities.

Because the net benefit from producing washers is largest, the simplex method recommends adding the maximum possible amount of washers (αW) to the firm's original production program in equation (5A.7) and, of course, deducting amounts of present activities that make equivalent demands on inputs, or $\alpha (5U_{L2} + 100U_{RM} + 4U_M)$. The solution then changes from equation (5A.7) to read

$$\alpha W - \alpha(5 U_{L2} + 100 U_{RM} + 4 U_M) + 350 U_{L1} + 125U_{L2} + 3{,}000U_{RM} + 200U_M = C \quad (5A.10)$$

or

$$\alpha W + 350U_{L1} + (125 - 5\alpha) U_{L2} + (3{,}000 - 100\alpha)U_{RM} + (200 - 4\alpha) U_M = C. \quad (5A.11)$$

Since no activity can occur at a negative level, the maximum possible value for α is 25, a value that completely eliminates the unemployment of labor #2. The new program becomes

Second feasible solution.
$$25W + 350U_{L1} + 500U_{RM} + 100U_M = C \qquad (5A.12)$$

By producing 25 washers, and leaving 350 units of labor #1, 500 units of raw materials, and 100 units of machine time unemployed, the firm would exactly "utilize" all available inputs. The value of this program, compared to the original one in equation (5A.7), equals 25 × $280, or $7,000. In Figure 5A.4, this computational step has moved the firm from point 0 to point P.

Looking for Further Improvement. Once more, the simplex method recommends a test of the net benefit to be gained from adding to any of the six types of activities. Table 5A.5 shows step 3 of the simplex method.

TABLE 5A.5

Simplex Method: Step 3

Activity That Might be Added to Program (1)	Equivalent Activities in Present Program (2)	Marginal Benefit of Addition (value of 1) (3)	Marginal Cost of Addition (value of 2) (4)	Marginal Net Benefit of Addition (5) = (3) − (4)
R	$10U_{L1} + 50\ U_{RM} + 5U_M$	$200	0	+$200
W	W	$280	$280	0
U_{L1}	U_{L1}	0	0	0
U_{L2}	$.2W - 20U_{RM} - .8\ U_M$	0	$56	−$56
U_{RM}	U_{RM}	0	0	0
U_M	U_M	0	0	0

Table 5A.5 tells us that we could further improve upon our goal by adding the production of as many refrigerators as possible (βR) to the latest production program in equation (5A.12), while deducting amounts of present activities that make equivalent demands on inputs, or $\beta(10\ U_{L1} + 50\ U_{RM} + 5\ U_M)$. The solution then changes from equation (5A.12) to read

$$\beta R - \beta(10\ U_{L1} + 50\ U_{RM} + 5\ U_M) + 25W + \\ 350\ U_{L1} + 500\ U_{RM} + 100\ U_M = C \qquad (5A.13)$$

or

$$\beta R + 25W + (350 - 10\beta)U_{L1} + (500 - \\ 50\beta)U_{RM} + (100 - 5\beta)U_M = C. \qquad (5A.14)$$

The maximum possible value for β (that preserves nonnegative parameters) is 10, a value that completely eliminates the unemployment of raw materials. The new program becomes

Third feasible solution.
$$10R + 25W + 250U_{L1} + 50U_M = C \qquad (5A.15)$$

Once more, the firm would exactly utilize available inputs; its output would be worth (10 × $200) + (25 × $280), or $9,000. In Figure 5A.4, this computational step has moved the firm from point P to point Q.

Further Improvement Still. The next test of possible further gain is provided by Table 5A.6, which shows step 4 of the method.

TABLE 5A.6

Simplex Method: Step 4

Activity That Might be Added to Program (1)	Equivalent Activities in Present Program (2)	Marginal Benefit of Addition (value of 1) (3)	Marginal Cost of Addition (value of 2) (4)	Marginal Net Benefit of Addition (5) = (3) − (4)
R	R	$200	$200	0
W	W	$280	$280	0
U_{L1}	U_{L1}	0	0	0
U_{L2}	$-.4R + .2W + 4U_{L1} + 1.2U_M$	0	−$24	+$24
U_{RM}	$.02R - .2U_{L1} - .1U_M$	0	$4	− $4
U_M	U_M	0	0	0

Surprisingly, this table tells our firm that it could do even better by keeping more of its labor #2 unemployed. If we add the maximum possible amount of labor #2 unemployment (γU_{L2}) to the last production program (equation 5A.15), while deducting amounts of present activities that make equivalent demands on inputs, or $\gamma(-.4R + .2W + 4U_{L1} + 1.2U_M)$, the solution changes to

$$\gamma U_{L2} - \gamma(-.4R + .2W + 4U_{L1} + 1.2U_M) + 10R + 25W + 250U_{L1} + 50U_M = C \quad (5A.16)$$

or

$$(10 + .4\gamma)R + (25 - .2\gamma)W + (250 - 4\gamma)U_{L1} + \gamma U_{L2} + (50 - 1.2\gamma)U_M = C. \quad (5A.17)$$

The maximum possible value for γ is 41.67, a value that completely eliminates the unemployment of machine time. The new program becomes

Fourth feasible solution.
$$26.67R + 16.67W + 83.33U_{L1} + 41.67U_{L2} = C. \quad (5A.18)$$

The firm's output now is worth (26.67 × $200) + (16.67 × $280), or $10,000. In Figure 5A.4, this computational step has moved the firm from point Q to the optimum at M. The simplex method provides a way of testing this fact.

Confirming the Optimum Solution. Consider Table 5A.7, constructed by the same procedure as the last three tables. Column (5) of this table indicates that no marginal net benefits can be reaped by any further changing of the firm's production program. Just as our graphical analysis did earlier, the algebraic procedure points to the production of 26.67 refrigerators and 16.67 washers as the revenue-maximizing program. Any other conceivable program would produce less revenue. Given the fixed inputs and, therefore, costs, any other program would produce lower profit.

TABLE 5A.7

Simplex Method: Step 5

Activity That Might be Added to Program (1)	Equivalent Activities in Present Program (2)	Marginal Benefit of Addition (value of 1) (3)	Marginal Cost of Addition (value of 2) (4)	Marginal Net Benefit of Addition (5) = (3) − (4)
R	R	$200	$200	0
W	W	$280	$280	0
U_{L1}	U_{L1}	0	0	0
U_{L2}	U_{L2}	0	0	0
U_{RM}	$-.0133R + .0166W + .133U_{L1} - .0833U_{L2}$	0	$2	−$2
U_M	$.33R - .166W$	0	$20	−$20

↑ Shadow prices of inputs

Indeed, equation (5A.18) tells us that our firm would be wise to leave 83.33 unit of labor #1 and 41.67 units of labor #2 totally unused. These are, of course, the amounts left when 26.67 × 10, or 266.67, units of labor #1 (needed for refrigerator production) are subtracted from 350 total units available and when 16.67 × 5, or 83.33, units of labor #2 (needed for washer production) are subtracted from the 125 unit total available. (Can you show that, under the optimum program, the totals of raw materials and machine time are fully utilized and how they are distributed between the production of the two products?)

Shadow Prices. The boxed numbers in column (4) of Table 5A.7 provide crucial information about the value to our firm of additional units of input. The numbers highlighted there are called **shadow prices**; these are the implicit valuations that always emerge as a by-product of solving a linear programming problem algebraically. Note that an extra unit of labor #1, as well as of labor #2, is designated as worthless to the firm (because the firm already has more of each than it can profitably use). On the other hand, the firm's production is now limited by the full use of all available raw materials and machine time, and these bottleneck inputs have positive shadow prices of $2 and $20, respectively. These shadow prices indicate that the firm could raise the value of its output by $2 (or $20) if it could get another unit of raw materials (or machine time). Such shadow prices are extremely important pieces of information for economic decision makers because they indicate the true scarcity of inputs or outputs (which may not be indicated by their actual market prices).

Note: The value of inputs used, calculated at their shadow prices, always equals the value of the output they produce. In our case, the shadow price of both types of labor is zero, but the total value of raw materials comes to 3,000 units × $2, or $6,000, and the total value of machine time comes to 200 units × $20, or $4,000. These values, of course, add to the $10,000 value of output.

The Manifold Uses of Linear Programming

Any reader who has worked through the forgoing example with pencil and paper will appreciate the help computers can provide to linear programming. Indeed, the advent of the computer has made linear programming a highly effective management tool with wide applicability.

Determining the Best Product Mix. Oil companies with limited crude oil supplies and refinery capacity use the technique to determine their product mix among diesel fuel, heating oil, kerosene, lubricants, and gasoline with different octane ratings. Forest product companies employ linear programming to determine the best combination of lumber, plywood, and paper that can be made from given supplies of logs and a fixed milling capacity. Tomato processors use it to determine their output mix among canned whole tomatoes, chili sauce, ketchup, stewed tomatoes, tomato juice, tomato paste, and soup. Police departments, law firms, and the Internal Revenue Service use linear programming to figure out the most efficient allocation of their staffs.

Determining the Best Input Mix. The technique is often directed toward achieving a *given* goal with the best combination of inputs. Producers of cake use linear programming to determine the ideal makeup of their ingredient lists; farmers use it to select the best combinations of cows in their herds in order to produce a given quality of milk. Hospitals want to supply their patients with minimum quantities of required calories, minerals, proteins, and vitamins (contained in different quantities in different foods), and linear programming helps them to do so at minimum cost. Directors of college dining halls have the same goal, as do managers of cattle feedlots! All kinds of firms use the technique to split their advertising budgets among such media as billboards, newspapers, magazines, radio, and TV in order to reach, with minimum cost, a specified number of customers of given age and income.

Determining the Best Transportation System. Innumerable users of the linear programming technique are concerned with the most efficient routing of products over space. In the United States, the technique was first developed to help with the complicated transportation tasks of the U.S. Air Force. Yet the procedure is just as useful for the efficient scheduling of everyday deliveries by private firms that may have many production, warehousing, and sales facilities in different parts of the country and wish to achieve a given set of deliveries at minimum cost.

Other Uses. The technique has found uses far beyond the management problems of individual firms or government departments. Scientists have used linear programming to model the movements of seasonal labor forces in Africa, the flood control, power production, and irrigation aspects of the Ganges-Brahmaputra river system in India, the national economic planning of the Soviet Union, and even the survival and diffusion of insect colonies.

Key Terms

isorevenue line
linear programming
objective function
shadow prices
simplex method

BIOGRAPHY 5A.1

Leonid V. Kantorovich

Leonid Vitalyevich Kantorovich (1912–   ) was born in St. Petersburg, Russia. His career as a mathematical genius advanced rapidly. At the age of 14, he enrolled at Leningrad University; by the time he was 22, he was a full professor of mathematics. During World War II, he worked at Leningrad's Naval Engineering School; in 1949, he received the Stalin Prize for his work in pure mathematics on functional analysis and computer development. In 1958, he was elected to the prestigious Soviet Academy of Sciences and received the 1965 Lenin Prize for his work in mathematical economics (shared with the chief architects of the mathematical revolution in Soviet economics, V. S. Nemchinov and V. V. Novozhilov). In the 1970s, Kantorovich headed the Moscow Institute for the Management of the National Economy. While there, in 1975, he received the Nobel Memorial Prize in Economic Science (jointly with T. C. Koopmans). He is now working at the All-Union Scientific Research Institute for Systems Research.

In the 1930s, the Central Plywood Trust approached Kantorovich with a problem. The producers of plywood were rotating logs in stripping machines that cut off a continuous thin sheet of material. These sheets were then laminated to make plywood. There were many types of logs and many machines with different productivities. How could one match logs to machines so as to process the largest possible volume per unit of time? Kantorovich responded by inventing linear programming. His 1939 Leningrad paper, "Mathematical Methods of Organizing and Planning Production," was the first publication ever to appear on the subject.

By 1943, Kantorovich realized that the national economic plan itself could be viewed as a grandiose linear programming problem. He extended linear programming from a tool for solving short-run planning problems of the firm to one for solving the short-run planning problems of the nation. He noted that the optimum solution of the national planning problem revealed shadow prices for all goods and resources, which could be used to make crucial decisions. His calculation of positive prices for the use of scarce capital and natural resources, however, clashed with Stalinist ideology (which insisted that such resources be priced at zero under socialism to reflect their common ownership by all the people). As a result of this clash Kantorovich's book on *The Best Use of Economic Resources* was not published until 1959, well after Stalin's death. Since then, Kantorovich has extended his earlier model of optimal short-run national planning to long-run planning as well.

Additional Readings

Baumol, William J. "Activity Analysis in One Lesson," *The American Economic Review,* December 1958, pp. 837–73; and *Economic Theory and Operations Analysis,* 4th ed. Englewood Cliffs, N.J.: Prentice-Hall, 1977.

Beals, Ralph E., et al. "Rationality and Migration in Ghana," *Review of Economics and Statistics,* November 1967, pp. 480–86.

    Uses linear programming to estimate implicit wage rates for seasonal migrants.

Bland, Robert G. "The Allocation of Resources by Linear Programming," *Scientific American,* June 1981, pp. 126–44.

    A superb exposition of the current state of the art.

Dantzig, George M. *A Procedure for Maximizing a Linear Function Subject to Linear Inequalities.* Washington, D.C.: Headquarters U.S. Air Force, 1948.

    The original statement of the simplex method.

Dantzig, George M. *Linear Programming and Extensions.* Princeton: Princeton University Press, 1963.

Dorfman, Robert. "Mathematical, or Linear, Programming: A Nonmathematical Exposition," *The American Economic Review,* December 1953, pp. 797–825.

Dorfman, Robert; Samuelson, Paul A., and Solow, Robert M. *Linear Programming and Economic Analysis.* New York: McGraw-Hill, 1958.

Kantorovich, Leonid V. "Mathematical Methods of Organizing and Planning Production," *Management Science,* July 1960, pp. 366–422; idem. *The Best Use of Economic Resources.* Cambridge: Harvard University Press, 1965; idem. "Essays in Optimal Planning." *Problems of Economics,* August–September–October 1976, pp. 3–251.

    Translations of the original 1939 article, of his 1943 work not published until 1959, and of 18 essays, with an introduction on the life and work of Kantorovich by Leon Smolinski. (In 1975, Kantorovich received the Nobel Prize in economics for this work.)

Koopmans, Tjalling C. *Activity Analysis of Production and Allocation.* Chicago: Cowles Commission, 1951; idem. *Three Essays on the State of Economic Science.* New York: McGraw-Hill, 1957; idem. *Scientific Papers of Tjalling C. Koopmans.* New York: Springer, 1970.

    Major works by the American co-winner of the 1975 Nobel Prize honoring the developers of linear programming.

Stigler, George J. "The Cost of Subsistence," *Journal of Farm Economics,* May 1945, pp. 303–14.

    A paper famous as a forerunner of the linear programming technique, written by the 1982 winner of the Nobel Prize in economics.

Ward, Benjamin. "Linear Programming and Soviet Planning." In John P. Hardt, ed. *Mathematics and Computers in Soviet Economic Planning.* New Haven: Yale University Press, 1967, chap. 3.

Wilson, Edward O. "The Ergonomics of Caste in the Social Insects," *The American Economic Review,* December 1978, pp. 25–35.

    A fascinating application of linear programming to the survival of insect colonies.

CHAPTER 6
Perfect Competition

MULTIPLE-CHOICE QUESTIONS

Circle the letter of the *one* answer that you think is correct or closest to correct.

1. A market is said to be in short-run equilibrium
   a. when the quantity supplied is absolutely fixed, regardless of price.
   b. when the number of firms is fixed, but they can vary quantity supplied by changing the utilization rate of given plants.
   c. when the price elasticity of supply is zero.
   d. whenever (a) and (c) are true.

2. In a perfectly competitive market, a price above the equilibrium level
   a. produces a shortage.
   b. induces competition among buyers and sellers.
   c. produces a surplus.
   d. results in both (b) and (c).

3. If, at a zero price, the quantity demanded a good falls short of the quantity supplied,
   a. the good is scarce.
   b. the good is free.
   c. a shortage exists.
   d. both (a) and (c) are true.

When answering questions 4–5, refer to the graph below.

4. A price of $7 in this perfectly competitive market implies
   a. a shortage of AD and pressure for price to rise.
   b. imminent competition among buyers but not sellers.
   c. an imminent decrease in quantity demanded from $0D$ to $0C$ and increase in quantity supplied from $0A$ to $0C$.
   d. all of the above.

5. A price of $13 in this perfectly competitive market implies
   a. a surplus of AE and pressure for price to fall.
   b. that the good involved is not scarce.
   c. imminent competition among sellers but not buyers.
   d. none of the above.

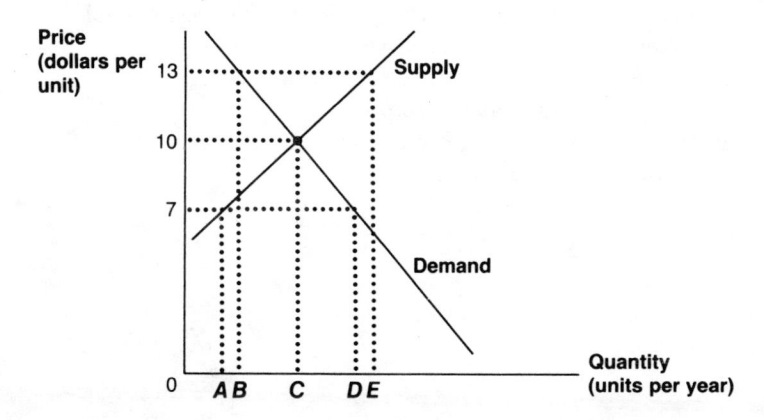

6. If a household in a perfectly competitive market were to boycott the market to protest too high a price, the equilibrium price would
   a. fall.
   b. not change.
   c. rise.
   d. first fall, but then rise again (once the household started buying at the lower price).

7. In the typical demand-supply diagram of a perfectly competitive market,
   a. a vertical line drawn through the equilibrium point shows how traders split their gains.
   b. a horizontal line drawn through the equilibrium point discriminates between actual trades made and potential trades rejected.
   c. the quantity that would be traded, should price ever be above equilibrium, can be read off on the demand line.
   d. the quantity that would be traded, should price ever be below equilibrium, can be read off on the demand line.

8. When a perfectly competitive industry expands,
   a. the typical firm has probably been making profits.
   b. the typical firm's minimum level of average total cost may be falling (as a result of favorable technological or pecuniary externalities).
   c. the typical firm's minimum level of average total cost may be rising (as a result of external diseconomies).
   d. all of the above are true.

9. In a perfectly competitive market, the function of losses is
   a. to discourage the use of expensive inputs and raise their supplies.
   b. to discourage firms from an industry so that output price can be raised to minimum average total cost.
   c. to encourage constant returns to scale so that output price can be maintained.
   d. to encourage firms to expand their scales so that costs can be reduced.

10. Once the owners of firms in a perfectly competitive industry earn from their own resources, used in their own firms, neither more nor less than they could earn in their best outside alternatives,
    a. the industry's product price is the normal price.
    b. the industry will begin to contract.
    c. the industry's costs will begin to rise.
    d. both (b) and (c) will occur.

11. When supply and demand in a perfectly competitive market intersect at a price of $20, while the typical firm's lowest possible long-run average total cost is $20 as well,
    a. the equilibrium price equals the normal price.
    b. the typical firm makes zero profit.
    c. the industry is in long-run equilibrium.
    d. all of the above are true.

12. When the normal price of an industry's product is unchanged after the industry has ceased to expand or contract,
    a. the law of diminishing returns does not apply.
    b. the industry is a constant-cost industry.
    c. the industry's long-run supply is a horizontal line.
    d. both (b) and (c) are true.

13. When technological and pecuniary externalities are absent from an industry,
    a. it is a constant-cost industry.
    b. its product price cannot change.
    c. its firms must be subject to constant returns to scale.
    d. both (a) and (b) are true.

14. When the normal price of an industry's product is higher after the industry has ceased to expand or lower after it has ceased to contract,
    a. each firm must be subject to increasing returns to scale.
    b. each firm must be subject to decreasing returns to scale.
    c. the industry is an increasing-cost industry.
    d. both (b) and (c) are true.

15. In the long run, a given increase in market demand
    a. leads to the same increase in the volume of production, regardless of whether the industry is one of constant, increasing, or decreasing cost.
    b. leads to a larger increase in the volume of production in a constant-cost industry than in an increasing-cost industry.
    c. leads to a smaller increase in the volume of production in a decreasing-cost industry than in an increasing-cost industry.
    d. leads to a larger increase in the volume of production in a constant-cost industry than in a decreasing-cost industry.

16. In the long run, a decrease in market demand will
    a. lower price in a constant-cost industry.
    b. lower price in a decreasing-cost industry.
    c. lower price in an increasing-cost industry.
    d. do all of the above.

17. Any cobweb cycle will be damped when
    a. the slope of the lagged supply line equals the absolute value of the slope of the demand line.
    b. the slope of the lagged supply line is larger than the absolute value of the slope of the demand line.
    c. the slope of the lagged demand line equals the absolute value of the slope of the supply line.
    d. the slope of the lagged demand line is larger than the absolute value of the slope of the supply line.

18. Marshall noted that the measurement of the consumers' surplus by a triangular area under a straight demand line for a good would be
    a. exaggerated or understated unless the good had a zero income elasticity of demand.
    b. exaggerated for inferior goods.
    c. understated for normal goods.
    d. all of the above.

When answering questions 19–20, consider the graph below, in which a consumer's optimum moves from C to E as the price of bread falls, given income 0A:

19. The price compensating variation (or the maximum amount of income the consumer would pay for the privilege of buying any desired quantity of a good at a lower price) equals
    a. AK.
    b. AF.
    c. FK.
    d. 0F.

20. The price-equivalent variation equals
    a. AK.
    b. AF.
    c. FK.
    d. 0F.

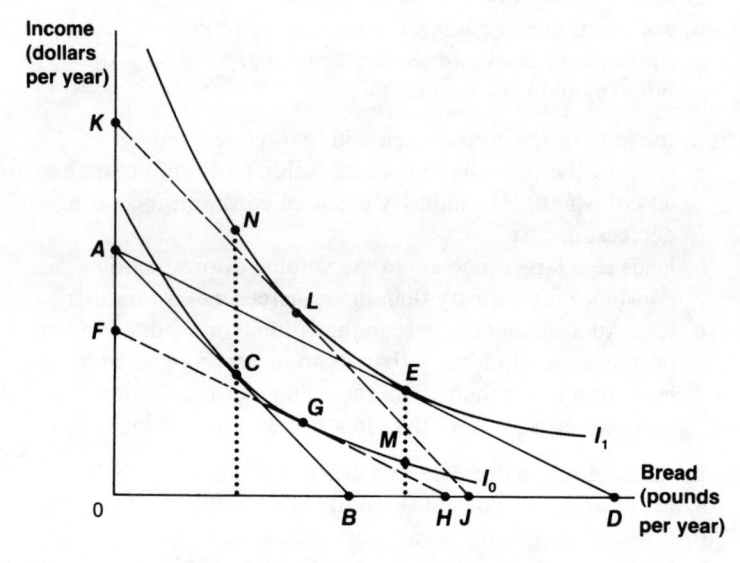

21. The quantity-equivalent variation involves
    a. a hypothetical payment by a consumer (equivalent to the value of a higher quantity of a good).
    b. a hypothetical change in quantity (equivalent in its effect on a consumer of an opposite change in price).
    c. a hypothetical payment to a consumer (as in the case of a quantity-compensating variation).
    d. none of the above.

When answering questions 22–24, consider the graph on page 99.

22. In the short run, the imposition of an excise tax of HF in this perfectly competitive market
    a. lowers the price to buyers from 0C to 0B.
    b. reduces the consumers' surplus by CDFG.
    c. reduces the producers' surplus by BCKH.
    d. reduces the producers' surplus by IHKL.

23. In the short run, the imposition of an excise tax of HF in this perfectly competitive market
    a. lowers quantity traded from 0L to 0I.
    b. creates a shortage of GK.
    c. creates a deadweight welfare loss of DEF.
    d. yields government revenue of 0DFI.

24. In the short run, after an excise tax of *HF* has been imposed on this perfectly competitive market
    a. the consumers' surplus equals *DEF*.
    b. the producers' surplus equals *ABH*.
    c. the government collects revenue of *BDFH*.
    d. all of the above are true.

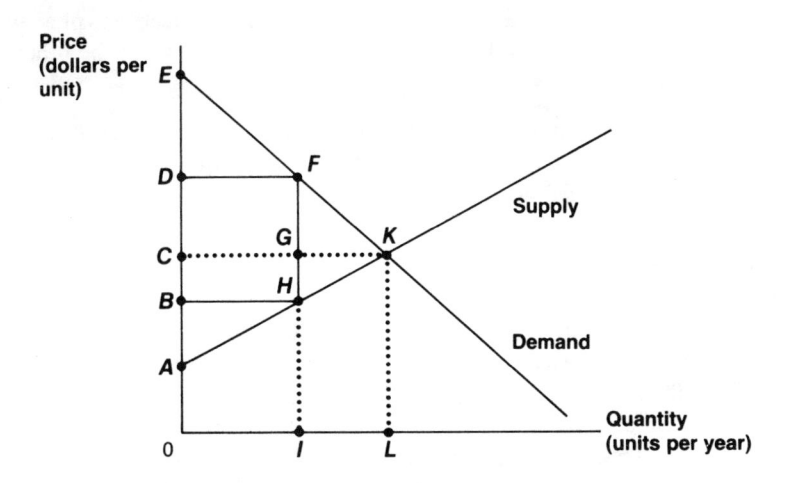

25. If a perfectly competitive industry expands as a result of the imposition of an import quota, the price of its product will ultimately
    a. return to the prequota level if the industry is one of constant cost.
    b. lie above the prequota level if the industry is one of decreasing cost.
    c. lie below the prequota level if the industry is one of increasing cost.
    d. do all of the above.

TRUE-FALSE QUESTIONS

In each space below, write a *T* if the statement is true and an *F* if the statement is false.

_____ 1. In a perfectly competitive market, a shortage causes a fall in demand, while a surplus causes a rise in supply.

_____ 2. In a perfectly competitive market, any shortage raises the equilibrium price.

_____ 3. In a perfectly competitive market, any shortage raises the industry's normal price.

_____ 4. A perfectly competitive industry will neither expand nor contract when the typical firm makes zero profit.

_____ 5. When the long-run supply of every firm in an industry is a horizontal line, the industry cannot be one of increasing or decreasing cost.

_____ 6. When a decreasing-cost industry contracts, the cost curves of individual firms shift up.

_____ 7. In the long run, a decrease in market demand will raise price in a decreasing-cost industry.

_____ 8. The *price-equivalent variation* is the minimum amount of income a consumer would accept for relinquishing the opportunity of buying any desired quantity of a good at a lower price.

_____ 9. The imposition of an excise tax on a perfectly competitive market yields government revenue equal to the difference between the loss of consumers' and producers' surplus on the one hand and a deadweight welfare loss on the other hand.

_____ 10. The imposition of an import quota can never result, ultimately, in consumers getting a larger quantity of the affected good at a lower price.

PROBLEMS

1. Suppose you knew, prior to an auction, that the only copy of a rare postage stamp was being offered for sale at a minimum price of $400,000.
   a. If you also knew demand to equal $Q = 2.5 - \frac{5}{3} P$ (with quantity Q expressed in single units and price P in millions of dollars), which equilibrium price would you predict? Find the answer graphically with the help of the following graph.

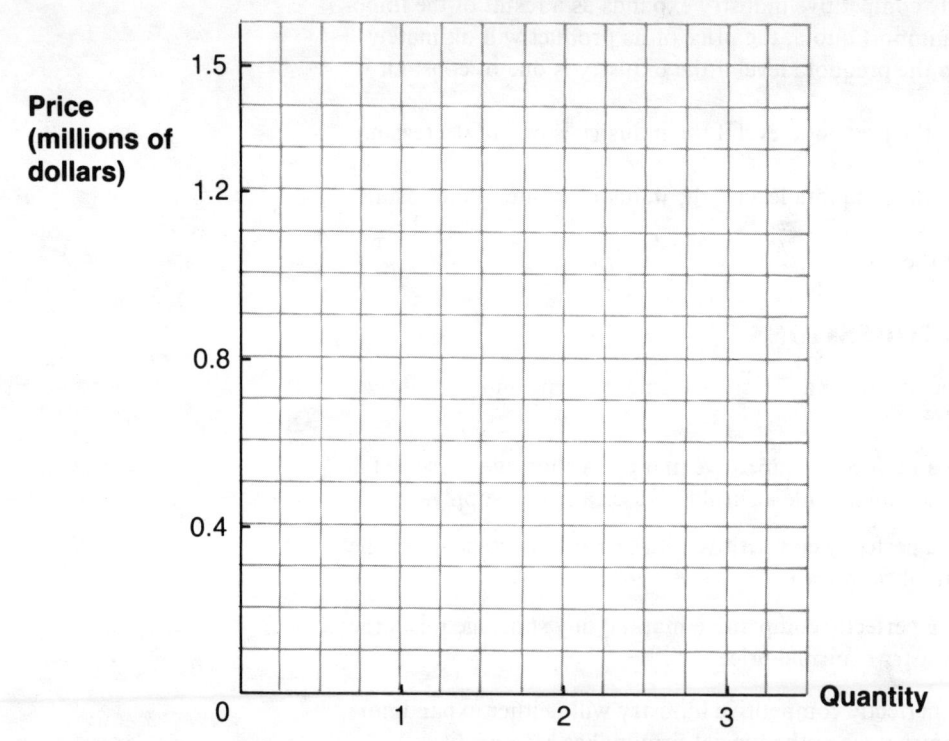

   b. How would your answer have differed, given the same demand, if two or three such stamps had been offered at any price?

c. Can you see why people have bought up all the giant crabs and tarantulas used in a horror movie (or all the Superman costumes used in a television series or all the robots from a sci-fi film) just to destroy most of them right away?

2. Now consider this hypothetical demand for Van Gogh paintings: $Q = 250 - \frac{50}{3}P$, where Q is stated in single units and P in millions of dollars. Let supply be $Q = -300 + 75P$ up to a maximum of $Q = 250$.
   a. What will the equilibrium price and quantity be in a perfectly competitive market? Find the answer graphically with the help of the following graph.

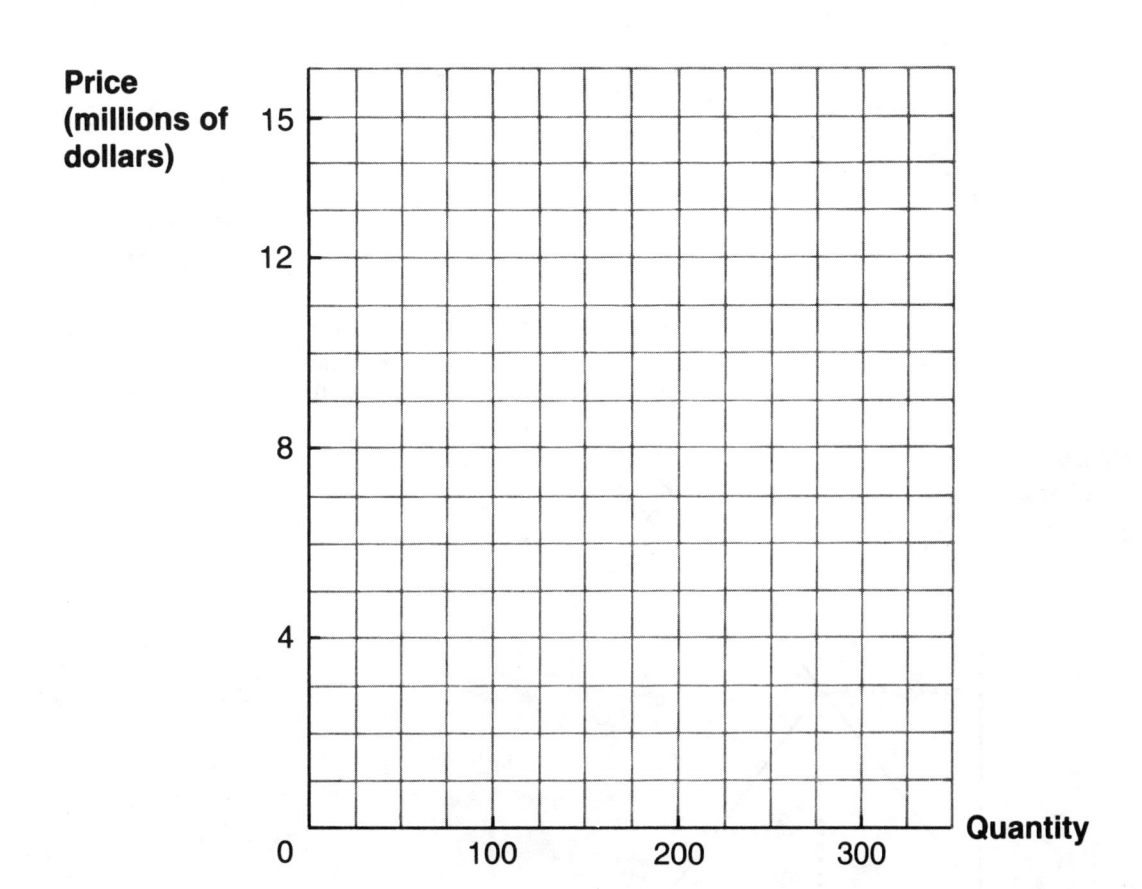

   b. Why will the price not be $12 million per painting? Why not $2 million?

3. Suppose that the typical firm in a perfectly competitive industry were correctly depicted by the next graph.
   a. Label the curves shown.
   b. Indicate the firm's profit-maximizing output level and the size of its profit (or loss).
   c. Identify the firm's short-run supply curve.

d. Is the firm's industry in long-run equilibrium? If yes, why? If not, what will happen?

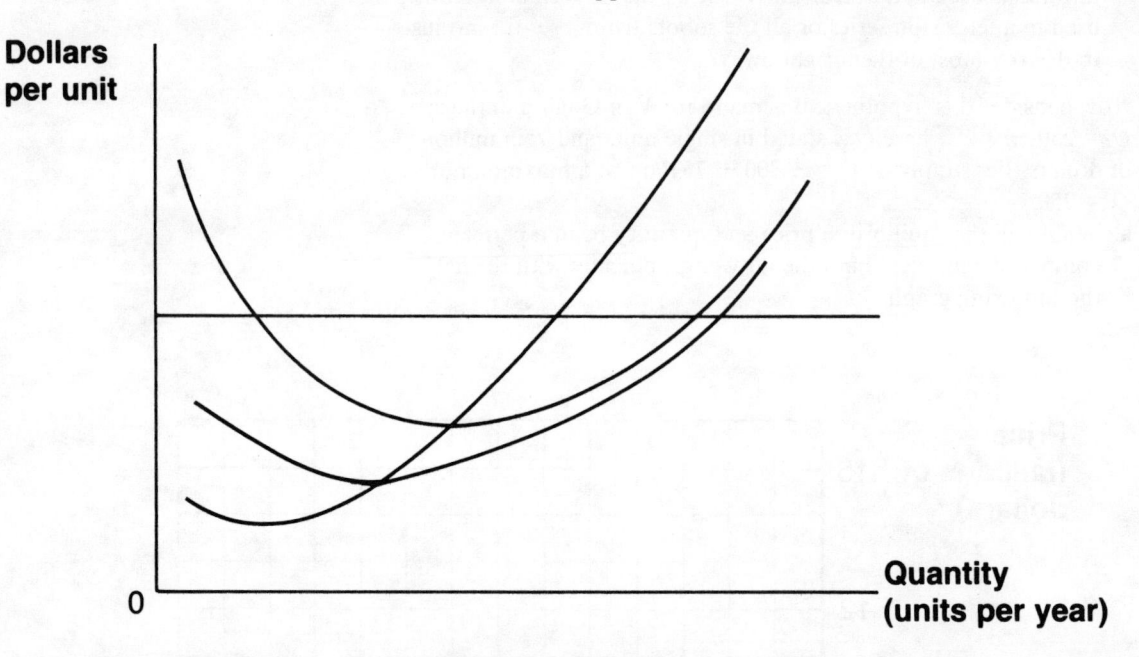

4. Consider the following graph, which depicts *E* at a short-run equilibrium in a perfectly competitive market. Suppose that the industry were in long-run equilibrium as well.

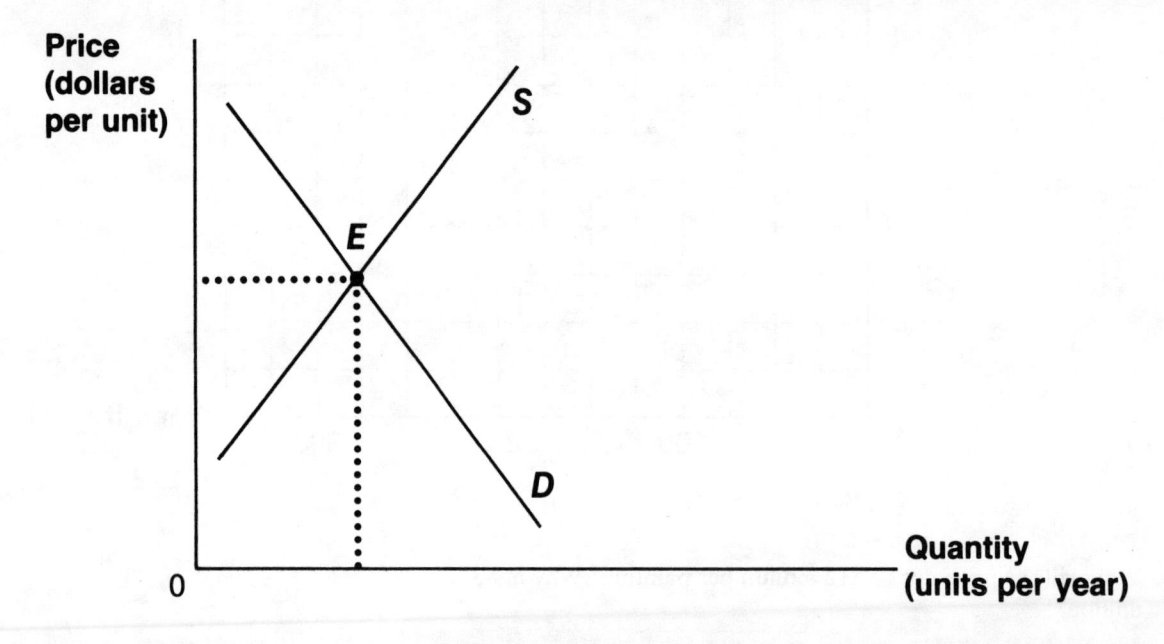

a. What would now happen if the industry were a constant-cost industry and demand fell? Show your answer graphically.
b. What would now happen if the industry were an increasing- or decreasing-cost industry and demand rose? Show your answer graphically.
c. Identify the long-run industry supply curves.

5. a. On the following graph, draw a lagged supply line that is bound to produce a damped cobweb cycle, given the momentary equilibrium at E.

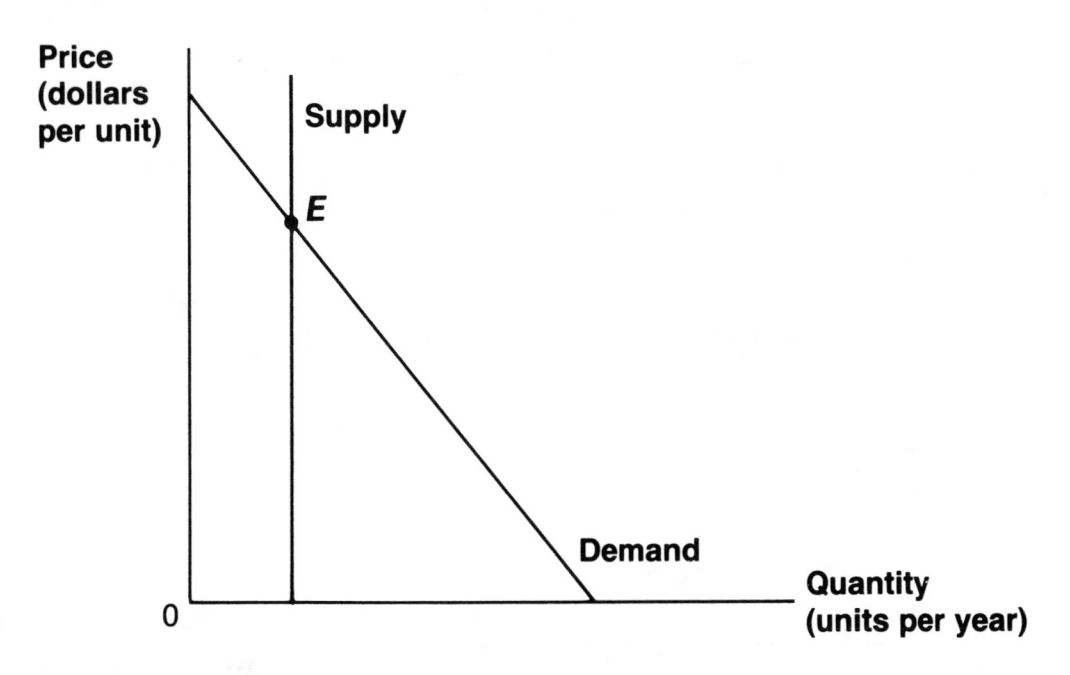

b. Prove that your line does indeed lead to a damped cycle. What was the key to drawing it correctly?

6. a. In the following graph, identify the consumers' and producers' surplus. Then show how each would change, if the government paid producers a $1 subsidy per unit of product.

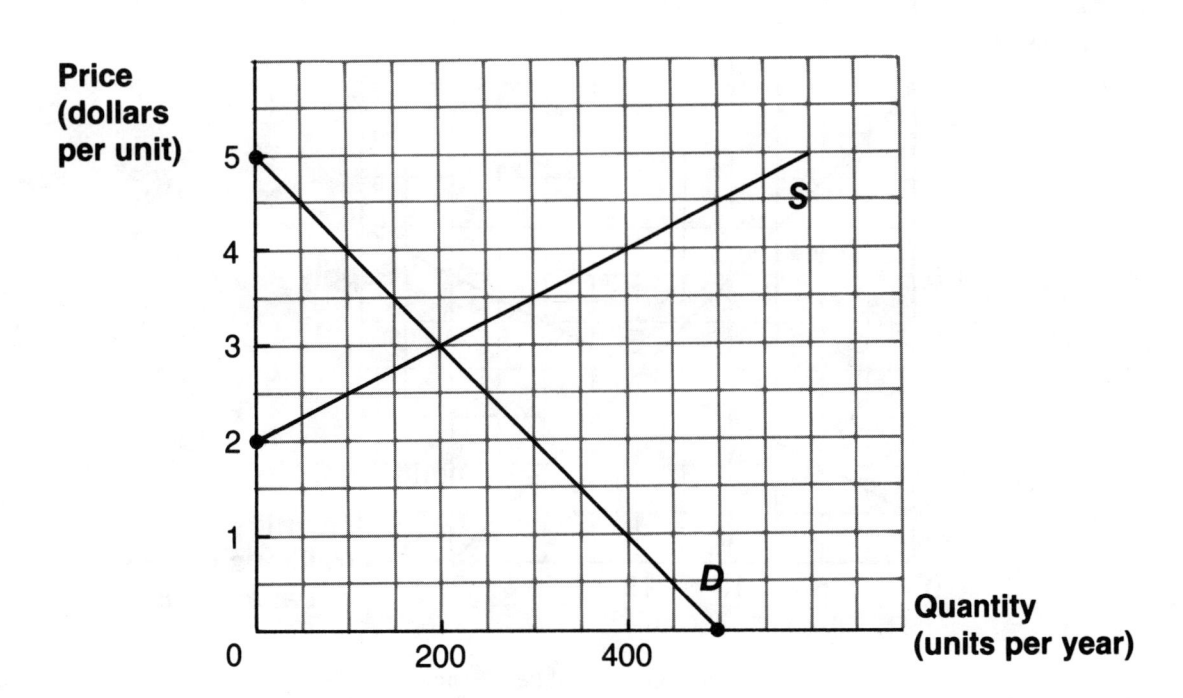

b. How big would be the government's payment?
c. If the industry were in long-run equilibrium prior to the subsidy, what would now happen?

7. The following graph pictures a consumer with income $0C$ in equilibrium at E on indifference curve I_0. Show how Hicks would measure the consumer's surplus from a halving of the price of bread as
   a. a quantity-compensating variation.
   b. a price-equivalent variation.

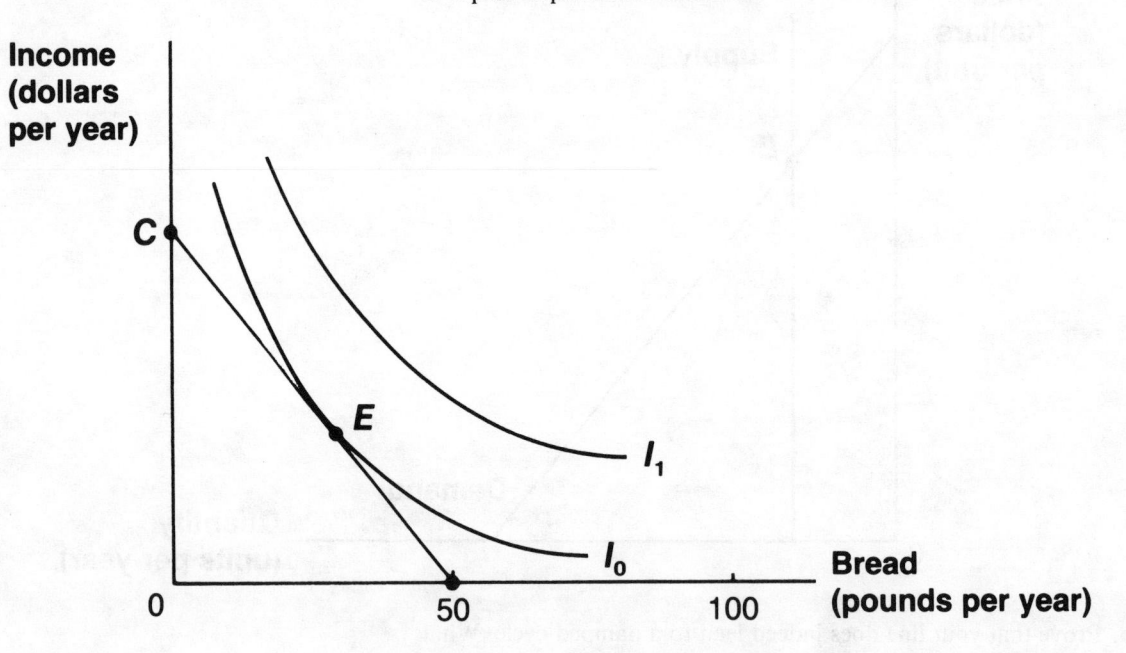

8. The following graph pictures the banana market in equilibrium at E.

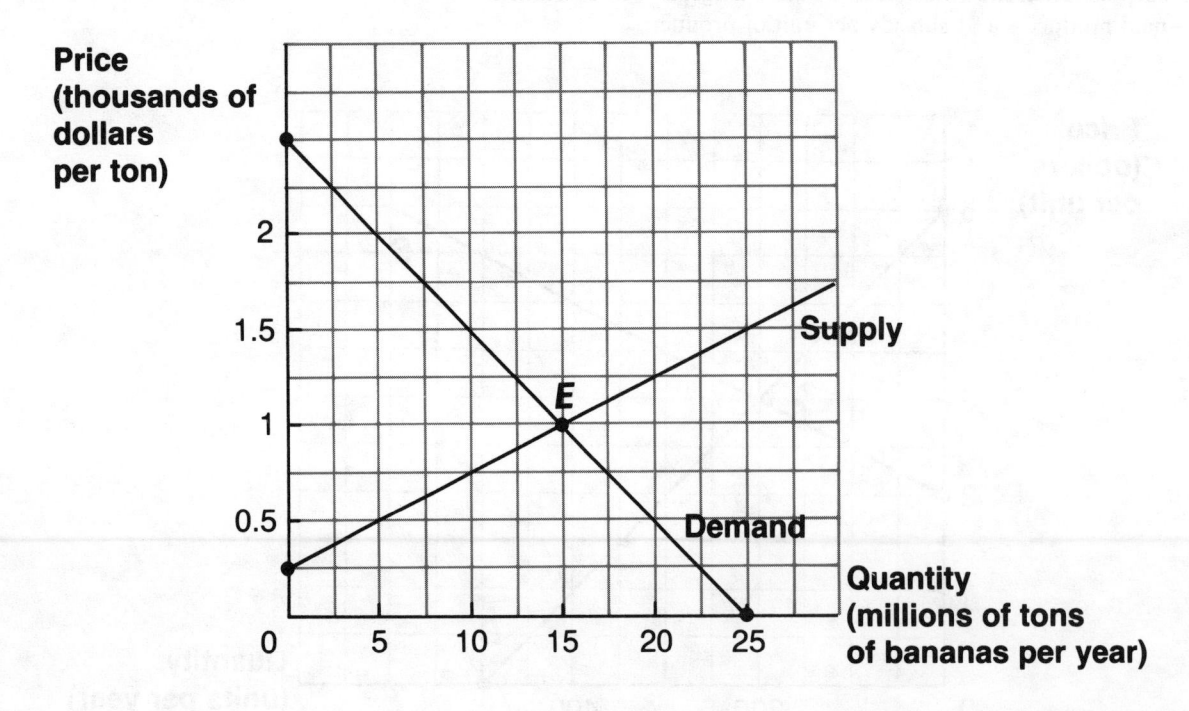

   a. What would be the short-run effect on price, trading volume, and the consumers' and producers' surplus if the government imposed an import quota of 5 million tons per year? (Assume that bananas cannot be produced domestically).
   b. Would there be a deadweight loss?
   c. What would be the size of governmental revenue?

9. For each of the following demand and supply functions, arithmetically determine the competitive equilibrium price and quantity, the effect of a price ceiling of $P = 3$, and of a price floor of $P = 10$.
   a. $Q_D = 10 - 3P$ and $Q_S = 5P - 1$
   b. $Q_D = 40 - 7P$ and $Q_S = 2P - 5$

ANSWERS

Multiple-Choice Questions

1. b	2. c	3. b	4. d	5. c	6. b	7. c	8. d	9. b
10. a	11. d	12. d	13. a	14. c	15. b	16. c	17. b	18. a
19. b	20. a	21. d	22. c	23. a	24. d	25. a		

True-False Questions

1. F (A shortage causes competition among frustrated would-be buyers. This raises price and thus lowers *quantity demanded*, while raising quantity supplied. A surplus causes competition among frustrated would-be sellers. This lowers price and thus lowers *quantity supplied*, while raising quantity demanded.)
2. F (It raises the *actual* price toward the *unchanged* equilibrium price.)
3. F (A shortage raises the industry's actual price.)
4. T
5. F (Text Figure 6.5, "Long-Run Industry Supply Curves," proves the opposite. Those horizontal long-run supply curves of every firm can be shifted up or down as the industry expands as a result of unfavorable or favorable technological and pecuniary externalities.)
6. T
7. T
8. T
9. T
10. F (See the last paragraph in the Chapter 6 text section, "Application 5: Import Quotas.")

Problems

1. a. The supply would then look like line S_1. Demand would look like line AB: At a price of $1.5 million, Q would equal $2.5 - \frac{5}{3}(1.5) = 0$, as shown by point A. At a zero price, Q would equal $2.5 - \frac{5}{3}(0) = 2.5$, as shown by point B. Thus the equilibrium price could be predicted at $900,000 (point C). Note: In 1980, the 1856 British Guiana 1-cent stamp was auctioned at

$850,000, the highest price ever for a postage stamp. A group of business executives had bought it in 1970 for $280,000 as a hedge against inflation.

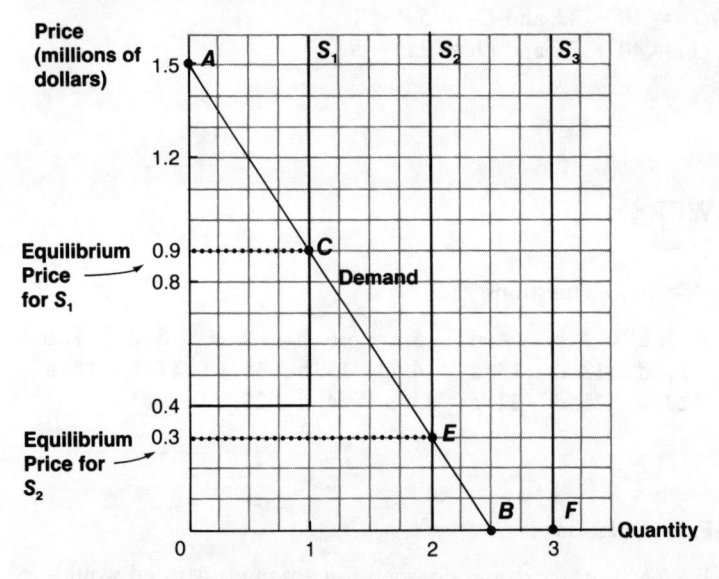

b. Supply would have looked like S_2 or S_3. Thus the equilibrium price would have been $300,000 (point E) or zero (point F).

c. Of course. Given the kind of demand shown here, the sale of 3 stamps would bring zero revenue, that of 2 stamps twice $300,000, that of 1 stamp $900,000. Note: It is rumored that a business executive once bought a second British Guiana 1-cent stamp for $60,000, then promptly set it on fire.

2. a. The demand curve is AB: At a price of $15 million, $Q = 250 - \frac{50}{3}(15) = 0$ (point A). At a price of zero, $Q = 250 - \frac{50}{3}(0) = 250$ (point B). The supply curve is CFG and beyond: At a price of $4 million, $Q = -300 + 75(4) = 0$ (point C). At a price of $6 million, $Q = -300 + 75(6) = 150$ (point E). But supply can never exceed 250 units; hence vertical section FG. Thus equilibrium price is $6 million; the equilibrium quantity is 150 (corresponding to point E). Note: In 1980, an actual Van Gogh (*The Poet's Garden*) sold for $5.2 million; in 1987, another Van Gogh (*Sunflowers*) sold for $36.3 million and a third one (*Irises*) for $53.9 million.

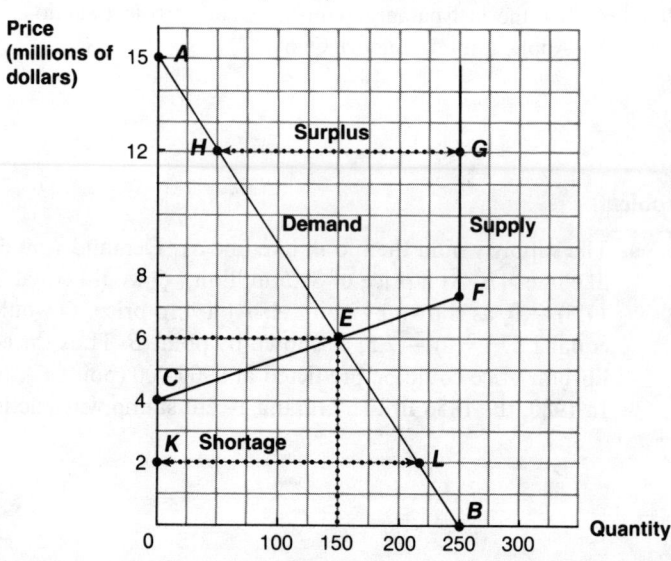

b. At $12 million, there would be a surplus of *HG*. At $2 million, there would be a shortage of *KL*.

3. a.

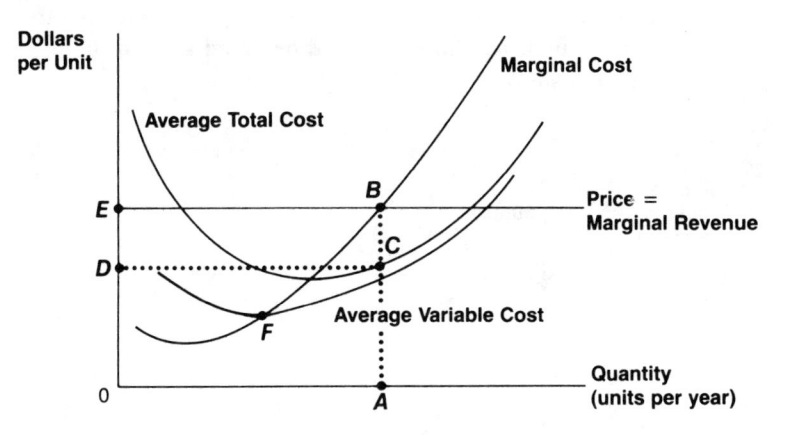

[Graph: Dollars per Unit vs Quantity (units per year). Shows Marginal Cost, Average Total Cost, Average Variable Cost curves, and horizontal Price = Marginal Revenue line at level E. Points B (intersection of MC and price line), C (on ATC below B), D (ATC level), F (minimum of AVC), A (on quantity axis below B/C).]

b. The profit-maximizing output level is $0A$; the size of profit is *BCDE*.
c. The marginal cost line above *F*.
d. No. Profit made by the typical firm leads to industry expansion. See text Figure 6.3, "A Profitable Industry Expands."

4. a. If demand fell to D_1, a new short-run equilibrium at *A* would eventually give way to a new long-run equilibrium at *B*, because the industry would contract, shifting *S* left until new supply S_1 intersected new demand D_1 at the original price. If demand fell to D_2, however, a new short-run equilibrium at *C* would theoretically give way to a new long-run equilibrium at *F*, because the industry would contract, shifting *S* left until new supply S_2 intersected new demand D_2 at the original price. In fact, the industry would cease to exist because nobody would be willing to pay a price high enough to cover the minimum average total cost of producing its product.

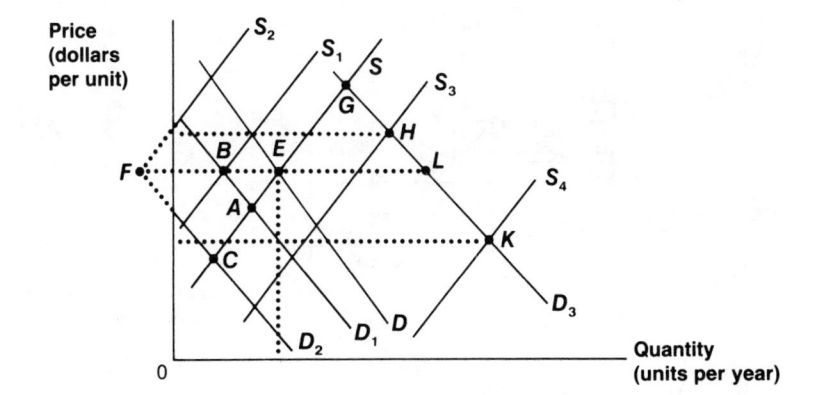

[Graph: Price (dollars per unit) vs Quantity (units per year). Shows supply curves S_2, S_1, S, S_3, S_4 and demand curves D_2, D_1, D, D_3 with labeled intersection points F, B, E, G, H, L, A, C, K.]

b. If demand rose to D_3, a new short-run equilibrium would be established at *G*. Eventually, this would give way to a new long-run equilibrium at *H* or *K*, because the industry would expand, shifting *S* right until new supply S_3 (increasing-cost industry) or S_4 (decreasing-cost industry) intersected new demand D_3 at a higher (or lower) price equal to the higher (or lower) minimum average total cost of firms.

c. For the *constant-cost-industry* case, the horizontal line through *BEL*; for the *increasing-cost-industry* case, a rising line through *E* and *H* (not drawn); for the *decreasing-cost-industry* case, a falling line through *E* and *K* (not drawn). *Note:* If the industry had been one of constant cost when demand rose to D_3, the ultimate equilibrium would be found at *L* on supply line S_5 (not shown).

5. a.

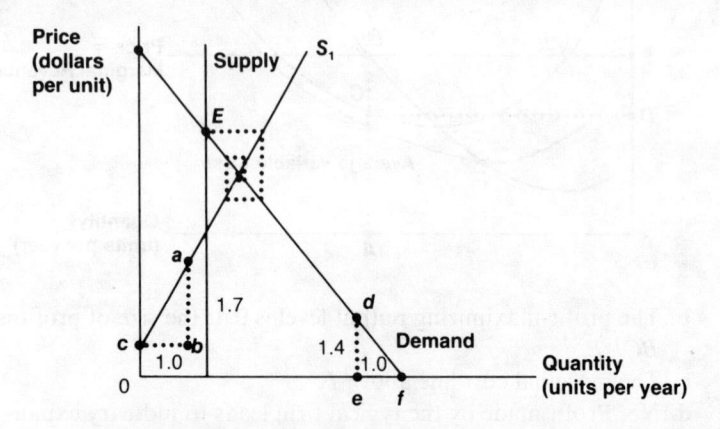

b. The dotted line to the right of *E* proves the point. The key lies in drawing S_1 more steeply than *D* (comparing absolute slope values only). Consider triangles *def* and *abc* and note how the slope of *D* equals |1.4|, while that of S_1 equals 1.7.

6. a. The consumers' surplus is *acd* originally; the producers' surplus is *acb*. The subsidy shifts supply line *S* vertically down by $1. This shift yields a new equilibrium at *e*. Thus the consumers' surplus rises to *ehd*; the producers' rises to *ehf* (plus the subsidy).

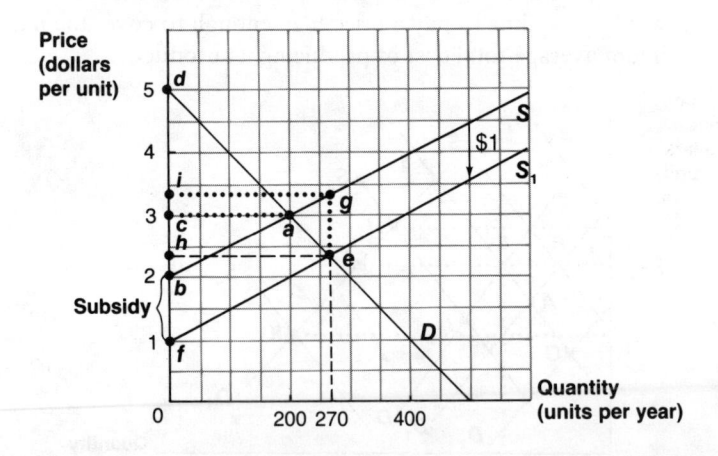

b. $270 per year (area *egih*)

c. The price to the seller would have become higher (0*i* instead of 0*c*). The typical firm would be making profits. The industry would expand. Line *S* (and with it parallel S_1) would shift to the right. In case of a *constant-cost-industry*, this expansion would cease when the price to the buyer (now 0*h*) plus subsidy (*hi*) just equalled 0*c* (the original price and minimum average total cost). In the case of an *increasing-cost industry*, the expansion would cease earlier: when the price to the buyer plus subsidy (now totalling $0i$) still exceeded the original price $0c$ (but then equalled the higher minimum average total cost of firms). In the case of a *decreasing-cost industry,* the expansion would go further: until the price to the buyer plus subsidy was below the original price (but then equalled the lower minimum average total cost of firms). In all cases, consumers would be getting an even larger quantity and an even lower price than is implied by intersection e.

7. The halving of the bread price would tilt the consumer's budget line to the right as shown, and the consumer would reach a higher welfare at A on indifference curve I_1.

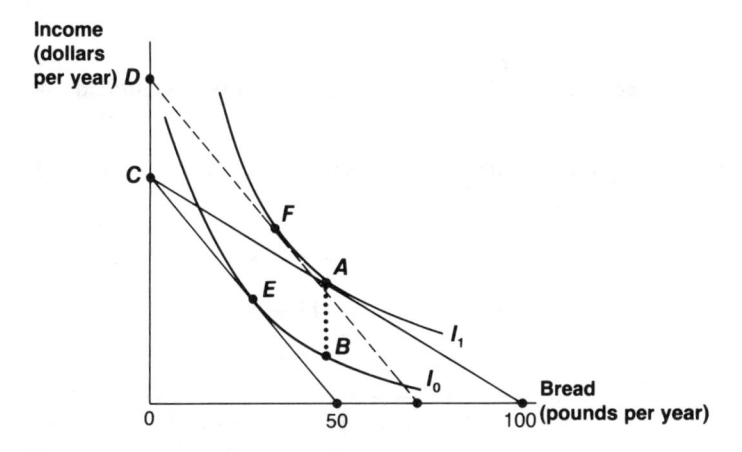

a. The quantity-compensating variation (the maximum amount of income the consumer would pay for the privilege of buying bread at the lower price, while being constrained to buying the quantity the consumer would buy at the lower price in the absence of compensation) equals AB. After paying this amount, the consumer could at B consume a combination that yields as much satisfaction as initial combination E. (Both B and E lie on indifference curve I_0.)

b. The price-equivalent variation (the minimum amount of income the consumer would accept for relinquishing the opportunity of buying any desired quantity of bread at a lower price) equals CD. After receiving this amount, the consumer could reach F along the dashed budget line parallel to the original one; F provides as much satisfaction as A (both lie on indifference curve I_1).

8. a. The import quota would change the supply line to ABC and beyond. Thus the equilibrium point would move from E to C. Price would rise from \$1,000 to \$2,000 per ton. Trading volume would fall from 15 million to 5 million tons per year. The consumers' surplus would fall from EFG to CFH. The producers' surplus would rise from AEG to $ABCH$ (a loss of BEK due to lower trading volume being overcome by a gain, at the consumers' expense, of $GKCH,$ due to the higher price).

[Graph: Price (thousands of dollars per ton) vs Quantity (millions of tons of bananas per year). Supply with Quota vertical line. Points F, H, C at top; G, K, E at P=1; A, B near bottom. Equilibrium E at Q=15, P=1.]

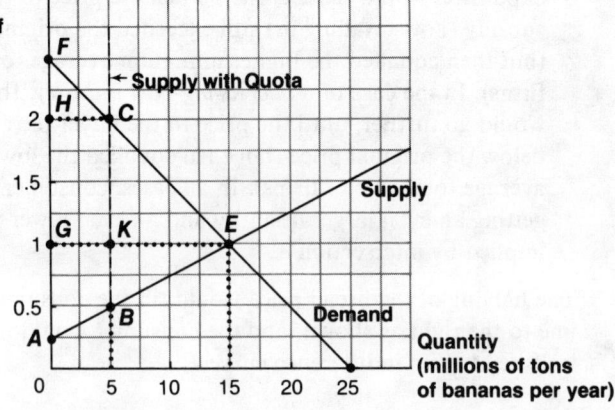

b. Yes, equal to *BEC*.
c. Zero, unless the government auctions off the limited quota rights.

9. a. Equating demand and supply, the equilibrium price and quantity are found:

$$10 - 3P = 5P - 1$$
$$11 = 8P$$
$$P = 1.375$$

$$Q_D = 10 - 3P = 10 - 4.125 = 5.875$$

check: $Q_S = 5P - 1 = 6.875 - 1 = 5.875$

At $P_{max} = 3$, $Q_D = 10 - 3P = 10 - 9 = 1$ and $Q_S = 5P - 1 = 15 - 1 = 14$. At such a price, a surplus of 13 would exist. However, the equilibrium price is lower, hence legal; the above-equilibrium ceiling is ineffective. At $P_{min} = 10$, $Q_D = 10 - 3P = 10 - 30 = -20$, which means zero, while $Q_S = 5P - 1 = 50 - 1 = 49$. A surplus of 49 exists. The lower equilibrium price is illegal; the price floor is effective.

b. Equating demand and supply, the equilibrium price and quantity are found:

$$40 - 7P = 2P - 5$$
$$45 = 9P$$
$$P = 5$$

$$Q_D = 40 - 7P = 40 - 35 = 5$$

check: $Q_S = 2P - 5 = 10 - 5 = 5$

At $P_{max} = 3$, $Q_D = 40 - 7P = 40 - 21 = 19$ and $Q_S = 2P - 5 = 6 - 5 = 1$. A shortage of 18 exists. The higher equilibrium price is illegal; the price ceiling is effective. At $P_{min} = 10$, $Q_D = 40 - 7P = 40 - 70 = -30$, which means zero, while $Q_S = 2P - 5 = 20 - 5 = 15$. A surplus of 15 exists. The lower equilibrium price is illegal; the price floor is effective.

BIOGRAPHY 6.1

Alfred Marshall

Alfred Marshall (1842–1924) was born in Clapham, a suburb of London, England. His tyrannical father, a Bank of England cashier who wrote a tract on *Man's Rights and Woman's Duties,* forced him to study Hebrew and the classics in preparation for the ministry. He also ordered him to stay away from chess and mathematics, but to no avail. When Alfred received a scholarship for the study of theology at Oxford, he rebelled against his father. A kindly uncle helped him finance his study of mathematics at Cambridge; eventually, Marshall became one of Cambridge's great professors of political economy.

Very much like Adam Smith, Marshall was a profoundly learned man, overflowing with ideas covering such diverse fields as biology, economics, history, mathematics, and philosophy. But Marshall's father had left his mark. Alfred was always afraid of speaking too soon and was never in a hurry to rush into print. He took infinite care and insisted on the highest standards of accuracy and truth and on the full mastery of his material. The bulk of his work is contained in his *Principles of Economics* (1890), *Industry and Trade* (1919), and *Money, Credit, and Commerce* (1923).

By 1890, when he published his *Principles* (fully 20 years after first sharing its contents with his students), he had already laid the foundation, through years of teaching, for what is now called the neoclassical school. His book's success was instant and complete. In the English-speaking world, it was the leading text for decades, going through eight editions in Marshall's lifetime. Generations of economists were brought up under the pervasive influence of his thought.

Even today, many copies of Marshall's *Principles* are sold every year. This is not surprising because almost the entire corpus of modern microeconomics can be traced to some suggestion by Marshall. He introduced such concepts as the short run and the long run; he was a master of *partial analysis,* the analysis of phenomena in relatively small sectors of the economy within which events can be assumed not to call forth repercussions in social aggregates. He made *consumers'* and *producers' surplus* (first found in Dupuit), *price elasticity of demand* (found in embryonic form in Cournot), the *marginal productivity theory of distribution* (found in von Thünen), and *external* (pecuniary and technological) and *internal* (returns to scale) *economies* permanent additions to the economist's set of tools. Marshall laid the groundwork for the statistical measurement of demand and cost functions. He introduced diagrammatic analysis into the discipline, the impact of which any comparison of modern economics texts with pre-Marshallian ones can quickly show. (Marshall himself, however, buried his diagrams in footnotes and appendices.) Most of all, of course, Marshall is known for his analysis of supply and demand. While Ricardo had elucidated the cost side of market phenomena and Jevons the utility side, Marshall put them together in his famous "scissors diagram" (Figure 6.2). About the old controversy, he writes:[1]

---

[1] Alfred Marshall, *Principles of Economics,* 8th ed. (London: Macmillan, 1922), p. 348.

We might as reasonably dispute whether it is the upper or the under blade of a pair of scissors that cuts a piece of paper, as whether value is governed by utility or cost of production. It is true that when one blade is held still, and the cutting is effected by moving the other, we may say with careless brevity that the cutting is done by the second; but the statement is not strictly accurate, and it is to be excused only so long as it claims to be merely a popular and not a strictly scientific account of what happens.

Marshall hoped that all the manifold tools which he had developed would be put to work in the great task of conquering scarcity: "Political economy follows the actions of individuals and of nations as they seek, by separate or collective endeavour, to increase the material means of their well-being and to turn their resources to the best account."[2] But, like Adam Smith, he was also aware of the costs of economic growth:

When the necessaries of life are once provided, everyone should seek to increase the beauty of things in his possession rather than their number. . . . An improvement in the artistic character of furniture and clothing train the higher faculties of those who make them, and is a source of growing happiness to those who use them. . . . The world would go much better if everyone would buy fewer and simpler things, and would take trouble in selecting them for their real beauty.

---

[2] Alfred Marshall, *Elements of Economics of Industry* (London: Macmillan, 1920), pp. 1, 83, and 84. For additional readings, *see:*

Pigou, A. C. *Memorials of Alfred Marshall.* London: Macmillan, 1925.
    Memorials by Edgeworth, Keynes, Pigou and others; selections from Marshall's writings and letters; a complete bibliography of his works.

Schumpeter, Joseph A. *Ten Great Economists.* New York: Oxford University Press, 1951, chap. 4.
    On Alfred Marshall.

Shove, G. F., et al. "The Centenary of the Birth of Alfred Marshall," *The Economic Journal,* December 1942, pp. 289–349.

CHAPTER 7

Monopoly and Cartels

MULTIPLE-CHOICE QUESTIONS

Circle the letter of the *one* answer that you think is correct or closest to correct.

1. The ability of a seller to raise the price of something that is for sale above the perfectly competitive level is called
   a. monopoly power.
   b. cartel power.
   c. a natural monopoly.
   d. a key resource.

When answering questions 2–3, refer to the graph below.

2. Given the market demand pictured here, a firm with cost curves such as those above
   a. is a natural monopoly.
   b. will never run a plant at the optimal rate.
   c. will not construct the optimum plant.
   d. is correctly described by all of the above.

3. Given the market demand pictured here, as well as cost curves such as those above,
   a. only three firms can profitably coexist in the industry.
   b. a monopoly with cost curve $SRATC_{41}$ will produce 240 units of output.

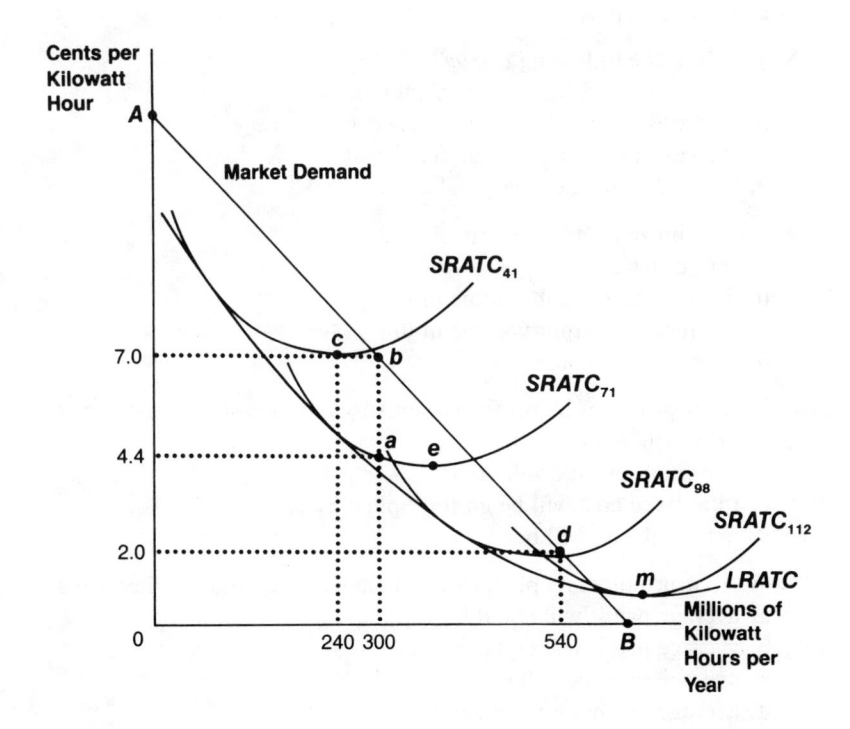

c. a monopoly with cost curve $SRATC_{41}$ will charge a price of 7 cents per unit of output.
   d. none of the above is true.

4. Which of the following is *not* a common source of monopoly?
   a. Decreasing returns to scale.
   b. Exclusive franchises.
   c. Exclusive ownership of key resources.
   d. Copyrights.

5. A monopoly that can sell either 500 units of output at $2 per unit or 700 units at $1 per unit has a marginal revenue
   a. of +$700.
   b. of −$1.50.
   c. of −$300.
   d. that cannot be calculated from the above data alone.

6. A monopoly could maximize its total revenue (if it wished) by
   a. minimizing its total cost.
   b. maximizing its profit.
   c. simultaneously doing (a) and (b).
   d. setting a price corresponding to a unitary price elasticity of demand for its product.

7. A monopoly makes profit (not necessarily maximum profit) only as long as
   a. the slope of its total-revenue curve exceeds that of its total-cost curve.
   b. the height of its average-revenue curve exceeds that of its average-total-cost curve.
   c. the height of its marginal-revenue curve exceeds that of its marginal-cost curve.
   d. either (a) or (c) is true.

When answering questions 8–12, refer to the graph on the top of page 115 about a monopoly firm.

8. Which of the following is *true*?
   a. Curve *A* indicates average total cost.
   b. Curve *B* indicates average total cost.
   c. Curve *C* indicates average fixed cost.
   d. Curve *D* indicates marginal cost.

9. To maximize profit, the firm should
   a. shut down at once.
   b. produce an output volume of 0*d*.
   c. produce an output volume of 0*m*.
   d. charge a price of *su*.

10. When it produces its profit-maximizing output volume, the firm's
    a. price will be *km*.
    b. marginal revenue will be *fi*.
    c. total fixed cost will be *gh* multiplied by 0*i*.
    d. marginal cost will be *xz*.

11. When it produces its profit-maximizing output volume, the firm's
    a. average profit will equal *kl*.
    b. average profit will equal *bc*.
    c. average revenue will equal *ei*.
    d. average revenue will equal *np*.

12. When it produces its profit-maximizing output volume, the firm's
    a. output will equal 0r.
    b. total cost will equal lm multiplied by 0m.
    c. total cost will equal tu multiplied by 0u.
    d. total cost will equal vw multiplied by 0w.

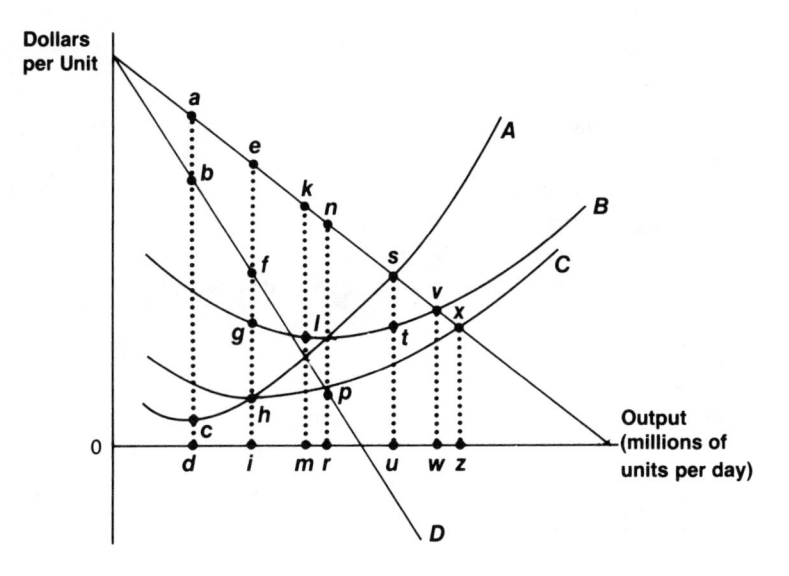

When answering questions 13–14, refer to the following graph about a monopoly.

13. Which of the following is *false*? When it maximizes profit, the firm
    a. charges a price of $3 per unit.
    b. has an average variable cost of $1.50 per unit.
    c. has an average variable cost that cannot possibly be calculated from the information given above.
    d. has an average fixed cost of $1.50 per unit.

14. To maximize profit, the firm will produce an output volume
    a. such that marginal cost is less than $3 per unit.
    b. such that average revenue exceeds average total cost.
    c. such that marginal revenue is negative.
    d. of 40 million units per day.

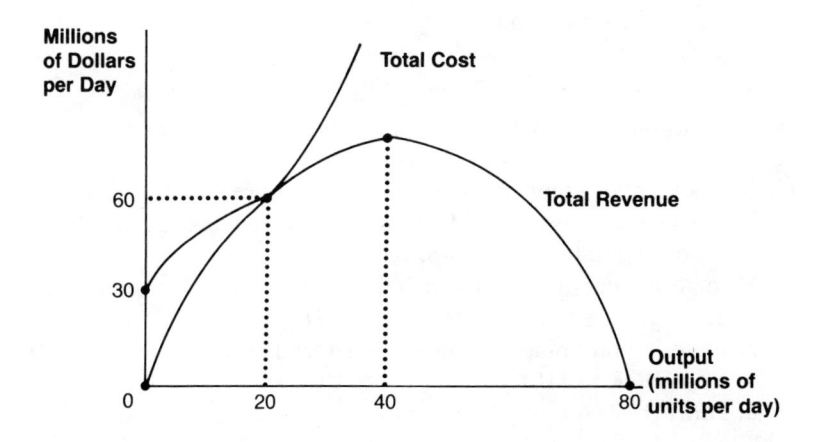

When answering questions 15–16, refer to the following graph about a profit-maximizing monopoly.

15. Assuming that the slope at *a* equals that at *b*, this firm will in the short run
   a. produce 100 million units per day.
   b. charge a price of 50 cents per unit.
   c. do both (a) and (b).
   d. do neither (a) nor (b).

16. Assuming that the slope at *a* equals that at *b*, this firm will in the long run
   a. charge a price of $2 per unit.
   b. charge a price of 40 cents per unit.
   c. have an average variable cost of zero.
   d. have an average variable cost of 60 cents per unit.

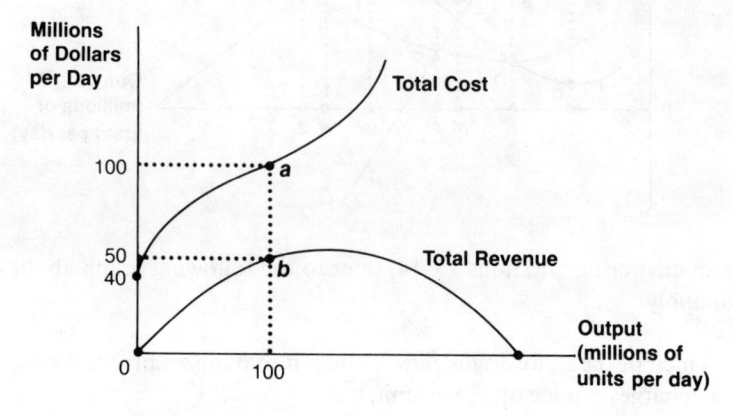

17. In the long run, a profit-maximizing monopoly produces an output volume that
   a. equates long-run marginal cost with long-run marginal revenue.
   b. equates long-run average total cost with long-run average revenue.
   c. assures permanent positive profit.
   d. is correctly described by both (a) and (c).

18. The Lerner index
   a. measures monopoly power as $\frac{P - MC}{MC}$.
   b. equals 1 for monopolies.
   c. equals zero for perfectly competitive firms.
   d. is correctly described by all of the above.

When answering questions 19–21, refer to the graph on page 117.

19. In an otherwise competitive lemon market, price could be raised from its competitive level $0B$ to the higher cartel level $0C$ by
   a. simply legislating the new price.
   b. imposing import or marketing quotas that tilted the supply curve around point E from EGI to ED.
   c. imposing rationing that tilted the demand curve around point D from DGK to DEF.
   d. either (b) or (c).

20. If the government stood ready to buy, at a price of $0C$, any lemons that producers could not sell to private buyers,
   a. private buyers would buy quantity $0H$.
   b. private buyers would spend $\$0BGH$.
   c. the government would spend $\$FDIL$ of the taxpayers' money.
   d. all of the above would be true.

21. If the governemnt offered to the producers of lemons subsidies that guaranteed receipts of $0C$ per bushel but refused to buy any lemons from these producers,
   a. the market price of lemons would come to equal $0A$.
   b. government subsidy payments of $ACIK$ would become necessary.
   c. producers would produce quantity $0L$.
   d. all of the above would occur.

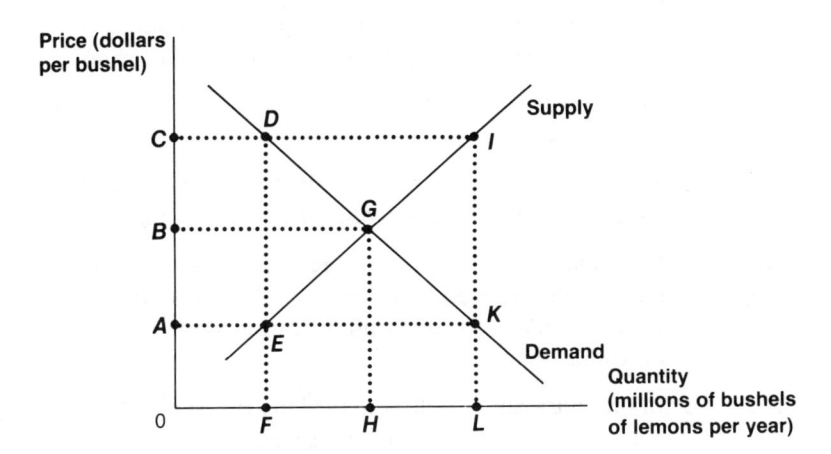

22. Textbook publishers usually prefer a higher sales price for books than do authors who are paid a fixed percentage of list price because
   a. profit cannot be maximized when the price elasticity of demand is zero (and that is what such authors want).
   b. profit cannot be maximized when marginal revenue is zero (and that is what such authors want).
   c. profit cannot be maximized when marginal revenue is maximized (and that is what such authors want).
   d. of all of the above.

23. A profit-maximizing third-degree price discriminator will
   a. charge a higher price in that market which has a lower price elasticity of demand at any given price.
   b. charge a lower price in that market which has a lower price elasticity of demand at any given price.
   c. charge a higher price in that market which has a lower price elasticity of demand at any given quantity.
   d. do none of the above.

24. A monopoly's demand function is $Q_D = 500 - 50P$. This implies that the demand line's vertical intercept is
   a. .02.
   b. 10.
   c. 50.
   d. 500.

25. A monopoly's total cost function is $TC = 120Q - Q^2 + .02Q^3$. This implies
   a. that the firm's average variable cost is
      $AVC = 120 - Q + .02Q^2$.
   b. that the firm's fixed cost is zero.
   c. both (a) and (b).
   d. that the firm's profit is maximized.

TRUE-FALSE QUESTIONS

In each space below, write a *T* if the statement is true and an *F* if the statement is false.

_____ 1. In perfectly competitive markets, sufficiently cunning manipulators of prices can always get more goods for themselves even when the overall quantity of goods is unchanged or shrinking: at the expense of other people.

_____ 2. Monopolies are called price takers because they confront their buyers with take-it-or-leave-it offers.

_____ 3. For a monopoly with zero marginal cost at all output levels, revenue maximization implies profit maximization.

_____ 4. If the manager of a monopoly refused to produce another unit of output for $1, although it could be sold for $2, the manager would obviously not be caring about profit maximization.

_____ 5. In the short run as well as in the long run, a profit-maximizing monopoly will charge a price in excess of marginal cost.

_____ 6. Domestic agricultural cartels are often buttressed by import quotas that set minimum physical limits on the amounts of imported goods.

_____ 7. Markup pricing is consistent with the marginal-cost-equals-marginal-revenue rule of profit maximization as long as marginal cost is constant over a wide range of output (and thus equals average variable cost).

_____ 8. Third-degree price discrimination is also called perfect price discrimination.

_____ 9. The imposition of a per-unit tax on a monopoly's output will not change the monopoly's output.

_____ 10. The imposition of a 40 percent profits tax on a monopoly will not change the monopoly's price.

PROBLEMS

1. The following graph pertains to a natural monopoly.
   a. Draw in a set of three possible curves of short-run average total cost, including those pertaining to the present plant (the short-run marginal cost curve of which is printed in the graph) and to the optimum plant.
   b. Now add the two missing short-run marginal-cost curves as well as a possible curve of long-run marginal cost.
   c. Determine graphically how much the firm will produce in the short run and what price it will charge, given the short-run marginal-cost curve printed in the graph.
   d. Would you change your answer to (c) with respect to the long run? Explain.

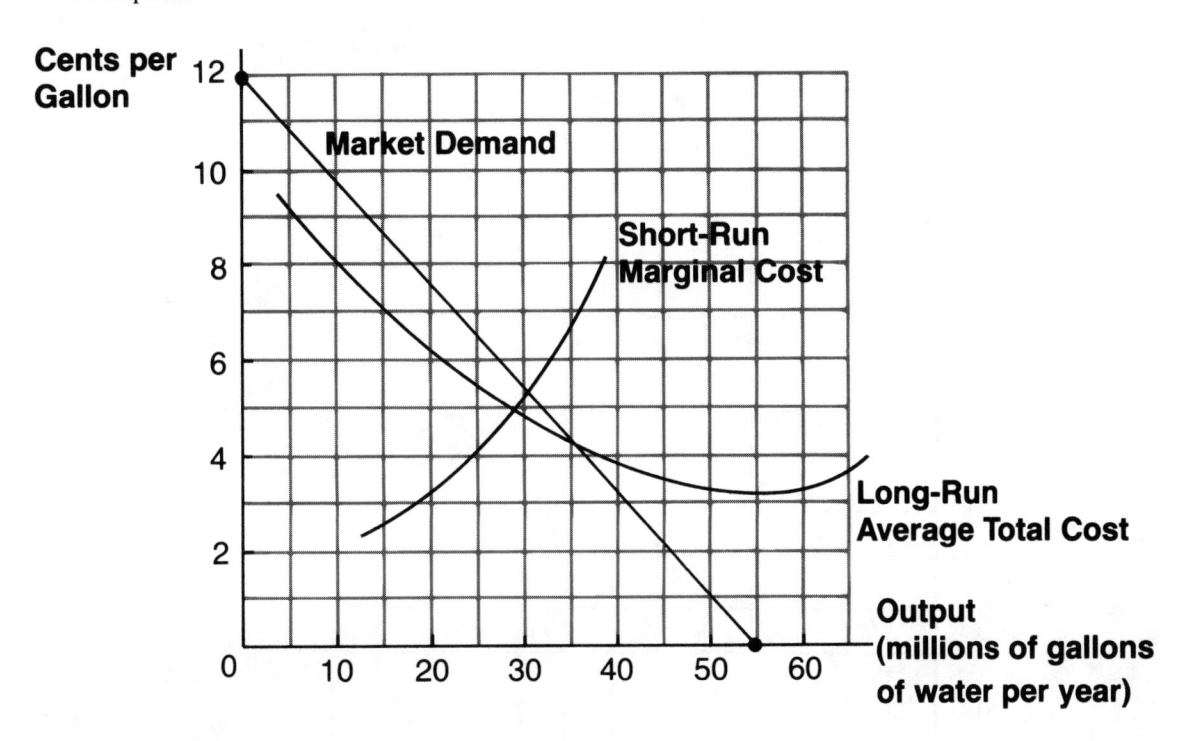

2. Consider this market-demand schedule facing a monopoly.

Price (dollars per unit)	Quantity (units per year)	Revenue Total	Revenue Marginal
(A) 500	0		
(B) 400	12		
(C) 300	24		
(D) 200	36		
(E) 100	48		
(F) 0	60		

a. Enter total and marginal revenue in the above table.
b. Plot your demand and marginal-revenue data in the following graph (plot marginal revenue at midpoints of associated quantity ranges).

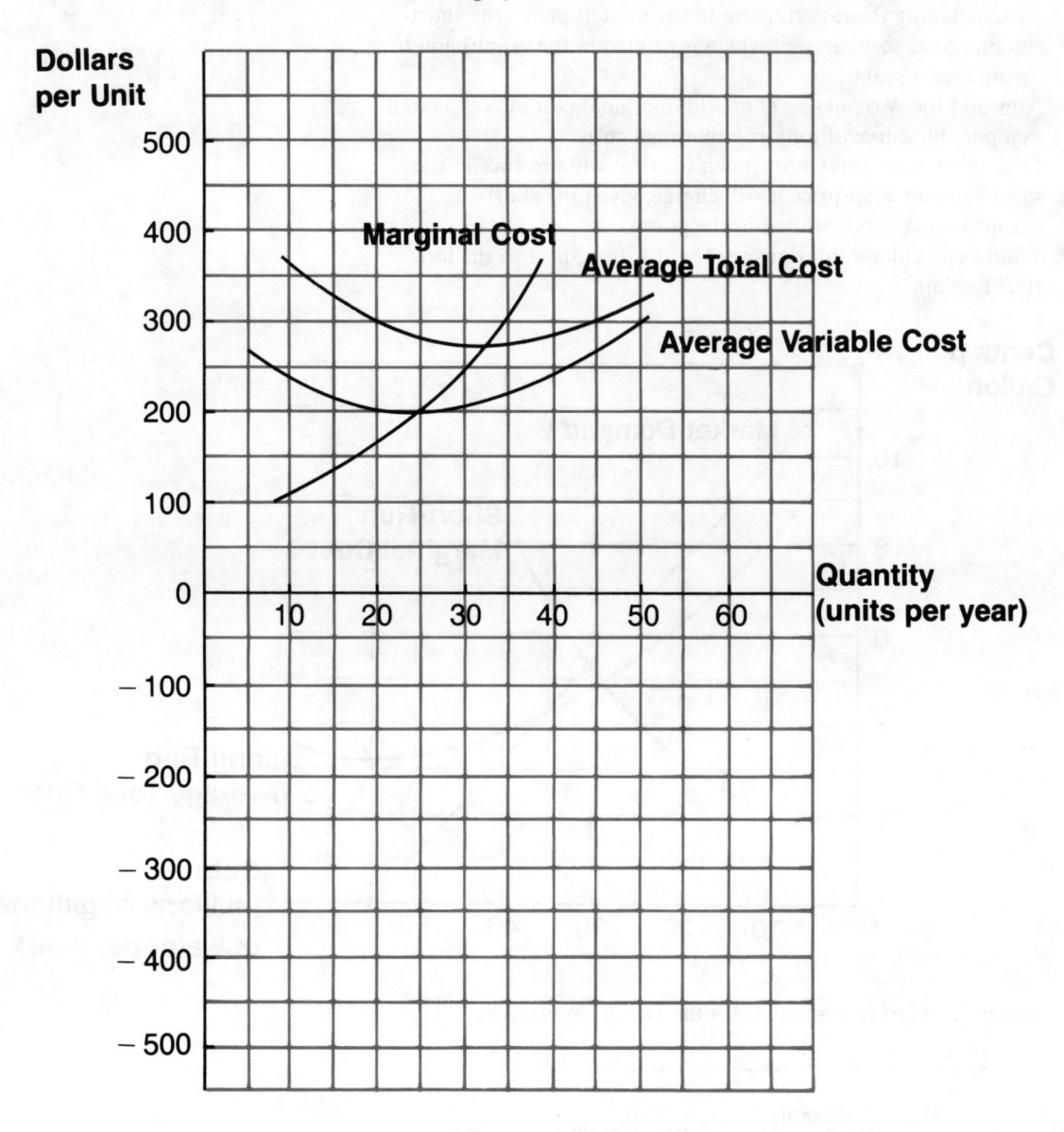

c. Graphically, determine the firm's profit-maximizing output level and price.
d. Determine also the firm's totals for revenue, cost, variable cost, fixed cost, and profit.
e. Identify the firm's supply curve.

3. **a.** Using the data from problem 2, plot the firm's curves of total revenue and fixed cost in the following graph.
   b. Identify in this graph the profit-maximizing levels that were computed above for the variables noted in problem 2(d).
   c. By how much would fixed cost have to rise before the firm's profit was eliminated? How much would the firm then produce in the short run? In the long run?
   d. Determine the maximum lump-sum tax a government could impose upon this firm in the short run and in the long run without affecting its level of production and price.
   e. Determine the maximum percent-of-profit tax a government could impose upon this firm in the short run and in the long run without affecting its level of production and price.
   f. What would happen if the government imposed upon this firm a tax per unit of output?

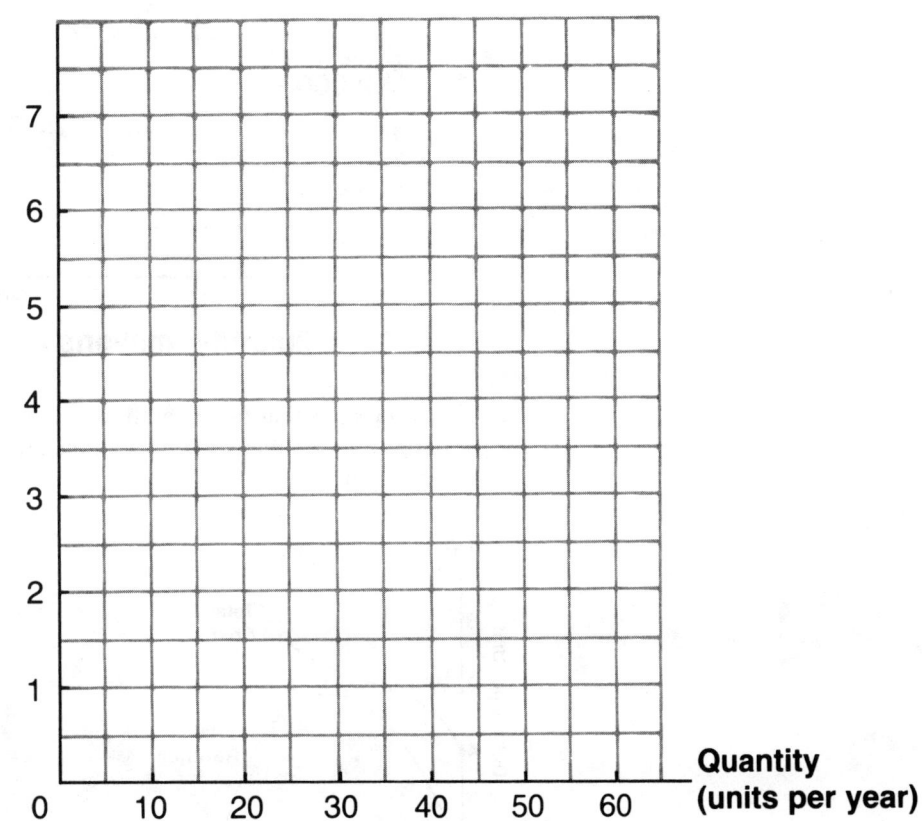

4. The following graph pictures a perfectly competitive market for raw tobacco.
   a. If you were a government official who wanted to guarantee tobacco farmers a minimum price of $3,000 per ton but did not want the government to have to buy up any surplus tobacco, how much of a subsidy would you have to be ready to pay? Illustrate your answer graphically.
   b. How else might you achieve your minimum-price-goal?

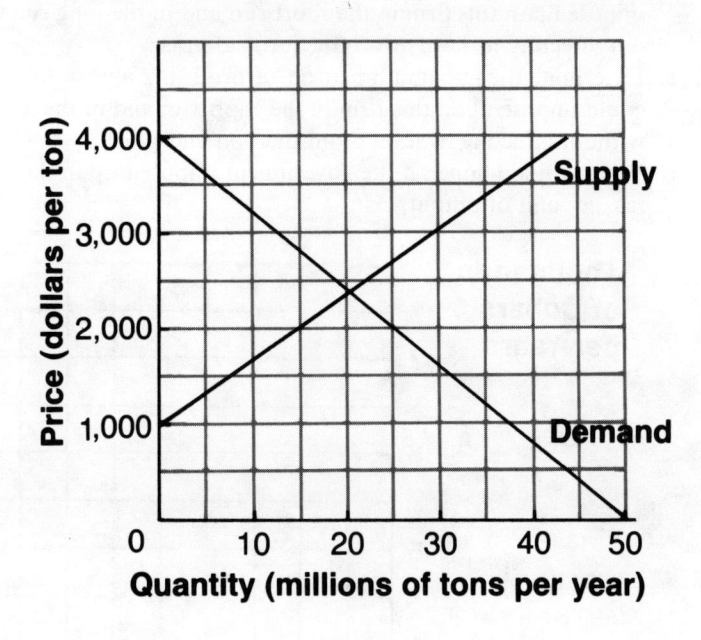

5. The two graphs that follow both refer to the same firm. A number of things are drastically wrong with the graphs. How many errors can you find?

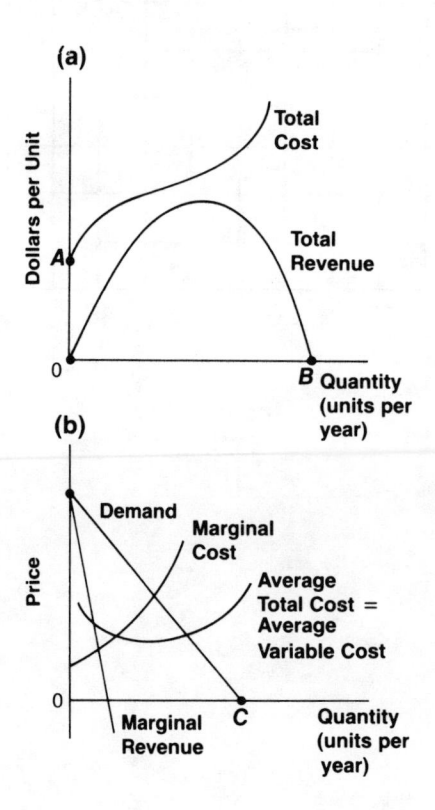

6. Consider the following graph. Assume that it refers to a monopoly that practiced perfect price discrimination.
   a. Determine graphically how much this firm would produce if it cared to maximize profit. Which price would it charge?
   b. Graphically indicate the sizes of the firm's total profit and total cost.

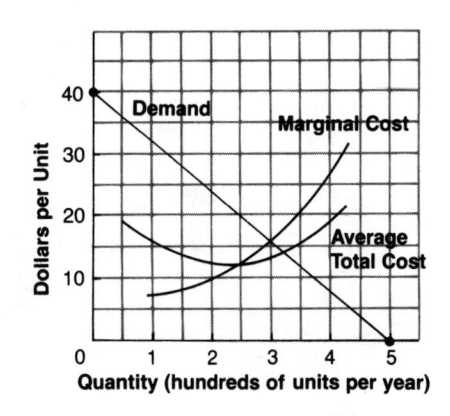

7. Consider the following graph. Assume that it refers to a monopoly that practiced second-degree price discrimination (according to the price schedule indicated).
   a. Determine graphically how much this firm would produce if it cared to maximize profit. Which price would it charge?
   b. Graphically indicate the sizes of the firm's total profit, total cost, total fixed cost, and total variable cost (if you can).

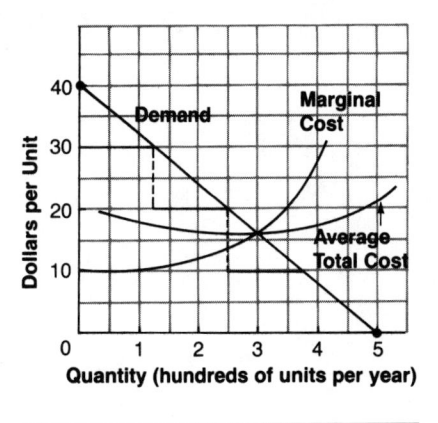

8. The table shows two straight-line demand schedules facing a monopoly in two separated markets, A and B.
   a. Plot the two demand curves in the following graph.

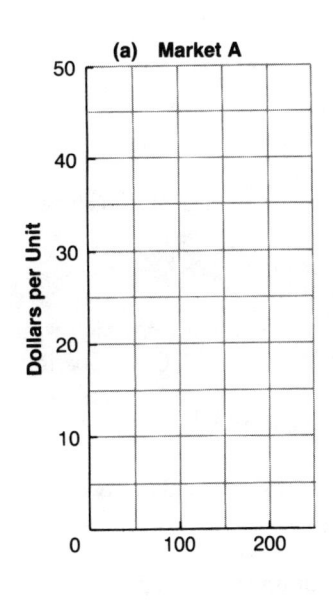

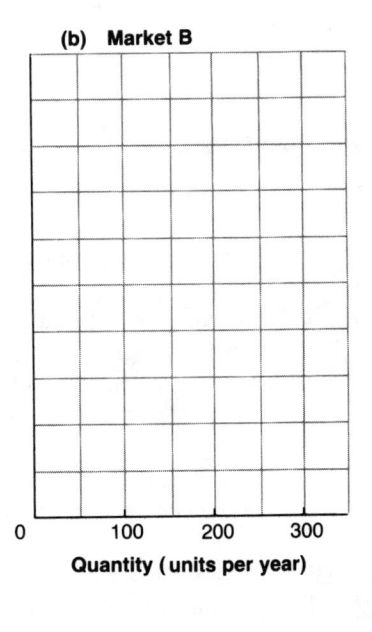

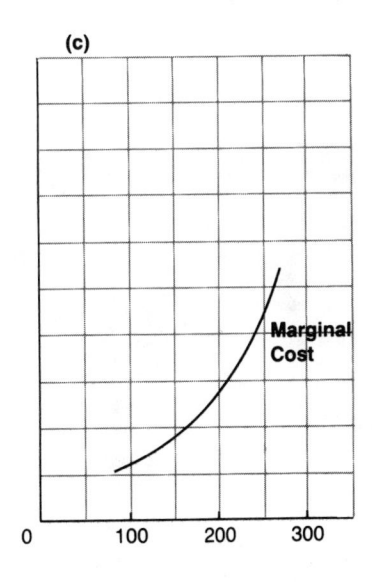

Price	0	10	20	30	40	50
Sales in A:	200	160	120	80	40	0
Sales in B:	300	150	0	0	0	0

   b. Plot the two marginal revenue curves.
   c. Determine the best third-degree price discrimination strategy, using all the information now available in graphs (a), (b), and (c).

9. Consider the following graph. Using the markup formula discussed in the text ("Application 3: Markup Pricing"), calculate the percentage markup on the firm's average variable cost of production that would be equivalent to equating marginal cost with marginal revenue.

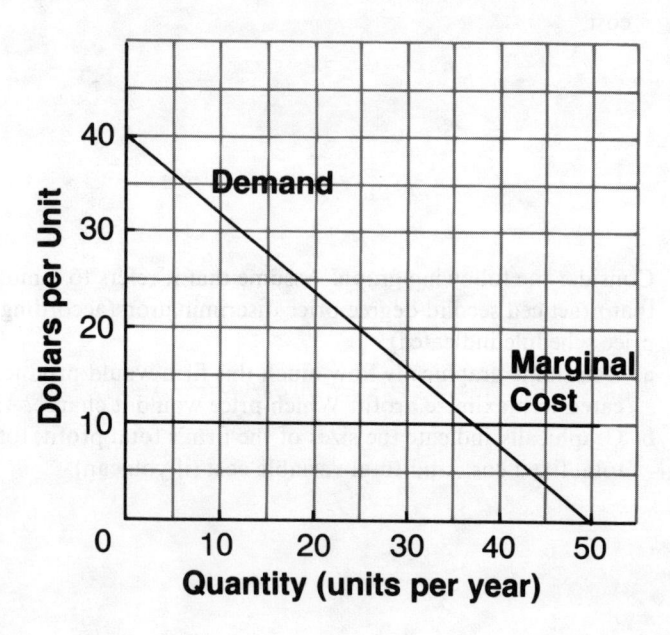

ANSWERS

Multiple-Choice Questions

1. a	2. d	3. d	4. a	5. b	6. d	7. b	8. b	9. c
10. a	11. a	12. b	13. c	14. a	15. d	16. c	17. a	18. c
19. b	20. c	21. d	22. b	23. a	24. b	25. c		

True-False Questions

1. F (Such price manipulators can accomplish this in *imperfect* markets.)
2. F (Perfect competitors are price takers.)
3. T
4. F ($1 would be the marginal cost, $2 the price, which would imply a marginal revenue of *less than $2*. Quite possibly, marginal revenue would be even less than $1, making increased production foolish, indeed.)
5. T
6. F (The import quotas set *maximum* physical limits on amounts of imported goods.)
7. T
8. F (*First*-degree price discrimination is so called.)
9. F (Output will fall because the marginal-cost curve will shift up.)
10. T

Problems

1. **a.** Consider $SRATC_1$. Then consider $SRATC_2$ (pertaining to the present plant and being intersected at its minimum a by the printed short-run marginal-cost curve). Finally, consider $SRATC_3$ (pertaining to the optimum plant, the capacity output of which at b corresponds to the minimum of long-run average total cost).

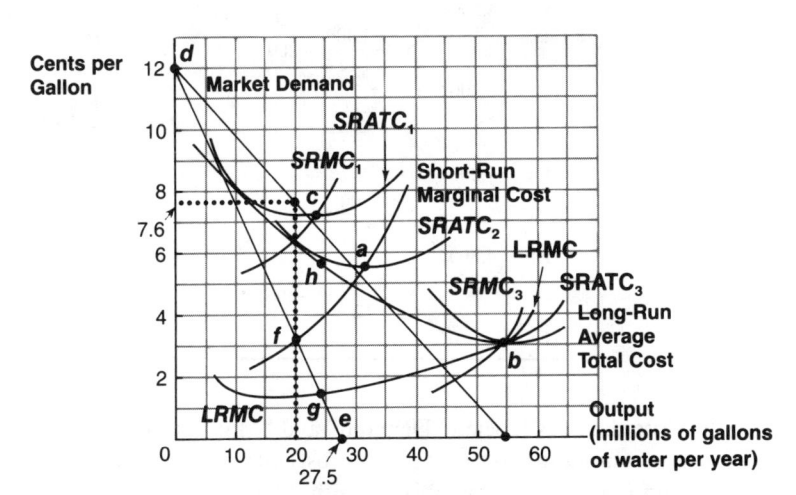

    b. Consider $SRMC_1$ and $SRMC_3$ and note that they must from below go through minimum points c and b of their associated short-run average-total-cost curves. Then consider $LRMC$ and note that it must lie below $LRATC$ as long as the latter is falling and above it when it is rising. Thus it must go from below through minimum $LRATC$ point b.

    c. Draw in marginal revenue de, lying at each price halfway between market demand and the vertical axis. Note the marginal revenue line's intersection f with printed short-run marginal cost. The firm will produce the corresponding output of 20 million gallons and charge the corresponding price of 7.6 cents per gallon.

    d. Yes. In the long run, this firm would choose a larger output volume that equated at g long-run marginal cost with (let us assume) long-run marginal revenue. The firm would then also construct a plant larger than the one depicted by $SRATC_2$; it would operate this plant below capacity at a point, such as h (vertically above g on the long-run average-total-cost curve). Note text Figure 7.6, "The Monopoly in the Long Run," wherein points A and G correspond to points g and h just discussed.

2. a.

	Revenue	
	Total (dollars per year)	Marginal (dollars per unit)
(A)	0	
		$\frac{+4{,}800}{+12} = 400$
(B)	4,800	
		$\frac{+2{,}400}{+12} = 200$
(C)	7,200	
		$\frac{+0}{+12} = 0$
(D)	7,200	
		$\frac{-2{,}400}{+12} = -200$
(E)	4,800	
		$\frac{-4{,}800}{+12} = -400$
(F)	0	

b. Note the lines labeled "Demand" and "Marginal Revenue."

c. Corresponding to the marginal cost/marginal revenue intersection b, the profit-maximizing output level is 20 units per year; the price is $333.33 per unit (Note: The demand data imply the equation $Q = 60 - \frac{60}{500}P$. Thus a Q of 20 goes with a P of precisely 333.33.)

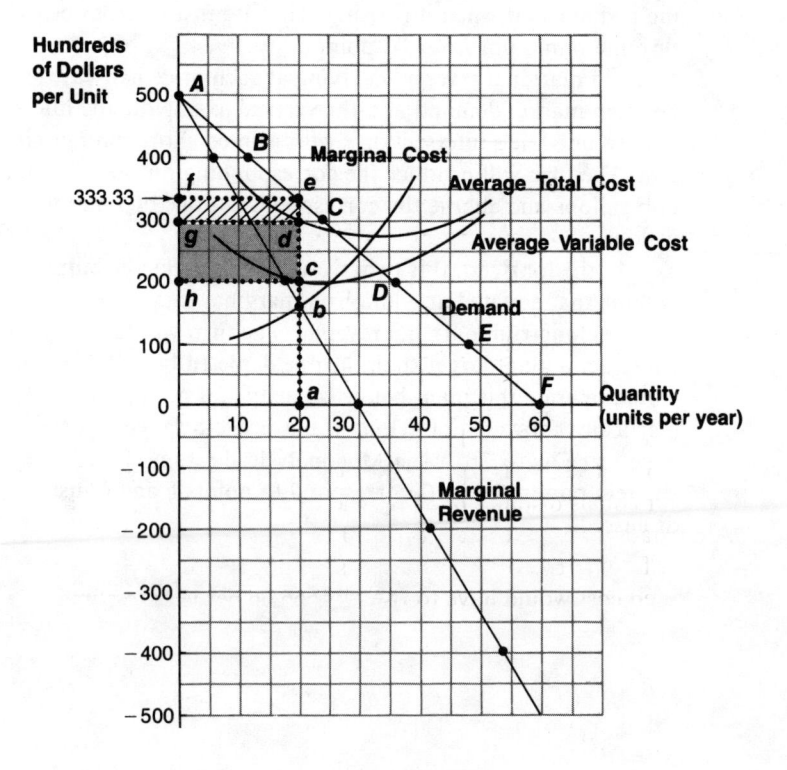

d. *Total revenue* = $333.33(20) = $6,666.66 per year (area 0*aef* in the graph). *Total cost* = $300(20) = $6,000 per year (area 0*adg* in the graph). *Variable cost* = $200(20) = $4,000 per year (unshaded area 0*ach* in the graph). *Fixed Cost* = $100(20) = $2,000 per year (shaded area *hcdg* in the graph). *Profit* = $33.33(20) = $666.66 per year (crosshatched area *gdef* in the graph).

e. A monopoly does not have a supply curve that tells us how much the firm would supply at various market-determined prices. The monopoly itself *sets* the price and in such a way as to make people demand the very quantity (here 20 units per year) that the firm has decided would be profit-maximizing because $MC = MR$ (here at *b*). If anything, a monopoly has only a supply *point* (here *e*).

3. a. Note the lines labeled "Total Revenue" and "Fixed Cost." Note: Total revenue is maximized (point *M*) at $7,500 per year and 30 units of output. This quantity can be read off in the graph for the answer to 2(b) where marginal revenue is zero at $Q = 30$ (and zero marginal revenue implies maximum total revenue). A quantity of 30 corresponds, in turn, to a price of $250 per unit and thus to a total revenue of $250(30) = $7,500 per year.

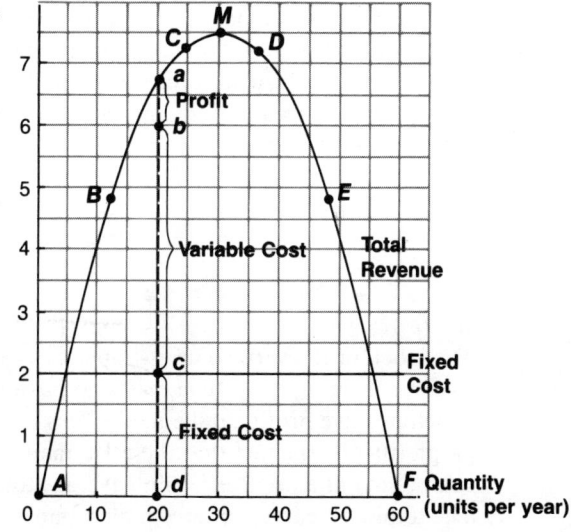

b. Note the vertical dashed line at $Q = 20$: Total revenue of $6,666.66 per year equals distance *ad*. Total cost of $6,000 per year equals distance *bd*. Variable cost of $4,000 per year equals distance. *bc*. Fixed cost of $2,000 per year equals distance *cd*. Profit of $666.66 per year equals distance *ab*.

c. Fixed cost would have to rise by $666.66 per year to eliminate profit. The firm would then produce the same amount of $Q = 20$, regardless of whether we considered the short run or long run.

d. In the short run, the government could impose a maximum lump-sum tax of $2,666.66 per year equal to profit plus fixed cost. The firm would then still be able to cover its variable cost, but would make a loss of $2,000 per year, the same loss it would make if it shut down at once. Such a loss-creating tax would put an end to the business in the long run (when fixed cost could be escaped). Thus the maximum long-run tax only equals profit, or $666.66 per year.

e. In the short run, for the reasons just cited under 3(d), the maximum percent-of-profit tax per year could be $\frac{\$2,666.66}{\$666.66} \cdot 100 = 400$ percent. In the long run, it could only be 100 percent.

f. The firm's marginal cost curve in the graph for the answer to 2(b) would shift up vertically by the amount of this per-unit tax. Thus intersection b would move up and left along the marginal-revenue line. The firm would select a new profit-maximizing position, involving a lower output and higher price.

4. a. At $3,000 per ton, farmers would supply quantity $dc = 29$ million tons per year. These could be sold privately (point b) only at $1,680 per ton. (Note: the demand line implies the equation

$$Q = 50 - \frac{50}{4,000} P.$$

If $Q = 29$, P is $1,680.) Thus government would have to pay a subsidy of $3,000 - 1,680 = \$1,320$ per ton (distance bc), or a total of $38.28 billion per year (shaded area $abcd$).

b. You might restrict supply to make it go through point e (using import quotas, acreage quotas, marketing quotas, and the like). You might instead force up demand to make it go through point c (mandating minimum purchases beyond what people would voluntarily buy).

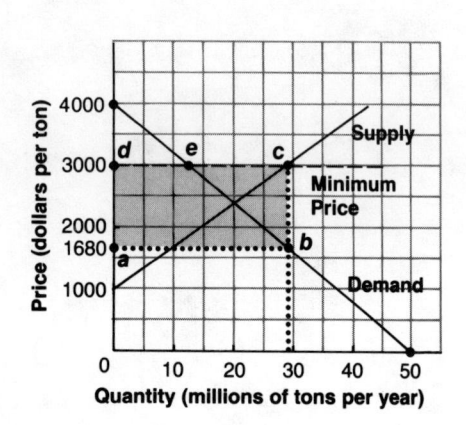

5. Graph (a):
   a. The label for the vertical axis should be "Dollars per Year."
   b. The curves imply that there exists no possible output level at which positive profit can be made. This is contradicted by graph (b)—by the fact that there are many levels of output that can be sold at a price exceeding average total cost.
   c. The total-cost curve's intercept at A implies the existence of fixed cost equal to $0A$. Yet graph (b)'s claim that average total cost is identical with average variable cost implies zero fixed cost.
   d. The total-revenue curve's maximum should correspond to the midpoint of the lower graph's straight demand line (where elasticity $= |1|$). It doesn't.
   e. The total-revenue curve's intercept with the horizontal axis at B should correspond to the demand line's intercept at C. It is impossible for total revenue to be positive for quantities beyond C once price is zero.
   f. As quantity rises, total cost is rising: at first at a decreasing rate, than at an increasing rate. This implies at first falling, then rising marginal cost. Yet marginal cost in the lower graph is steadily rising.

5. Graph (b):
   a. The label for the vertical axis should be "Dollars per Unit" because not only revenue per unit (which is price), but also various costs per unit are measured in the graph.
   b. The marginal-cost curve must logically intersect the average-total-cost curve at the latter's minimum. It doesn't.
   c. The marginal-revenue curve must lie halfway between the straight demand line and the vertical axis at every price. It doesn't.

6. a. The firm would produce 300 units per year, because marginal revenue (in this case identical with the demand line) intersects marginal cost at *a*.
   The firm would charge a different price for every unit sold: $39.92 for the first unit (remember $Q = 500 - \frac{500}{40}P$) and $16 for the 300th unit.
   b. Total cost would equal 0*cbd*; total profit *dbae*.

7. a. The firm would equate stair-step marginal revenue with marginal cost at *a* and produce 250 units per year. It would charge $30 each for the first 125 units and $20 each for the second 125 units.
   b. The firm's total profit would equal the shaded area; the total cost would equal area 0*ebc* (*eb* being the average total cost of producing 250 units per year). Total fixed cost would equal *abcd*, because total variable cost would equal 0*ead*, the *sum* of marginal costs up to the chosen output level.

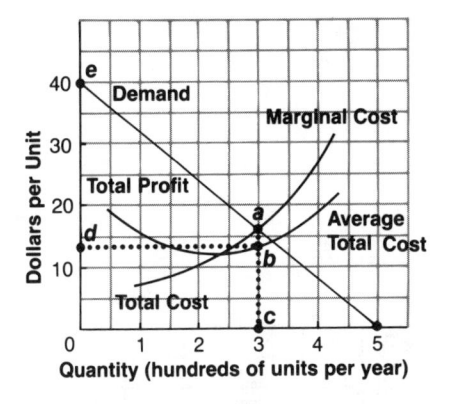

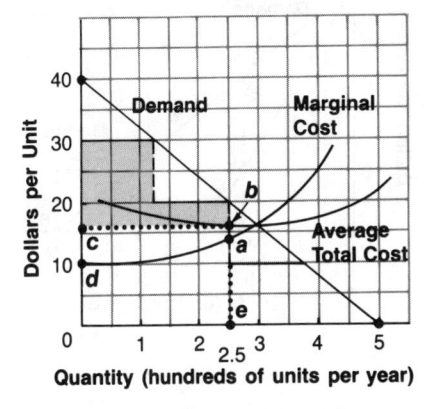

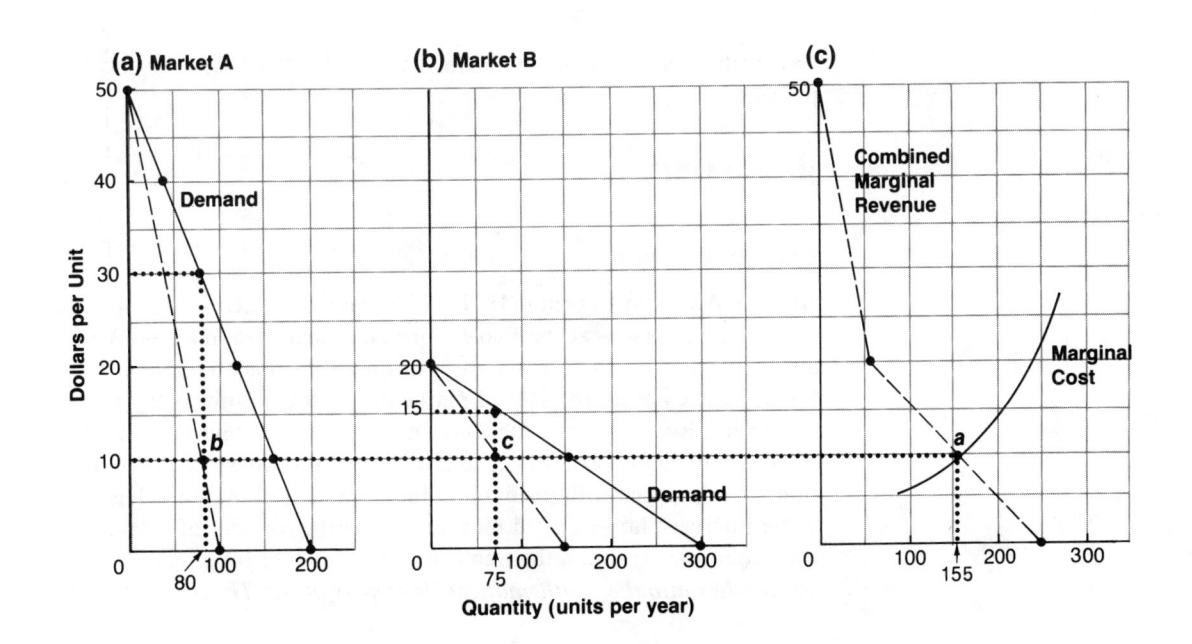

b. The marginal-revenue curves are the broken lines, halfway between demand and the vertical axis.

c. The firm would maximize profit by equating marginal cost with *combined* marginal revenue at *a* and selling 155 units per year. This implies, as point *b* tells us, selling 80 units at $30 each in Market A. It implies, as point *c* tells us, selling 75 units at $15 each in Market B. (The demand line of Market B implies the equation $Q = 300 - \frac{300}{20}P$. Thus $Q = 75$ goes with $P = 15$.) *Caution:* The horizontally combined marginal revenue only serves the purpose of finding the best output volume and hence the marginal cost (here $10) with which the price-discriminating firm then separately equates the marginal revenues of the two markets. The kinked line of combined marginal revenue should not be confused with the kinked demand curve discussed in text Figure 8.2, "Sweezy's Model," and in that chapter's Question 2. Those kinked demand curves refer to oligopolists who do *not* engage in price discrimination but charge uniform prices.

9. The marginal-revenue line can be drawn halfway between demand and the vertical axis, yielding line *ab*.

   Intersection *c* implies a profit-maximizing output of 18.75 units per year and a price of $25 per unit. For constant marginal cost, average variable cost is the same (here $10 per unit) and the markup formula then suggests a price of

$$P = AVC\left(\frac{|\epsilon|}{|\epsilon| - 1}\right).$$

Elasticity at point *d* equals *de/da* which, in turn, equals *fe/f0* or $\frac{31.25}{18.75} = \frac{5}{3}$. Thus the recommended price comes to

$$P = 10\left(\frac{\frac{5}{3}}{\frac{5}{3} - 1}\right) = 10\left(\frac{\frac{5}{3}}{\frac{2}{3}}\right) = 10\left(\frac{5}{2}\right) = 10(2.5) = 25.$$

This is the same price noted above. The markup, of course, equals 150 percent.

BIOGRAPHY 7.1

Antoine A. Cournot

Antoine Augustin Cournot (1801–1877) was born at Gray, France. He studied mathematics at the École Normale Supérieure in Paris. While a student, he worked as a secretary for one of Napoleon's generals. Later, he became a professor of mathematics at the University of Lyons and Rector, first at the Academy of Grenoble and then at the Academy of Dijon. The works that brought him widespread recognition are concerned with probability theory and epistemology. His wider interests, however, led him to become the founder of mathematical economics, through the publication, in 1838, of his greatest book, *Researches into the Mathematical Principles of the Theory of Wealth.*

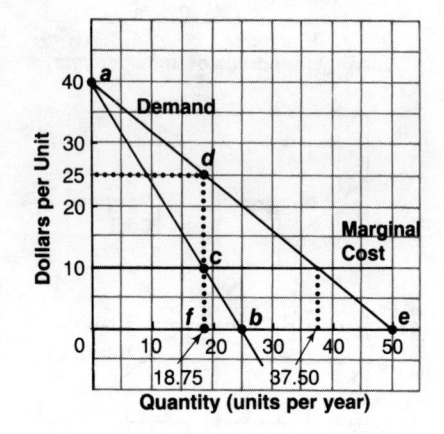

At the time, the book had no impact at all. Not a single copy was sold! This was, of course, disappointing to the author, who proceeded to simplify the presentation and produce two less mathematical versions of the book in 1863 (*Principles of the Theory of Wealth*) and 1876 (*Summary View of Economic Doctrines*) but to no avail. Not until William Jevons (see Biography 2.2) paid glowing tribute to Cournot two years after his death were his pioneering qualities recognized.

Cournot's great book has few equals in economics for sheer originality and boldness of conception. It contained the nucleus of Alfred Marshall's economics. Unlike any previous book, it developed a theory of monopoly, introducing for the first time demand, marginal-revenue, and total-revenue functions, contrasting these with total and marginal costs, and deriving clearly the profit-maximizing principle of marginal-revenue-equal-to-marginal-cost. Starting from monopoly, Cournot similarly explored the economics of two sellers (duopoly), few sellers (oligopoly) and, eventually, innumerable sellers. While Cournot took the partial-equilibrium approach to analysis, he was not blind to the desirability of studing the entire economic system at once, but he thought that such a general-equilibrium approach was beyond the reach even of mathematical analysis. Even so, he influenced Léon Walras, inventor of general equilibrium analysis (see Biography 13.2), no less than Alfred Marshall (see Biography 6.1). And Cournot produced his work in defiance of dispiriting conditions. During many years, he was troubled by an infirmity of the eyes that made continuous work impossible and eventually led to blindness.[1]

---

[1] For readings on Cournot, *see:*

Moore, Henry L. "The Personality of Antoine Augustin Cournot," *The Quarterly Journal of Economics,* May 1905, pp. 370–99; and Nichol, A. J. "Tragedies in the Life of Cournot," *Econometrica,* July 1938, pp. 193–97.

CHAPTER 8

Oligopoly and Monopolistic Competition

MULTIPLE-CHOICE QUESTIONS

Circle the letter of the *one* answer that you think is correct or closest to correct.

1. *Monopolistic competition* refers to a market structure in which
   a. the entry of new firms is difficult.
   b. differentiated products are being offered for sale.
   c. strategic behavior takes on crucial importance.
   d. all of the above are true.

2. The name of Cournot is connected with
   a. focal point price.
   b. the kinked demand curve.
   c. product group.
   d. the reaction curve.

When answering questions 3–5, refer to the accompanying graph about an oligopolistic firm.

3. The kinked demand curve shown here implies that the firm in question expects
   a. to sell relatively little itself (segment *CE*), while competitors sell relatively much (segment *AC*).
   b. that competitors will match price decreases but will not match price increases which this firm initiates.
   c. that competitors will never initiate price changes.
   d. all of the above.

4. A marginal revenue curve for this kinked demand curve
   a. must start at *A*, bisect *BC*, and end at *E*.
   b. must start at *A* and intersect the horizontal axis halfway between 0 and *E*.
   c. clearly lies above the horizontal axis for all quantities between 0 and *D*.
   d. cannot be derived without further information.

5. We can expect that this firm's
   a. output falls short of 0*D*.
   b. price exceeds 0*B*.
   c. marginal cost curve goes through segment *CD*.
   d. equilibrium is correctly described by (a) and (b) above.

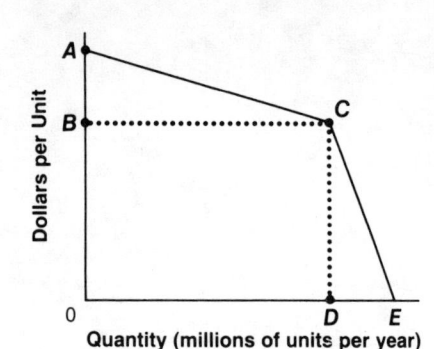

6. The kinked oligopolistic demand curve implies this about oligopolists:
   a. They may keep price unchanged in the face of moderate changes in marginal cost, but they would certainly not keep quantity supplied unchanged.
   b. They may keep price and quantity supplied unchanged even in the face of moderate changes in demand.
   c. They may keep price and quantity supplied unchanged even in the face of moderate changes in marginal cost.
   d. Both (b) and (c).

7. The kinked-oligopoly-demand-curve model
   a. has been confirmed empirically by Stigler and others.
   b. suggests that price should be more rigid in oligopolistic industries than in those dominated by monopoly.
   c. implies a strong bandwagon effect which overwhelms both the income and substitution effects of price changes.
   d. is correctly described by none of the above.

8. All participants in a basing point system
   a. quote prices equal to those charged by a specified firm at a specified place (the basing point).
   b. do (a) above, then add freight to the buyer's location from the basing point.
   c. do (a) above, then add freight to the buyer's location from the actual point of shipment.
   d. agree to base their prices on the point of intersection of their marginal revenue and marginal cost curves.

9. Product differentiation can refer to
   a. physical aspects of goods (color, flavor, etc.).
   b. legal aspects of goods (brand names, trademarks, etc.)
   c. conditions of sale (store location, business hours, etc.).
   d. all of the above.

10. According to Hotelling's paradox,
    a. competition among oligopolists tends to create monopoly.
    b. competition among oligopolists tends to create monopolistic competition.
    c. competition by means of product differentiation may lead to products that are hardly differentiated at all.
    d. competition by means of price may lead to uniform and rigid prices throughout the affected industry.

11. Persuasive advertising
    a. is designed to divert people's attention from facts to images and make them buy more as a result of imagined advantages.
    b. is easily disentangled from informative advertising that reduces people's ignorance about the existence of products and sellers, about product quality, about price, and more.
    c. represents pure social waste.
    d. is correctly described by all of the above.

12. If the firms in a monopolistically competitive "industry" made economic profit,
    a. they might earn this profit permanently.
    b. new firms would enter their "industry" until the profit was eliminated.
    c. the price elasticity of demand would have to be less than $|1|$.
    d. both (b) and (c) would be true.

When answering questions 13–15, refer to the following graph about a monopolistically competitive firm.

13. The firm's profit-maximizing output level equals
    a. $0I$.
    b. $0H$.
    c. $0G$.
    d. none of the above.

14. The firm's profit in the long run
    a. equals $BCDE$.
    b. equals $ABEF$.
    c. equals zero.
    d. cannot be determined from this graph.

15. When it produces output $0G$, the firm's
    a. fixed cost equals $ABEF$.
    b. variable cost cannot be determined.
    c. quasi rent equals $BCDE$.
    d. total revenue equals $BCDE$.

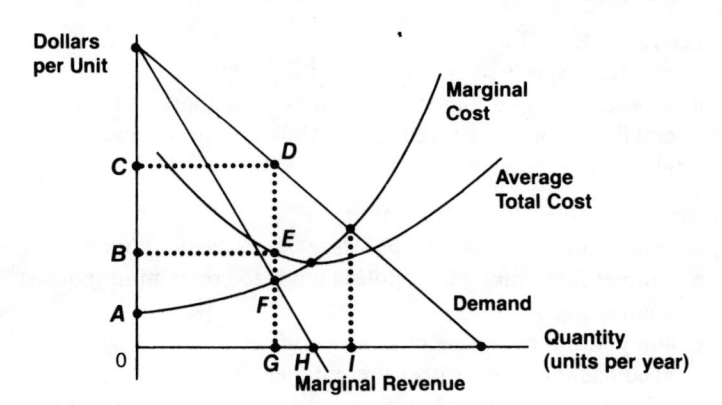

16. In long-run equilibrium, a monopolistically competitive firm will find
    a. marginal cost below average total cost.
    b. marginal cost equal to minimum average total cost.
    c. both (a) and (b).
    d. neither (a) nor (b).

17. A merger between firms that sell closely related products in the same geographic market is termed a
    a. horizontal merger.
    b. vertical merger.
    c. conglomerate merger.
    d. market-extension merger.

18. Most mergers during the second U.S. merger wave (1923–33) were
    a. vertical.
    b. horizontal.
    c. conglomerate.
    d. none of the above.

19. Conglomerate mergers
    a. occurred in the United States at a rapid rate in the 1950s and beyond.
    b. often serve to reduce risk through diversification.
    c. include product- and market-extension mergers.
    d. are correctly described by all of the above.

20. An eight-firm concentration ratio of 71 indicates that
    a. the eight largest firms in the industry account for 71 percent of industry sales.
    b. the 71 largest firms in the industry account for 8 percent of industry sales.
    c. the eight largest firms in the industry account for 29 percent of industry sales.
    d. the eight largest firms in the industry account for 29 percent of industry profits.

21. The market power of firms is likely to be understated by concentration ratios that
    a. refer to the nation as a whole.
    b. are based on too-narrow product groupings.
    c. ignore imports.
    d. do any of the above.

22. If an industry's four-firm concentration ratio equaled 44, its eight-firm concentration ratio
    a. would equal 88.
    b. could equal 88, but would not have to.
    c. would equal 56.
    d. could not possibly equal 56.

23. Typical examples of oligopolistic competition in the United States can be found in
    a. retail trade and services.
    b. mining and wholesale trade.
    c. most of manufacturing.
    d. both (b) and (c).

24. The U.S. economy is best described as
    a. a genuine monopoly economy, wherein buyers of anything are confronted by giant corporations or tightly knit cartels among a few (or even a multitude of) firms.
    b. a perfectly competitive economy, wherein buyers of anything are confronted by large numbers of independently acting sellers.
    c. an oligopoly economy, wherein buyers of anything are confronted by a small number of independently acting or informally (and, therefore, imperfectly) colluding sellers.
    d. none of the above.

25. Experience clearly shows that advertising
    a. raises costs and, therefore, prices.
    b. creates loyalty to brand names and, therefore, raises prices in the affected industry above the levels that would otherwise prevail.

c. leads to both (a) and (b).
d. can be an important weapon for those who want to challenge established firms by competing with them through lower prices.

TRUE-FALSE QUESTIONS

In each space below, write a *T* if the statement is true and an *F* if the statement is false.

_____ 1. When each of two firms in an industry makes output decisions on the assumption that its rival is supplying a fixed quantity that will not be adjusted in response to any output decision made by itself, a Cournot equilibrium is said to emerge.

_____ 2. In the United States, price wars among oligopolists have not occurred since the 1880s.

_____ 3. Cournot has argued persuasively that oligopolistic firms, seeking to avoid suicidal price wars, are likely to have sticky prices.

_____ 4. Maximum-fee schedules routinely circulated by professional societies among their members are examples of gentlemen's agreements.

_____ 5. Focal-point pricing is an important form of product differentiation.

_____ 6. The earning of economic profit by monopolistically competitive firms leads to the entry of new firms into their "industry."

_____ 7. The difference between a monopolistically competitive firm's capacity output and its profit-maximizing lower actual output in long-run equilibrium is called its *excess capacity*.

_____ 8. Mergers between firms that do not directly compete but are using related production processes are called *product-extension mergers*.

_____ 9. All conglomerate mergers are either product-extension mergers or market-extension mergers.

_____ 10. In many U.S. industries, concentration ratios are high, and the number of firms is small.

PROBLEMS

1. Consider the market demand pictured in graph (a) and assume there are two oligopolists who must somehow share it.
   a. In graph (b), graphically determine the Cournot equilibrium. Assume that each firm has zero marginal costs and believes
   1. that it can count on market demand minus whatever quantity the rival supplies and 2. that its rival will supply a fixed quantity that is not adjusted in response to any output decision of its own.

b. Assuming zero fixed costs, what would be the profit a monopoly could extract from these conditions?
c. What would be the combined profit of our two oligopolists?

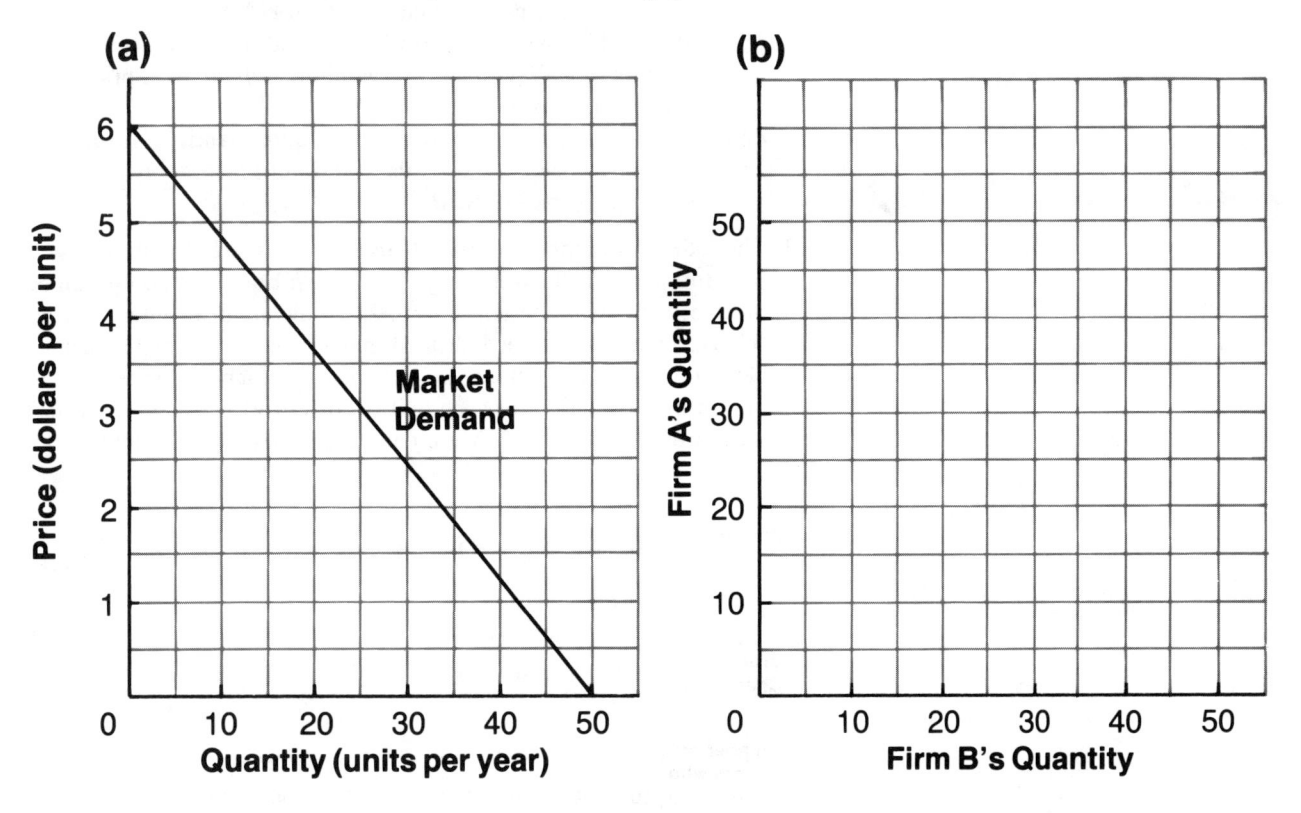

2. Consider an oligopolist's kinked demand curve, along with the cost curves shown in the following graph.

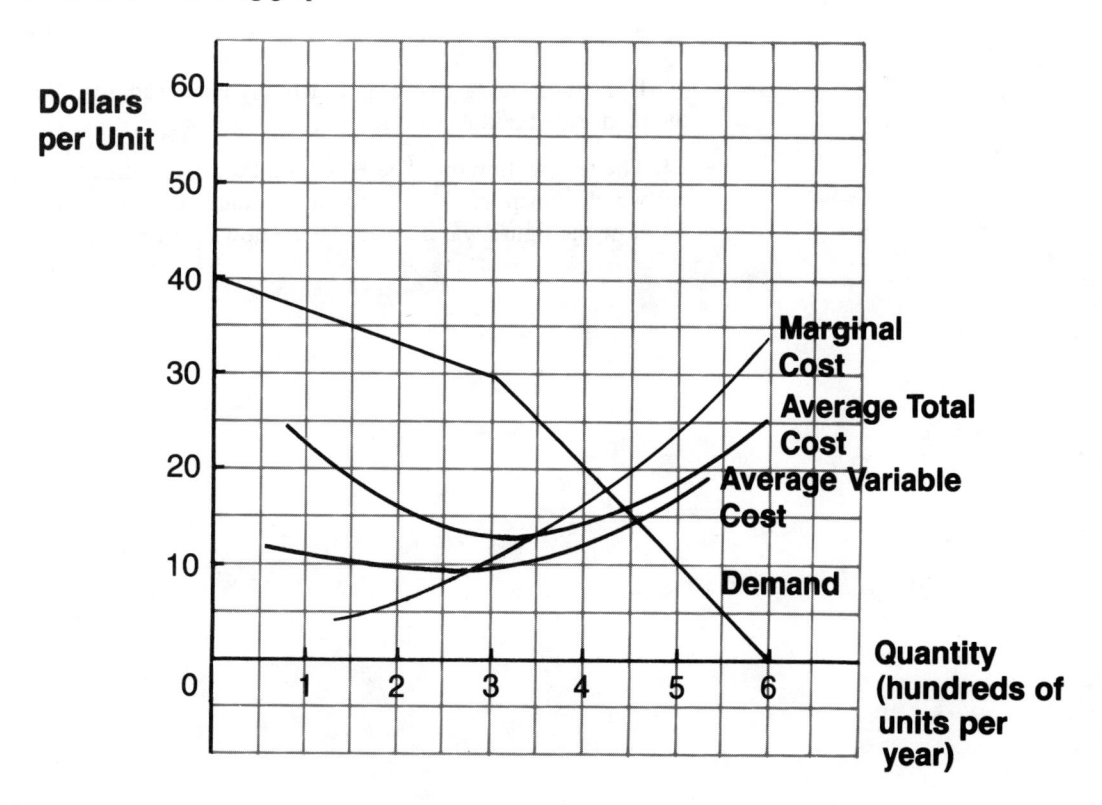

a. Graphically determine this profit-maximizing firm's totals of revenue, cost, fixed cost, variable cost, and profit.
b. By how much could the firm's marginal cost curve shift up or down before the firm would change its price?
c. What would happen to the firm's price and quantity if it suddenly came to believe that rivals would match *all* of its price changes (up or down)?
d. What would happen to the firm's price and quantity if it suddenly came to believe that rivals would not react to *any* of its price changes (up or down)?

3. The following graph is commonly used to illustrate Hotelling's paradox. Assume the following: 1. only two brands of cheese can be produced (by two different sellers); 2. people are distributed over "characteristics space" as indicated underneath the graph; 3. each person wants to contact a seller only once; 4. each person feels one unit of dissatisfaction for each unit of "distance" between his or her ideal brand (number 1, 2, 3, etc.) and the closest actual brand available.

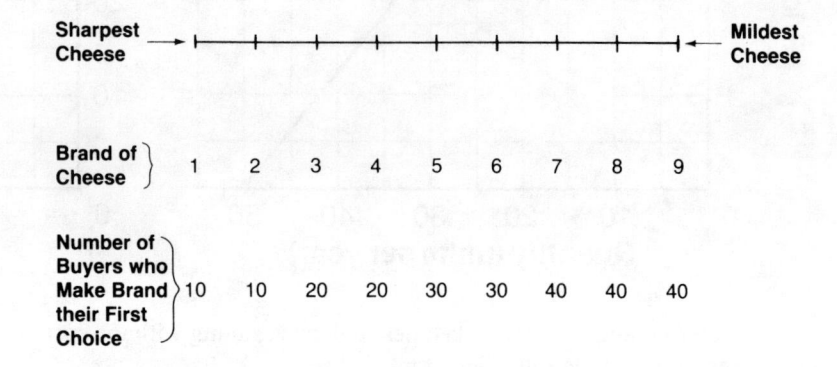

Sharpest Cheese ⟵|—|—|—|—|—|—|—|—|⟶ Mildest Cheese

Brand of Cheese: 1 2 3 4 5 6 7 8 9

Number of Buyers who Make Brand their First Choice: 10 10 20 20 30 30 40 40 40

a. Determine the two brands that would minimize overall dissatisfaction cost. What would this cost be?
b. Determine the two brands actually produced as well as the actual dissatisfaction cost.

4. The text section on "The Hotelling Paradox" considers a case in which transportation costs would equal $101 and asks you to show how the figure was derived. Do so now.

5. The following graph depicts the demand and cost conditions pertaining to a profit-maximizing, monopolistically competitive firm.

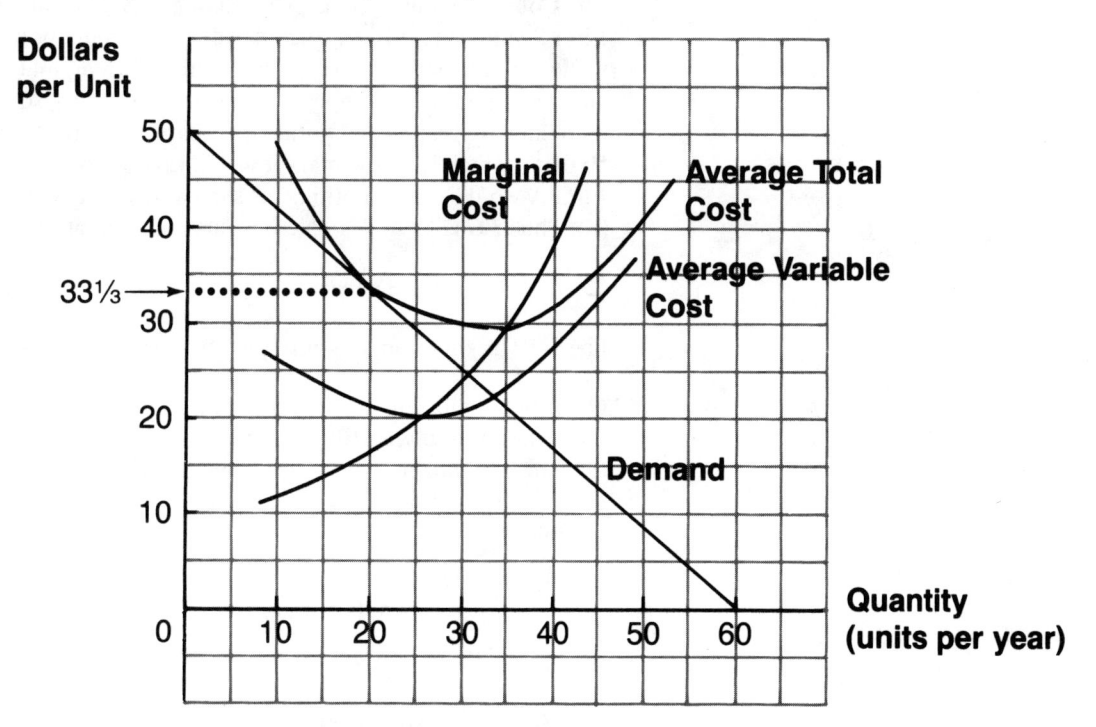

a. If the firm maximized profit, would it find itself in short-run or long-run equilibrium? Explain.
b. Graphically identify this profit-maximizing firm's totals of revenue, cost, fixed cost, variable cost, and profit.
c. Identify the extent of excess capacity under profit maximization.
d. Might a monopolistic competitor ever operate at capacity? Explain.

ANSWERS

Multiple-Choice Questions

1. b	2. d	3. b	4. c	5. c	6. c	7. b	8. b	9. d
10. c	11. a	12. b	13. c	14. c	15. a	16. a	17. a	18. a
19. d	20. a	21. a	22. b	23. d	24. d	25. d		

True-False Questions

1. T
2. F (The text provides numerous examples to the contrary.)
3. F (*Sweezy* argued thusly, not necessarily persuasively.)
4. F (*Minimum*-fee schedules are examples of such agreements.)
5. F (Pricing is not a form of product differentiation.)
6. T
7. T
8. T
9. F (While product-extension and market-extension mergers are conglomerate mergers, there are other types of conglomerate mergers as well.)
10. T

Problems

1. **a.** The Cournot equilibrium is pictured in graph (b). Note: If Firm B supplied nothing, Firm A would be a monopoly. It would equate its zero marginal cost with zero marginal revenue at E in graph (a) and supply 25 units per year as is also shown by point F in graph (b). On the other hand, if Firm B supplied 50 units (which is all the market would take at a zero price), Firm A would supply nothing, as shown by point G in graph (b). Thus FG is Firm A's reaction curve. (Its equation is $Q_A = 25 - \frac{25}{50} Q_B$). Firm B's reaction curve is derived similarly as line HI. (Its equation, of course, is $Q_B = 25 - \frac{25}{50} Q_A$). The Cournot equilibrium is found at point K: each firm would, in fact, end up supplying $16\frac{2}{3}$ units per year. If you cannot read this precise number in the graph, you can find it by solving the above two equations:

$$Q_A = 25 - .5Q_B = 25 - .5(25 - .5Q_A) =$$
$$25 - 12.5 + 0.25 Q_A.$$
$$.75 Q_A = 12.5$$
$$Q_A = 16\tfrac{2}{3}. \text{ Analogously, } Q_B = 16\tfrac{2}{3}.$$

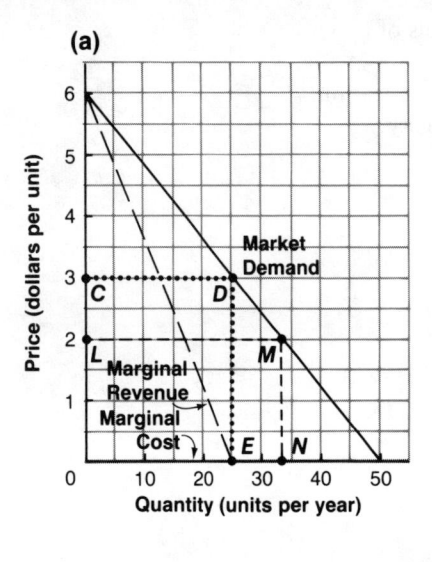

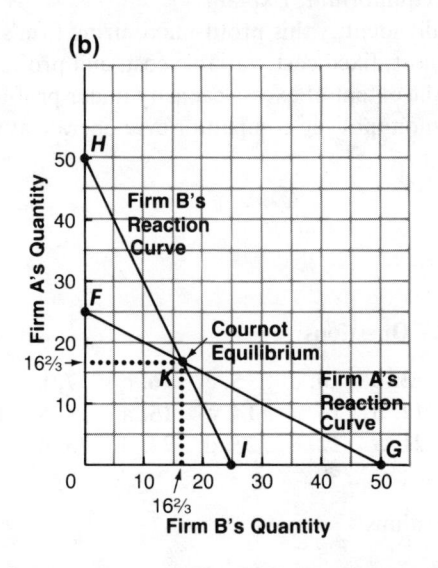

(a)

(b)

b. The monopoly profit would equal area $0CDE$ in graph (a), or $75 per year.

c. The combined output of the oligopolists would equal $33\frac{1}{3}$ units instead of 25 units per year. This would sell at $2 per unit. Thus their combined profit would equal area $0LMN$ in graph (a), or only $66\frac{2}{3}$ per year.

2. **a.** To answer the question, draw in the dashed marginal revenue line *abc* and beyond. For quantities up to 300 units per year, demand section *ad* is relevant, hence the associated marginal-revenue section is *ab*. For quantities above 300 units per year, demand section *de* is relevant, hence the associated marginal-revenue section (partially shown) lies below *c*. (Note that *c* precisely bisects distance 0*e*). The profit-maximizing price-quantity combination ($30 per unit and 300 units per year) corresponds to marginal cost/marginal revenue intersection *f*. The firm's total revenue equals $9,000 per year (area 0*gdc*). Total cost equals $3,900 per year (area 0*hic*), average total cost being $13 per unit (*ic*). Fixed cost equals $1,200 per year (area *khim*), average fixed cost being $4 per unit (*im*). Variable cost equals $2,700 per year (area 0*kmc*), average variable cost being $9 per unit (*mc*). Profit equals $5,100 per year (area *hgdi*), average profit being $17 per unit (*di*).

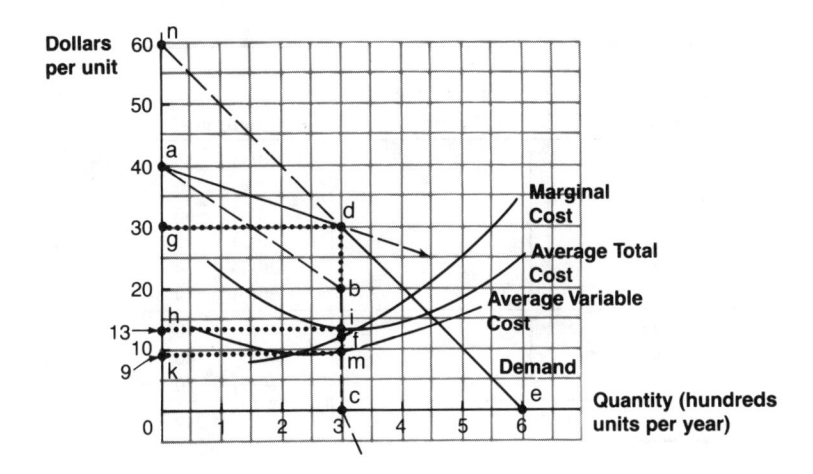

b. Up from *f* to *b*, down from *f* to *c*.

c. The relevant demand line would be the steeper line from *e* to *n*. The marginal revenue line would run from *n* to *c* and beyond and would intersect marginal cost to the left of *f*. Thus the firm would supply a lower quantity corresponding to this intersection, and it would charge a higher price (again corresponding to this intersection) on demand segment *nd*.

d. The relevant demand line would be the flatter line from *a* to *d* and beyond (note the arrow). The marginal revenue line would run from *a* to *b* and beyond. It would intersect marginal cost to the right of *f*. Thus the firm would supply a larger quantity corresponding to this intersection, and it would charge a lower price (again corresponding to this intersection) on the dashed demand segment to the right of *d*.

3. **a.** The line of indifference lies between brands #6 and #7, because there are 120 customers on either side of it. Overall dissatisfaction cost would be minimized by producing brand #8 and a mixture of #4 and #5, each being located in the center of the two halves of customers. The cost would equal 240 units, which can be calculated as follows:

The lovers of brand #	would buy brand #	and incur costs of
1		10(3.5) = 35
2		10(2.5) = 25
3	4/5	20(1.5) = 30
4		20(0.5) = 10
5		30(0.5) = 15
6		30(1.5) = 45
7		40(1) = 40
8	8	40(0) = 0
9		40(1) = 40
		Sum = 240

b. The brands actually produced would be right next to the line of indifference: #6 and #7. The cost would equal 340 units, which can be calculated as follows:

The lovers of brand #	would buy brand #	and incur costs of
1		10(5) = 50
2		10(4) = 40
3	6	20(3) = 60
4		20(2) = 40
5		30(1) = 30
6		30(0) = 0
7		40(0) = 0
8	7	40(1) = 40
9		40(2) = 80
		Sum = 340

4. The $101 cost was calculated as follows:

The lovers of brand #	would buy brand #	and incur costs of
1		1(4) = 4
2		1(3) = 3
3	5	1(2) = 2
4		1(1) = 1
5		1(0) = 0
6		1(0) = 0
7		1(1) = 1
8		1(2) = 2
9		1(3) = 3
10		1(4) = 4
11		1(5) = 5
12		1(6) = 6
13	6	1(7) = 7
14		1(8) = 8
15		1(9) = 9
16		1(10) = 10
17		1(11) = 11
18		1(12) = 12
19		1(13) = 13
		Sum = 101

5. **a.** By drawing in marginal-revenue line *ab* and beyond and by noting its intersection with marginal cost at *c*, the profit-maximizing price-quantity combination ($33⅓ per unit and 20 units per year) is found. For this quantity, average total cost equals price (point *d*), which points to the existence of long-run equilibrium.

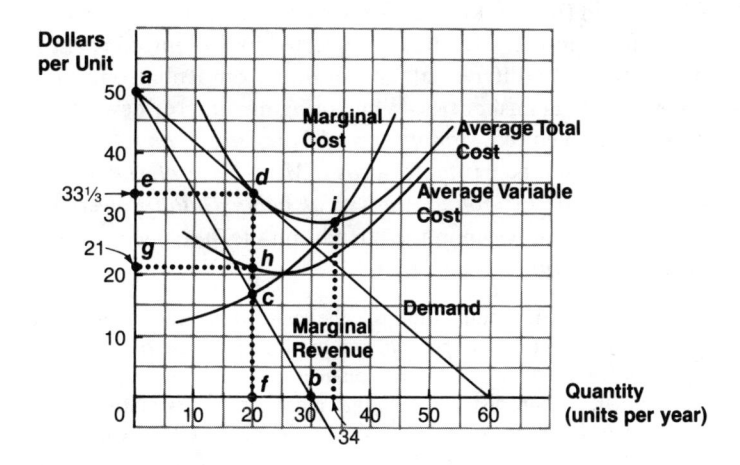

b. The firm's total revenue equals $666.67 per year (area 0*edf*). The total cost equals $666.67 per year (area 0*edf*), average total cost being $33.33 per unit *df*. Fixed cost equals $246.67 per year (area *gedh*), average fixed cost being $12.33 per unit (*dh*). Variable cost equals $420 per year (area 0*ghf*), average variable cost being $21 per unit (*hf*). Profit is zero.

c. Excess capacity equals the horizontal distance between *d* and *i*, or 14 units per year.

d. Yes, in the short run. Suppose the average-total-cost curve were lower such that its minimum point *i* coincided with the marginal cost/marginal revenue intersection *c*. (Average variable cost would then, of course, have to be lower still). Under such circumstances, the firm would be operating at capacity. It would also be earning a profit of *dc* per unit; this profit would lead to entry of new firms, a fall in demand for this firm, and, eventually, the type of long-run, no-profit, excess-capacity equilibrium pictured here and in panel (b) of text Figure 8.8. "The Monopolistic Competitor."

BIOGRAPHY 8.1

Edward H. Chamberlin

Edward Hastings Chamberlin (1899–1967) was born at La Conner, Washington. He studied at the Universities of Iowa, Michigan, and Harvard. He taught at Harvard during his entire career, where he was also editor of the *Quarterly Journal of Economics*.

Chamberlin's 1927 Ph.D. thesis, in which he fused the hitherto separate theories of monopoly and perfect competition, became his first book as *The Theory of Monopolistic Competition: A Reorientation of the Theory of Value* (1933). Its brilliant exposition and original contributions quickly swept the profession, which soon talked of the Chamberlinian revolution. Note: As Chamberlin used the term "monopolistic competition," it included what is now generally termed (differentiated) oligopoly as well as what is now called monopolistic competition. All these sellers have, of course, an absolute monopoly of their differentiated product but are subject to the competition of imperfect substitutes. Hence the title of the book.

As fate would have it, and as has happened before in the history of science, Joan Robinson of Cambridge, England, published a similar book six months later (see Biography 11.1). Rather grieved, Chamberlin (who in fact had stressed all along such aspects as product differentiation and advertising, neglected by Robinson) spent much of his life trying to differentiate his product from Robinson's and defending his work against critics. While bringing out ever new editions of his book, he also edited *Monopoly and Competition and Their Regulation* (1954), and he wrote *Towards a More General Theory of Value* (1957) and *The Economic Analysis of Labor Union Power* (1958).

Says Chamberlin in the 8th edition of *The Theory of Monopolistic Competition:*

> Monopolistic competition is a challenge to the traditional viewpoint of economics that competition and monopoly are alternatives and that individual prices are to be explained in terms of either the one or the other. By contrast, it is held that most economic situations are composites of both competition and monopoly and that, wherever this is the case, a false view is given by neglecting either one of the two forces. . . . This seems to be a very simple idea. . . . Its inherent reasonableness was never better expressed than by a student who observed to me . . . "Chapter IV is easy—you don't say anything in it."
>
> My own observation of Chapter IV, however, would be quite different. . . . It contains not a technique, but a way of looking at the economic system; and changing one's economic Weltanschauung is something very different from . . . adding new tools to one's kit. . . . This concept of blending of competition and monopoly is quite lacking in Mrs. Robinson's *Imperfect Competition*. . . . Imperfect and monopolistic competition have been commonly linked as different names for the same thing. Their elements of . . . dissimilarities [seem to be] hardly recognized. I submit . . . that there is no evidence . . . that Mrs. Robinson thinks of monopoly . . . and competition in any other way but as mutually exclusive. This difference in conception between us is in fact the key to an understanding of many other differences in treatment. (pp. 204–7).

In 1965, the American Economic Association honored Chamberlin by making him one of its Distinguished Fellows.

CHAPTER 9

Decision Theory and Game Theory

MULTIPLE-CHOICE QUESTIONS

1. Decision making under uncertainty
   a. occurs in a situation in which the ultimate outcome of a decision maker's choice depends on chance.
   b. occurs in a situation in which the ultimate outcome of a decision maker's choice depends on future events over which the decision maker has no control.
   c. can be aided greatly by decision theory.
   d. is correctly described by all of the above.

2. Which one of the following pairs corresponds to the action/event dichotomy that is central to modern decision theory?
   a. Choice/chance.
   b. Cost/payoff.
   c. Risk/uncertainty.
   d. Stock/flow.

3. The existence of a dominant action implies the existence of
   a. a dominant event.
   b. an inadmissible event.
   c. an inadmissible action.
   d. a positive payoff.

4. Which one of the following is not equivalent to two of the others?
   a. Action point.
   b. Decision fork.
   c. Decision point.
   d. Event point.

5. When nothing is known about the likelihood of occurrence of those alternative future events that are certain to affect the eventual outcome of a present decision, it is common practice to employ
   a. the maximin (or minimax) criterion.
   b. the minimax (or minimin) criterion.
   c. the minimax-regret criterion.
   d. any of the above.

6. Someone employing the maximin criterion
   a. finds the lowest possible cost associated with each possible action, identifies the lowest cost among these minima, and then chooses the action corresponding to this lowest minimum.
   b. finds the highest possible cost associated with each possible action, identifies the lowest cost among these maxima, and then chooses the action corresponding to this lowest maximum.
   c. finds the lowest possible benefit associated with each possible action, identifies the highest benefit among these minima, and then chooses the action corresponding to this highest minimum.

d. finds the highest possible benefit associated with each possible action, identifies the highest benefit among these maxima, and then chooses the action corresponding to this highest maximum.

7. The maximax (or minimin) criterion
   a. seeks to achieve the best of the best possible outcomes.
   b. seeks to achieve the best among the poorest possible outcomes.
   c. is ideally suited to the pessimist.
   d. is correctly described by (b) and (c).

8. To achieve the highest of the lowest possible profits, a decision maker would employ the criterion of
   a. maximax.
   b. maximin.
   c. minimin.
   d. minimax.

9. Nonprobabilistic decision-making criteria have been criticized for a variety of reasons, including
   a. undue reliance on extreme values.
   b. the index-number problem.
   c. the maximum-likelihood problem.
   d. all of the above.

10. Decision makers who employ the expected-monetary-value criterion identify as optimal
    a. the action with the largest expected monetary value.
    b. the action with the lowest expected monetary value.
    c. either (a) or (b), depending on the nature of their problem.
    d. the action with zero expected monetary value.

11. The Laplace criterion is identical with
    a. Bayes's postulate.
    b. the criterion of insufficient reason.
    c. the equal-likelihood criterion.
    d. all of the above.

12. Critics of probabilistic decision criteria that maximize monetary benefit or minimize monetary cost argue that one should rather devise criteria that help people
    a. minimize regret.
    b. maximize utility.
    c. maximax.
    d. minimin.

13. Based on the criterion of expected monetary value, the St. Petersburg game is
    a. a fair gamble.
    b. a more-than-fair gamble.
    c. an unfair gamble.
    d. no gamble at all.

14. Whenever a person considers the utility of a certain prospect of money to be higher than the expected utility of an uncertain prospect of equal expected monetary value, the person is
    a. risk-averse.
    b. risk-neutral.
    c. risk-seeking.
    d. one of the above, but one cannot tell which without a knowledge of the person's utility function.

15. Whenever a person considers the utility of a certain prospect of money to be lower than the expected utility of an uncertain prospect of equal expected monetary value, the person is said to be
    a. risk-averse.
    b. risk-neutral.
    c. risk-seeking.
    d. one of the above, but one cannot tell which without a knowledge of the person's utility function.

16. Any decision-making situation in which the payoff to people's choices depends not only on them and the "state of the world" but also on other people's choices is called
    a. a game.
    b. a mixed strategy.
    c. a saddle point.
    d. a zero-sum game.

TRUE-FALSE QUESTIONS

In each space below, write a *T* if the statement is true and an *F* if the statement is false.

_____ 1. A decison-making situation in the context of uncertainty is often summarized with the help of an action point.

_____ 2. A point in a decision tree at which the decision maker is in control is referred to as an *event point*.

_____ 3. The nature of a decision tree is well summarized by the phrase: "Choice plus chance equals outcome."

_____ 4. The maximin (or minimax) criterion is a decidedly conservative approach to decision making; it guarantees an outcome no worse than the best among the poorest possible outcomes.

_____ 5. The maximin (or minimax) criterion is ideally suited to the born optimist.

_____ 6. The lowest of the lowest possible costs can be achieved by employing the minimax criterion.

_____ 7. In modern decision theory, the payoff table reappears in the form of the regret table.

_____ 8. Critics argue that use of the minimax regret criterion amounts to "playing ostrich" because so much that might happen is being ignored.

_____ 9. The Laplace criterion is also referred to as the *maximum-likelihood criterion*.

_____ 10. In decision-tree analysis, branch pruning can occur only at event points.

_____ 11. When the expected monetary value of what is given up by an action exceeds the expected monetary value of what is received, a decision maker is said to face an unfair gamble.

_____ 12. According to Bernoulli, a gamble that is fair in monetary terms is also fair in utility terms.

_____ 13. Whenever a person considers the utility of a certain prospect of money to be equal to the expected utility of an uncertain prospect of equal expected monetary value, the person is said to minimize regret.

_____ 14. Saddle point is another word for action point.

PROBLEMS

1. An advertising manager must sign a contract now with an aerial photographer concerning an assignment one week hence. The manager must choose between two contracts. According to A, the photographer gets $500 regardless of the weather and regardless of whether the flight takes place. According to B, the photographer gets $800 if there are no clouds and all of the work is done as planned, $500 if there are scattered to broken clouds and only half of the work is done, and $200 if the sky is overcast and the assignment cannot be carried out. Determine the best contract from the manager's point of view under the criterion of
   a. minimax and
   b. minimin.

2. Rework Problem 1, but use the criterion of minimax regret.

3. An advertising executive must decide whether to invest the firm's profits in an expansion of the firm—that is, investing in the advertising industry—or whether to choose one of two promising alternatives—that is, investing in the air-charter or medical-equipment business. The annual real rates of return are forecast to be fairly high but also to vary substantially, depending on whether the next 20 years are predominantly years of inflation or recession. These rates of return under inflation or recession are believed to be, respectively, 12 or 6 percent in advertising, 20 or 2 percent in the air-charter business, and 8 or 7 percent in the medical-equipment business. Determine the best investment under the criterion of
   a. maximin and
   b. maximax.

4. Rework Problem 3, but use the criterion of minimax regret.

5. Reconsider Problem 1, but assume event probabilities are known as $p(E_1) = .5$, $p(E_2) = .3$, and $p(E_3) = .2$. Now solve the problem under the criterion of
   a. expected monetary value and
   b. expected utility (assuming the manager has a utility function of $U = \sqrt{\$}$).

6. Reconsider Problem 1, along with the event probabilities given in Problem 5. Now solve the problem under the criterion of expected opportunity loss. (*Hint:* Your answer to Problem 2 will be useful.)

7. A production manager believes that a given volume of output can be produced at different costs, depending on whether process A, B, or C is employed. However, it is also believed that future electric-power prices (which can be lower, unchanged, or higher) will influence the cost figures such that production costs (in thousands of dollars) with process A will be 40, 50, 60; with process B will be 55, 60, or 65; and with process C will be 20, 60, or 90. Determine the best production method under the criterion of
   a. minimax and
   b. minimin.

8. Solve Problem 7 again, but use the criterion of minimax regret.

9. A farm manager must decide whether to plant a frost-sensitive crop on March 15, April 15, or May 15. If there is no frost after planting, profits (in thousands of dollars) will be 100, 60, or 20 for the respective dates (because the crop catches premium prices if it gets to the market early). If there is frost, profits of -20, 60, or 80 are expected (not only because frost destroys crops, but also because greater scarcity at later dates drives up the value of smaller usable crops). Determine the best planting date under the criterion of
   a. maximin and
   b. maximax.

10. Solve Problem 9 again, but use the criterion of minimax regret.

11. Reconsider Problem 7, but assume event probabilities are known as $p(E_1) = .1$, $p(E_2) = .5$, and $p(E_3) = .4$. Now solve the problem under the criterion of
    a. expected monetary value and
    b. expected utility (assuming the manager has a utility function of $U = \sqrt{\$}$).

12. Reconsider Problem 7, along with the event probabilities given in Problem 11. Now solve the problem under the criterion of expected opportunity loss. (*Hint:* Your answer to Problem 8 will be useful.)

13. Solve the game problem illustrated by the accompanying table, assuming both firms use the maximin strategy.

| | | Matrix of each firm's annual gain (+) or loss (−) compared to original position (in millions of dollars) ||
| | | Firm 2 ||
		Publish new book this year	Publish new book next year
Firm 1	Publish new book this year	+5 / +2	+7 / +10
	Publish new book next year	+5 / +8	+10 / +3

ANSWERS

Multiple-Choice Questions

1. d
2. a
3. c
4. d
5. d
6. c
7. a
8. b
9. a
10. c
11. d
12. b
13. b
14. a
15. c
16. a

True-False Questions

1. F (It is summarized with the help of a payoff table or decision tree.)
2. F (It is variously referred to as an action point, decision point, decision fork, or even decision node, but never as an event point; at the latter, "nature" is in charge.)
3. T
4. T
5. F (It is ideal for the born pessimist.)
6. F (It requires the minimin criterion.)
7. F (This is pure nonsense; the two tables have nothing to do with each other.)
8. F (This is a common critique of the maximum-likelihood criterion.)
9. F (It is the equal-likelihood criterion.)
10. F (It can only occur at action points.)
11. T
12. F (It is unfair in utility terms.)
13. F (The person is said to be risk-neutral.)
14. F (The two are unrelated; see the text Glossary.)

Problems

1. See Table 9.A.

    a. The minimax action is A_1.

    b. The minimin action is A_2.

TABLE 9.A

Actions	E_1 = no clouds	E_2 = scattered to broken clouds	E_3 = overcast sky	Row Maximum	Row Minimum
A_1 = use contract A	500	500	500	500 minimax	500
A_2 = use contract B	800	500	200	800	500 minimin

2. See Table 9.B. The minimax-regret action is A_1 or A_2; it is a toss-up.

TABLE 9.B

	Events			
Actions	$E_1 =$ no clouds	$E_2 =$ scattered to broken clouds	$E_3 =$ overcast sky	Row Maximum
$A_1 =$ use contract A	$500 - 500 = 0$	$500 - 500 = 0$	$500 - 200 = 300$	300 ←
$A_2 =$ use contract B	$800 - 500 = 300$	$500 - 500 = 0$	$200 - 200 = 0$	300 ← minimax regret?

3. See Table 9.C.

a. The maximin action is A_3.

b. The maximax action is A_2.

TABLE 9.C

	Events		Row Minimum	Row Maximum
Actions	$E_1 =$ inflation	$E_2 =$ recession		
$A_1 =$ invest in advertising industry	12	6	6	12
$A_2 =$ invest in air-charter industry	20	2	2	⟨20⟩ maximax
$A_3 =$ invest in medical-equipment industry	8	7	⟨7⟩ maximin	8

4. See Table 9.D. The minimax-regret action is A_2.

TABLE 9.D

	Events		Row Maximum
Actions	$E_1 =$ inflation	$E_2 =$ recession	
$A_1 =$ invest in advertising industry	$20 - 12 = 8$	$7 - 6 = 1$	8
$A_2 =$ invest in air-charter industry	$20 - 20 = 0$	$7 - 2 = 5$	⟨5⟩ ← minimax regret
$A_3 =$ invest in medical-equipment industry	$20 - 8 = 12$	$7 - 7 = 0$	12

5. a. See Table 9.E. The expected-monetary-value action is A_1.

TABLE 9.E

	Events			
Actions	$E_1 =$ no clouds $p(E_1) = .5$	$E_2 =$ scattered to broken clouds $p(E_2) = .3$	$E_3 =$ overcast sky $p(E_3) = .2$	EMV
$A_1 =$ use contract A	500	500	500	⟨500⟩ ← optimum
$A_2 =$ use contract B	800	500	200	590

b. See Table 9.F. The expected-utility action is A_1.

TABLE 9.F

Actions	E_1 = no clouds $p(E_1) = .5$	E_2 = scattered to broken clouds $p(E_2) = .3$	E_3 = overcast sky $p(E_3) = .2$	EU
A_1 = use contract A	$\sqrt{500} = 22.36$	$\sqrt{500} = 22.36$	$\sqrt{500} = 22.36$	(22.36) optimum
A_2 = use contract B	$\sqrt{800} = 28.28$	$\sqrt{500} = 22.36$	$\sqrt{200} = 14.14$	23.68

Caution: Costs, not benefits are involved; thus, the *lowest EMV* is optimal in (a) and the *lowest EU* is optimal in (b).

6. See Table 9.G. The expected-opportunity-loss action is A_1. Note: The result is necessarily the same as for Problem 5a.

TABLE 9.G

Actions	E_1 = no clouds $p(E_1) = .5$	E_2 = scattered to broken clouds $p(E_2) = .3$	E_3 = overcast sky $p(E_3) = .2$	EOL
A_1 = use contract A	500 − 500 = 0	500 − 500 = 0	500 − 200 = 300	(60) optimum
A_2 = use contract B	800 − 500 = 300	500 − 500 = 0	200 − 200 = 0	150

7. See Table 9.H. We can ignore process B because it is worse than A for all events.

a. The minimax action is A_1.

b. The minimin action is A_2.

TABLE 9.H

Actions	E_1 = lower price	E_2 = same price	E_3 = higher price	Row Maximum	Row Minimum
A_1 = use process A	40	50	60	(60) minimax	40
A_2 = use process C	20	60	90	90	(20) minimin

8. See Table 9.I. The minimax-regret action is A_1.

TABLE 9.I

Actions	E_1 = lower price	E_2 = same price	E_3 = higher price	Row Maximum
A_1 = use process A	40 − 20 = 20	50 − 50 = 0	60 − 60 = 0	(20) minimax regret
A_2 = use process C	20 − 20 = 0	60 − 50 = 10	90 − 60 = 30	30

9. See Table 9.J.

   a. The maximin action is A_2.

   b. The maximax action is A_1.

   TABLE 9.J

	Events			
Actions	E_1 = no frost	E_2 = frost	Row Minimum	Row Maximum
A_1 = plant on March 15	100	−20	−20	(100) ← maximax
A_2 = plant on April 15	60	60	(60) ← maximin	60
A_3 = plant on May 15	20	80	20	80

10. See Table 9.K. The minimax-regret action is A_2.

    TABLE 9.K

	Events		
Actions	E_1 = no frost	E_2 = frost	Row Maximum
A_1 = plant on March 15	100 − 100 = 0	80 − (−20) = 100	100
A_2 = plant on April 15	100 − 60 = 40	80 − 60 = 20	(40) ← minimax regret
A_3 = plant on May 15	100 − 20 = 80	80 − 80 = 0	80

11. a. See Table 9.L. The expected-monetary-value action is A_1.

    TABLE 9.L

	Events			
Actions	E_1 = lower price $p(E_1)$ = .1	E_2 = same price $p(E_2)$ = .5	E_3 = higher price $p(E_3)$ = .4	EMV
A_1 = use process A	40	50	60	(53) ← optimum
A_2 = use process C	20	60	90	68

    b. See Table 9.M. The expected-utility action is A_1.

    TABLE 9.M

	Events			
Actions	E_1 = lower price $p(E_1)$ = .1	E_2 = same price $p(E_2)$ = .5	E_3 = higher price $p(E_3)$ = .4	EU
A_1 = use process A	$\sqrt{40}$ = 6.32	$\sqrt{50}$ = 7.07	$\sqrt{60}$ = 7.75	(7.27) ← optimum
A_2 = use process C	$\sqrt{20}$ = 4.47	$\sqrt{60}$ = 7.75	$\sqrt{90}$ = 9.49	8.12

Caution: Costs, not benefits are involved; thus, the *lowest EMV* is optimal in (a) and the *lowest EU* is optimal in (b).

12. See Table 9.N. The expected-opportunity-loss action is A_1. Note: The result is necessarily the same as for 11(a).

TABLE 9.N

	Events			
Actions	E_1 = lower price $p(E_1) = .1$	E_2 = same price $p(E_2) = .5$	E_3 = higher price $p(E_3) = .4$	EOL
A_1 = use process A	40 − 20 = 20	50 − 50 = 0	60 − 60 = 0	2 optimum
A_2 = use process C	20 − 20 = 0	60 − 50 = 10	90 − 60 = 30	17

13. Firm 1, comparing row minima +2 and +3, will publish next year; Firm 2, comparing column minima +5 and +7, will do likewise.

BIOGRAPHY 9.1

Frank H. Knight

Frank Hyneman Knight (1885–1972) was born in rural McLean County, Illinois. He studied philosophy, theology, and social science at Tennessee's Milligan College and later at Cornell University. He was a professor of economics at the University of Iowa and, from 1927 to his death, at the University of Chicago. Together with Friedrich von Hayek (see Biography 14.2) and Henry Simmons, he established a tradition known as the Chicago School of Economics, later to be carried on by Milton Friedman (see Biography 1.1) and George Stigler (see Biography 15.1). The School's advocacy of free enterprise and its rejection of government interference in markets are well reflected in two of Knight's books: *The Ethics of Competition and Other Essays* (1935) and *Freedom and Reform: Essays in Economics and Social Philosophy* (1947). In these collections of essays, Knight identifies the greatest enemies of the free market as those who argue for the free market only in order to defend their own special interests. He laments the destruction of religion by science and the fact that nothing has replaced this moral force. As people turn to government to solve their problems, Knight argues, the free enterprise system is being corrupted, with loss of freedom and dictatorship as the inevitable consequence.

Knight's most famous book, however, is *Risk, Uncertainty and Profit* (1921). This work stimulated rich advances in the study of uncertainty and earned Knight, in 1950, the presidency and, in 1957, the Walker medal of the American Economic Association.

In this book, Knight distinguishes decisions involving risk from those involving uncertainty. Risk decisions, he argues, are insurable (because the nature and probabilities of various future outcomes are objectively known); uncertainty decisions are not insurable (because the nature and probabilities of various future outcomes can only be estimated subjectively). Knight argues that *profit* is a reward for those who are willing to act in the face of uncertainty (as he defines it) and who are lucky enough to avoid loss.

BIOGRAPHY 9.2

Oskar Morgenstern

Oskar Morgenstern (1902–1977) was born in Görlitz, Germany. He studied at Vienna, where he earned his doctorate, at the Universities of London, Paris, and Rome, as well as at Harvard and Columbia. He became a professor of economics at the University of Vienna, where he also edited the famous *Zeitschrift für Nationalökonomie,* and was director of the Austrian Institute for Business Cycle Research. In 1938, like so many other European scholars, Morgenstern moved to the United States where he joined the faculty at Princeton University.

At the time, Morgenstern had already published a steady stream of papers, on business cycle theory, monetary policy, and international trade. His first major book, *Wirschaftsprognose* (1928) was a study of the theory and applications of economic forecasting. It was never translated, unlike his second work, on economic policy, *The Limits of Economics* (1937). Early on, Morgenstern showed a great ability to suggest important overlooked problems, as well as new ways to approach them. His most imaginative and ambitious contribution to economics grew out of his collaboration with John von Neumann: the *Theory of Games and Economic Behavior* (1944). The two authors adopted a thoroughly mathematical approach, using a kind of mathematics seldom seen in economics, with concepts drawn from set theory, group theory, and mathematical logic. The present appendix does no more than convey the flavor of their contribution, which extends to much more than two-person games and has spawned innumerable studies of both conflict and cooperation.

Morgenstern's work, however, did not cease with the theory of games. Other major books of his include *On the Accuracy of Economic Observations* (1950, revised 1963), a brilliant critique of the common types of statistics used in economic discourse and a "must" reading for every economist. This work was followed by the editing of *Economic Activity Analysis* (1954) and the writing of *International Financial Transactions and Business Cycles* (1959) and *The Question of National Defense* (1959). In his last book, written with Gerald Thompson, Morgenstern returned to his early interest in business cycle theory: *Mathematical Theory of Expanding and Contracting Economies* (1976).

Morgenstern was active in a large number of diverse undertakings which ranged from the Econometric Research Program at Princeton, the Rand Corporation, and Mathematica to the League of Nation's study on *Economic Stability in the Postwar World* to the Atomic Energy Commission and consultations at the White House. In 1976, the American Economic Association made Morgenstern one of its Distinguished Fellows.

Throughout his life, Morgenstern offered deep, but constructive criticism of economists. This criticism is illustrated in one of his last articles in which he commented on the inadequacy of textbooks on which young economists were being brought up.[1] "What these books have in common . . . ," he says, "is that few, if any unsolved theoretical (as distinct from applied) problems in economics are mentioned. . . . It is, therefore, all the more surprising that anyone should want to go into a science that seems to have no open theoretical problems left—a vastly different situation from that of physics or biology where even the layman knows that those worlds are filled with riddles. . . . however, the world of social phenomena . . . holds such a plenitude of difficult, important and unsolved theoretical problems."

Morgenstern mentioned 13 such problems, but this author won't spoil the fun for those readers who want to find out for themselves!

---

[1] Oskar Morgenstern, "Thirteen Critical Points in Contemporary Economic Theory: An Interpretation," *Journal of Economic Literature,* December 1972, pp. 1163-89.

CHAPTER 10

Insurance and Gambling, Search and Futures Markets

MULTIPLE-CHOICE QUESTIONS

Circle the letter of the *one* answer that you think is correct or closest to correct.

1. Primary uncertainty exists because
   a. certain facts about the present are known to some people but not to other people.
   b. certain facts about the future are known to some people but not to other people.
   c. certain future events, which are bound to affect the outcome of present decisions, have not yet occurred, and no one can possibly know what they will be like.
   d. of all of the above.

2. In contingent-claim markets, people
   a. trade rights to variable quantities of particular goods—the quantities being dependent on the occurrence of specified "states of the world."
   b. invariably reduce *risk*—the uncertainty-induced chance of variation in their welfare.
   c. commit themselves now to trade, at specified dates in the future, specified quantities and qualities of goods at specified prices.
   d. do both (a) and (b).

3. When people cannot foretell the specific outcome of an action because two or more outcomes are possible, but when they do know the types of possible outcomes and the objective probabilities of their occurrence, they find themselves in a situation of
   a. Knightian risk.
   b. Knightian uncertainty.
   c. contingency claim.
   d. insurance.

4. When irrigating an orchard yields profit of $120, $60, or $30, respectively, depending on whether nature provides no rain, some rain, or much rain (all of which events are considered equally likely) the expected monetary value of irrigation
   a. equals $60.
   b. equals $70.
   c. equals $210.
   d. depends on the actual weather and cannot be calculated.

5. Daniel Bernoulli solved the St. Petersburg paradox by
   a. postulating that people acting under uncertainty maximized expected utilities rather than expected monetary values.
   b. postulating that people acting under uncertainty maximized expected monetary values rather than expected utilities.
   c. pointing out that people were not stupid, but only gullible people would ever play the St. Petersburg game (who could believe that the game's infinite payoff could ever be made?).
   d. pointing out that the marginal utility of money, unlike that of goods, was rising the more money people had.

When answering questions 6–8, refer to the graph.

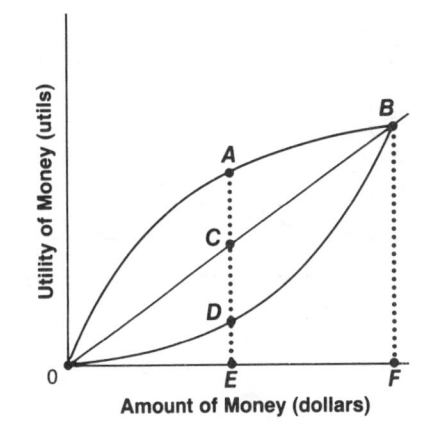

6. A person with utility function $0AB$
   a. is risk-averse.
   b. prefers the certainty of getting $0E$ of money to a 99 percent chance of getting $0F$ and a 1 percent chance of getting nothing.
   c. prefers an equal chance of getting $0F$ or nothing to the certainty of getting $0E$ (which is half of $0F$).
   d. is correctly described by all of the above.

7. A person with utility function $0CB$
   a. finds that more money is associated with a rising marginal utility thereof.
   b. is indifferent between an equal chance of getting nothing or $0F$ on the one hand and getting $0E$ (which is half of $0F$) with certainty on the other hand.
   c. is indifferent between getting $0E$ with certainty and getting nothing or $0F$ with any combination of probabilities.
   d. is not correctly described by any of the above.

8. A person with utility function $0DB$
   a. may well find the total utility of money always equalling the square root of the amount of money.
   b. is clearly risk-averse.
   c. prefers the certainty of $0E$ of money to the equal chance of getting nothing or $0F$.
   d. is not correctly described by any of the above.

9. If a person always considered the utility of a certain prospect of money as equal to the expected utility of an uncertain prospect of equal monetary value, the person would be called
   a. rational.
   b. risk-neutral.
   c. a born gambler.
   d. none of the above.

When answering questions 10–11, refer to the graph.

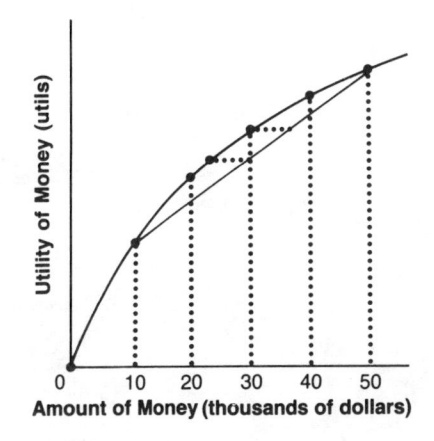

10. A person with the utility function pictured here, who was expecting $50,000 with certainty but was also subject to a 50-50 chance of having or not having a $40,000 loss, would pay, in order to be insured against the loss, a maximum of
    a. $10,000.
    b. $20,000.
    c. between $20,000 and $30,000.
    d. between $10,000 and $20,000.

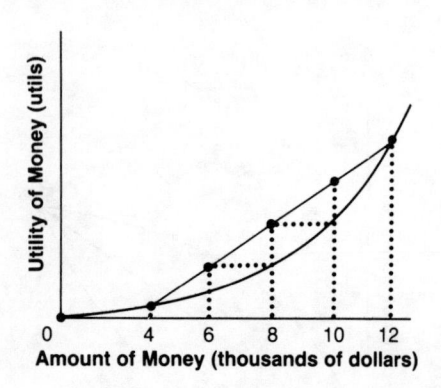

11. A person with the utility function pictured here, who was expecting $50,000 with certainty but was also subject to an 80-20 chance of having or not having a $40,000 loss, would pay, in order to be insured against the loss, a maximum of
    a. well over $32,000.
    b. $32,000.
    c. $18,000.
    d. $10,000.

When answering questions 12–13, refer to the graph.

12. A person with the utility function pictured here, who was expecting $8,000 with certainty and who could make a gamble equally likely to involve a $4,000 loss or gain, would pay, for the mere privilege of entering the gamble,
    a. as much as $2,000.
    b. as much as $4,000.
    c. as much as $8,000.
    d. not a cent.

13. A person with the utility function pictured here, who was expecting $8,000 with certainty and who could make a gamble with a 90 percent chance of losing $4,000 and a 10 percent chance of gaining $4,000, would pay, for the mere privilege of entering the gamble,
    a. $2,000.
    b. $3,200.
    c. nothing.
    d. $4,800.

14. A fair insurance contract
    a. will be rejected by a risk-neutral person.
    b. will be accepted by a risk averter.
    c. may be accepted by a risk seeker.
    d. is correctly described by all of the above.

15. A fair gamble
    a. may well be accepted by anybody, even a risk averter.
    b. will never be accepted by a risk averter.
    c. will never be accepted by a risk-neutral person.
    d. will only be accepted by a risk seeker.

16. A risk-neutral person will
    a. reject all insurance and gambling contracts.
    b. be indifferent about all insurance and gambling contracts, provided they are fair.
    c. accept all insurance and gambling contracts.
    d. be indifferent about all insurance and gambling contracts, regardless of their nature.

17. According to the Friedman-Savage hypothesis,
    a. people buy fair and not-too-unfair insurance simultaneously with taking fair and not-too-unfair gambles.
    b. the marginal utility of money declines (with increasing amounts of money) for relatively low amounts of it, then rises.
    c. the marginal utility of money declines (with increasing amounts of money) for relatively high amounts of it, having increased earlier.
    d. all of the above are true.

18. Search is an activity that
    a. can overcome event uncertainty.
    b. can never overcome market uncertainty.
    c. can be overdone.
    d. is correctly described by all of the above.

19. The used-car market is likely to abound with "lemons" because
    a. buyers find it too costly to search for the best price forever.
    b. sellers of low-quality used cars will find prices delightfully high.
    c. sellers of high-quality used cars will find prices unreasonably low.
    d. of both (b) and (c).

20. Speculators are people who
    a. prefer screening to signaling.
    b. attempt to have neither a long position nor a short position.
    c. wish to assume price risk.
    d. sell hedges.

21. Which of the following is *true*?
    a. Hedgers work the short side of the market, while speculators work the long side.
    b. Hedgers, like speculators, can work the short or long side of the market.
    c. Hedgers work the long side of the market, while speculators work the short side.
    d. None of the above.

22. If an insurance company offered fair insurance (based on the behavior of average people), the premiums would be viewed as
    a. fair by everybody.
    b. fair by nobody.
    c. more-than-fair by high-risk people.
    d. less-than-fair by high-risk people.

When answering questions 23–25, refer to the graph on page 160. Assume that it depicts the utility function of every member of a group of people seeking insurance.

23. A low-risk person, who is receiving $30,000 with certainty but is subject to a potential loss of $20,000 (with a probability of .1),
    a. has an expected monetary value of $28,000 without insurance as well as with fair insurance.
    b. has an expected utility of ab without insurance.
    c. would pay a maximum premium of dc for insurance.
    d. is correctly described by all of the above.

24. A high-risk person, who is receiving $30,000 with certainty but is subject to a potential loss of $20,000 (with a probability of .9),
    a. has an expected monetary value of $12,000 without insurance as well as with fair insurance.
    b. has an expected utility of ef with fair insurance.
    c. would pay a maximum premium of gh for insurance.
    d. is correctly described by all of the above.

25. An insurance company that knew that half of the people were low-risk and half were high-risk as defined above, but that could not identify which particular person belonged to which category,
    a. could issue fair insurance for the group as a whole by charging everyone $10,000.

b. would then drive away low-risk people because *i* (utility with averaged fair premiums) is below *c* (utility with maximum premium) and below *b* (utility without insurance).
   c. would then attract high-risk people because *i* is above *g* (utility with maximum premium) and above *k* (utility without insurance).
   d. would do all of the above.

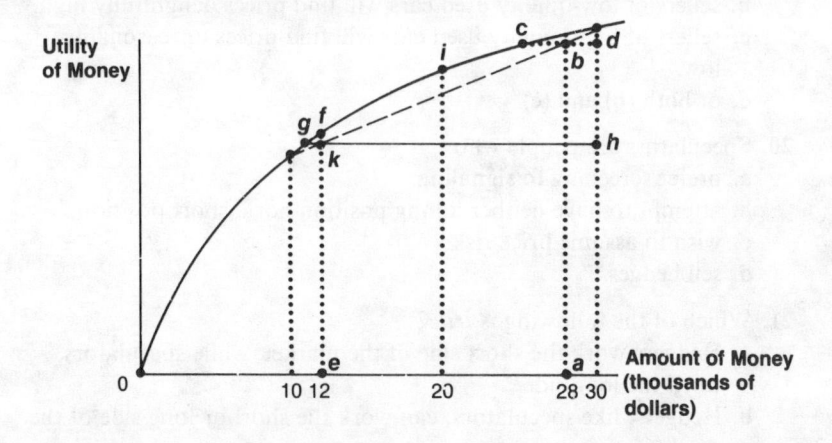

TRUE-FALSE QUESTIONS

In each space below, write a *T* if the statement is true and an *F* if the statement is false.

_____ 1. It is useful to distinguish between primary uncertainty and event uncertainty.

_____ 2. When people enter contingent-claim markets, they invariably increase *risk* (the uncertainty-induced chance of variation in their welfare).

_____ 3. The probabilities inherent in situations of Knightian risk are objective, while those in situations of Knightian uncertainty are subjective.

_____ 4. If irrigating an orchard yielded profit of $500, $400, or $100, respectively, depending on whether nature provided no rain, some rain, or much rain (with subjective probabilities of .1, .5, and .4), the expected monetary value of irrigation would equal $333.33.

_____ 5. If a person were willing to pay a maximum insurance premium of $5,000 in order to replace a gamble between a 30 percent chance of getting $1,000 and a 70 percent chance of getting $10,000 with the certainty of getting $10,000 (minus the premium), the person would be risk-neutral.

_____ 6. Insurance is called "fair" when its expected monetary value is positive.

_____ 7. Regardless of their attitudes toward risk, people may find insurance or gambling desirable or undesirable.

_____ 8. A risk averter will buy a fair insurance contract and even a not-too-unfair one but will never take a fair or unfair gamble.

____ 9. Screening is an activity designed to select high-quality sellers.

____ 10. The selling hedge (or short hedge) aims to provide price protection for producers, merchants, and warehousers while they hold inventories of commodities.

PROBLEMS

1. Consider the following payoff matrix, showing a California firm's expected 1991 net income (in millions of dollars):

Individual Actions \ States of the World	S_1 = Earthquake	S_2 = No Earthquake
A_1 = Heavy advertising	1	10
A_2 = No advertising	4	4
Subjective probabilities	.1	.9

   a. Calculate the expected monetary value of each action.
   b. Is there a gamble involved in choosing A_1 over A_2? Explain.
   c. If there is a gamble, is it fair? Explain.
   d. If the firm chose A_1, would it be risk-averse, risk-neutral, or risk-seeking? Explain.

2. Consider the following graph.

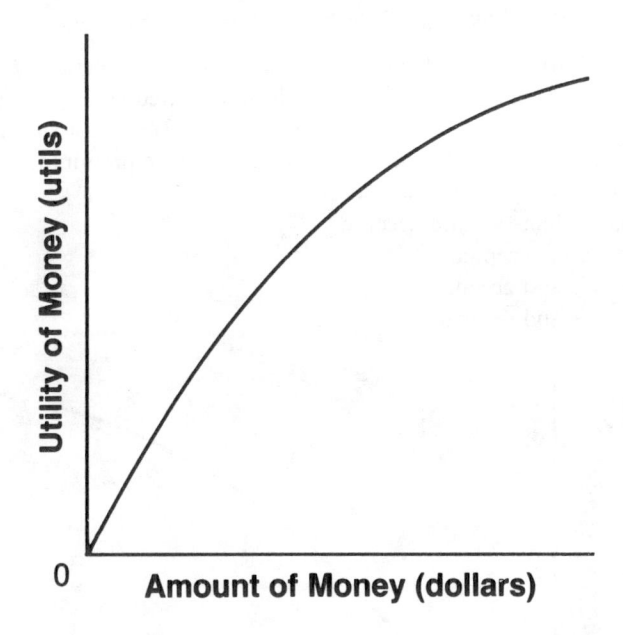

   a. What happens to this person's marginal utility of money as the amount of money rises?
   b. Explain your answer to (a) graphically.
   c. Is the person risk-averse, risk-neutral, or risk-seeking? Explain in the next graph.

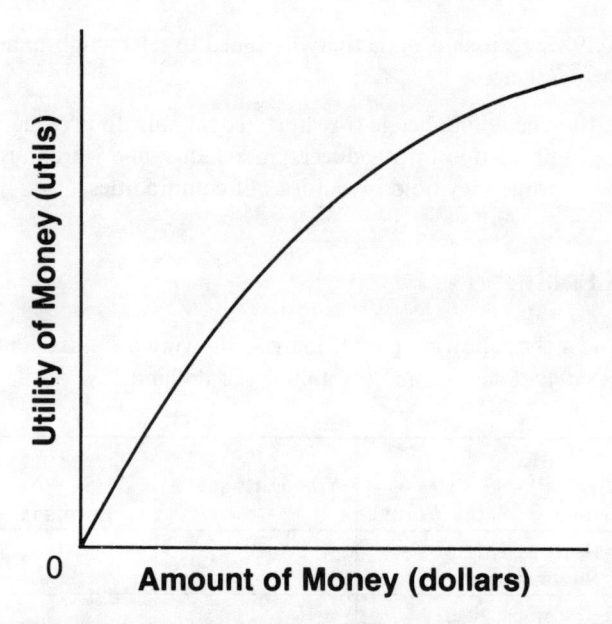

3. Calculate the expected monetary values of each of the following:
   a. a 90 percent chance of getting $10 and a 10 percent chance of getting nothing.
   b. a 50 percent chance each of getting $30 and $50.
   c. a 1 percent chance each of getting $100,000 or losing $90,000, and a 98 percent chance of getting nothing.

4. If the choice between getting $1,000 with certainty and getting either $10 or $10,000 were to be a fair gamble, what would the probabilities of getting $10 or $10,000 have to be?

5. The following graph depicts the utility-of-money function of a risk-averse person. Let this person be confronted with getting either $10 (probability .7) or $3 (probability .3). Now assure this person of the receipt of $10 (minus an insurance premium) through an insurance plan that would be
   a. more-than-fair and accepted.
   b. fair and accepted.
   c. unfair and accepted.
   d. unfair and rejected.

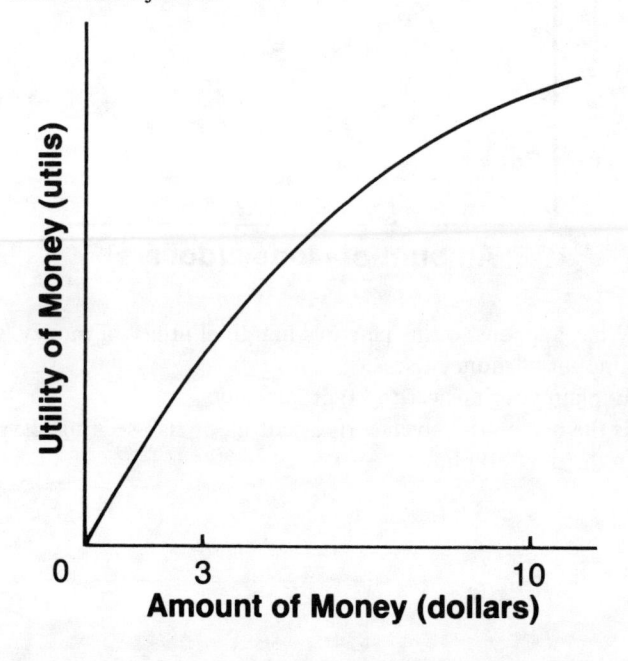

6. The following graph depicts the utility-of-money function of a risk-seeking person. Let this person be confronted with getting either $10 (probability .3) or $3 (probability .7). Would this person ever pay an insurance premium to be assured of the receipt of $10? If yes, how much would the person pay? If not, why not?

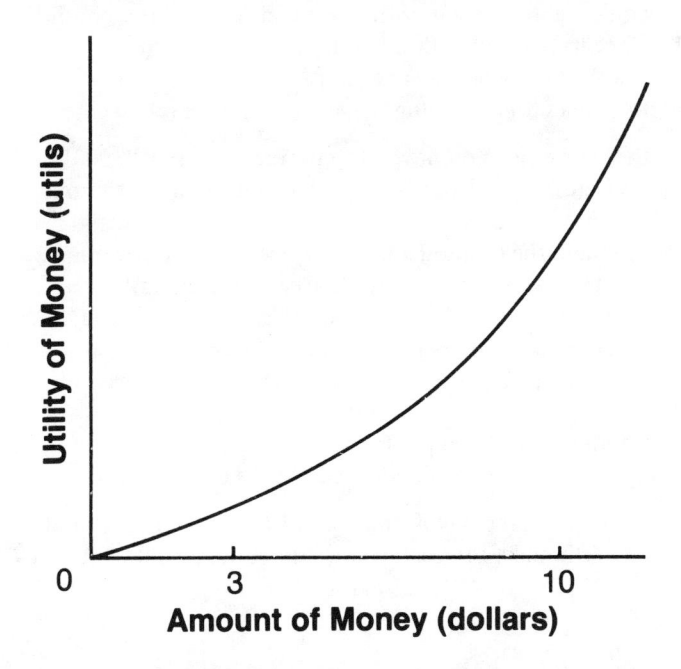

7. The following graph depicts a person's utility-of-money function. Show graphically whether such a person would prefer:
   a. a gamble between $1 (probability .5) and $4 (probability .5) or the certainty of the expected monetary value.
   b. a gamble between $1 (probability .9) and $9 (probability .1) or the certainty of the expected monetary value.
   c. a gamble between $1 (probability .1) and $9 (probability .9) or the certainty of the expected monetary value.
   d. a gamble between $9 (probability .4) and $12 (probability .6) or the certainty of the expected monetary value.

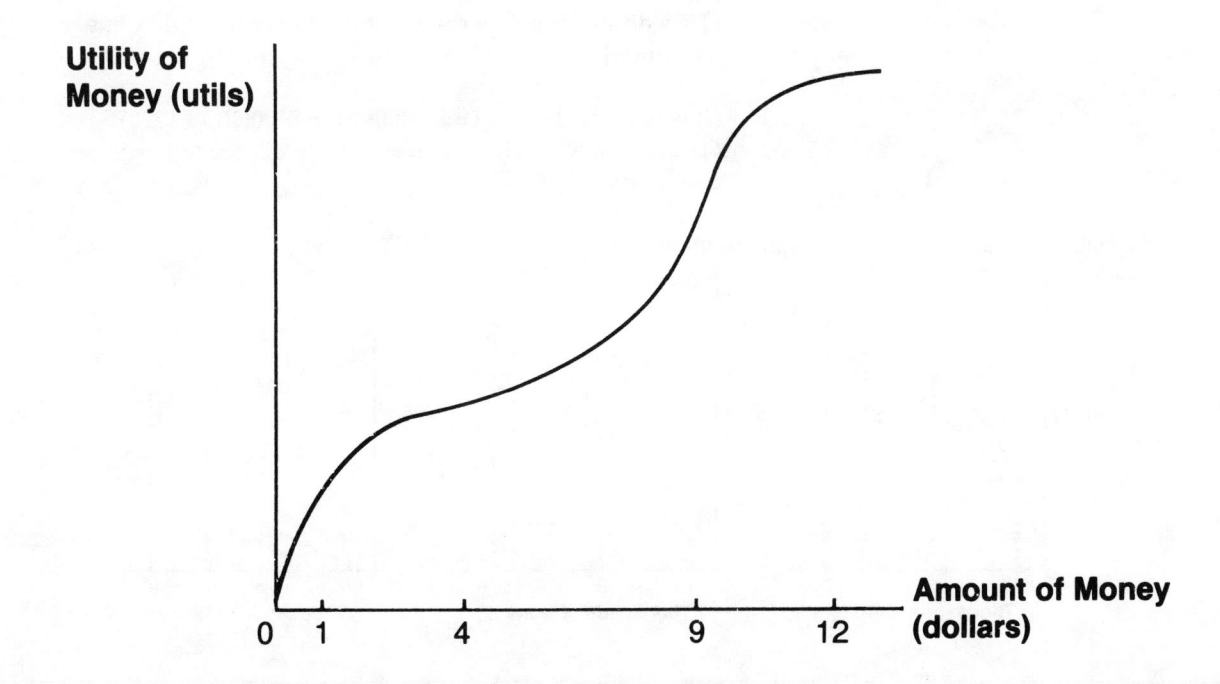

8. Consider the following: The current spot price of orange juice is $1 per unit; you expect to produce 5,000 units of it within a month. The current futures price (for 1 month from now) is $1.30 per unit.
   a. If you wanted to "lock in" the $1 price, what could you do with the help of the futures market?
   b. What would happen to you if the spot and futures prices both fell by 20 cents per unit within the month?
   c. What if both rose by 60 cents per unit?
   d. What if you had done nothing in case (b)? In case (c)?

9. Consider the following: You have just contracted to sell 10,000 units of orange juice in 6 months, requiring you to buy 5,000 tons of oranges eventually. Currently, the spot price of oranges is $2,000 per ton and the 6-months futures price is $2,200 per ton (assume that there is in fact a futures market for oranges).
   a. If you wanted to "lock in" the $2,000 price, what could you do with the help of the futures market?
   b. What would happen to you if the spot and futures prices both fell by $500 per ton within 6 months?
   c. What if both rose by $300 per ton?
   d. What if you had done nothing in case (b)? In case (c)?

10. Have another look at the payoff matrix in Problem 1. If you could get a perfect earthquake forecast, what, if anything, would you pay for it? Explain.

ANSWERS

Multiple-Choice Questions

1. c	2. a	3. a	4. b	5. a	6. a	7. b	8. d	9. b
10. c	11. a	12. a	13. c	14. b	15. b	16. b	17. d	18. c
19. d	20. c	21. b	22. c	23. d	24. d	25. d		

True-False Questions

1. F (The two concepts are identical.)
2. F (They do so when they gamble; they do not when they buy insurance.)
3. T
4. F (It would equal $500(.1) + $400(.5) + $100(.4) = $290)
5. F (The person would be risk-averse. Consider the following set of graphs:

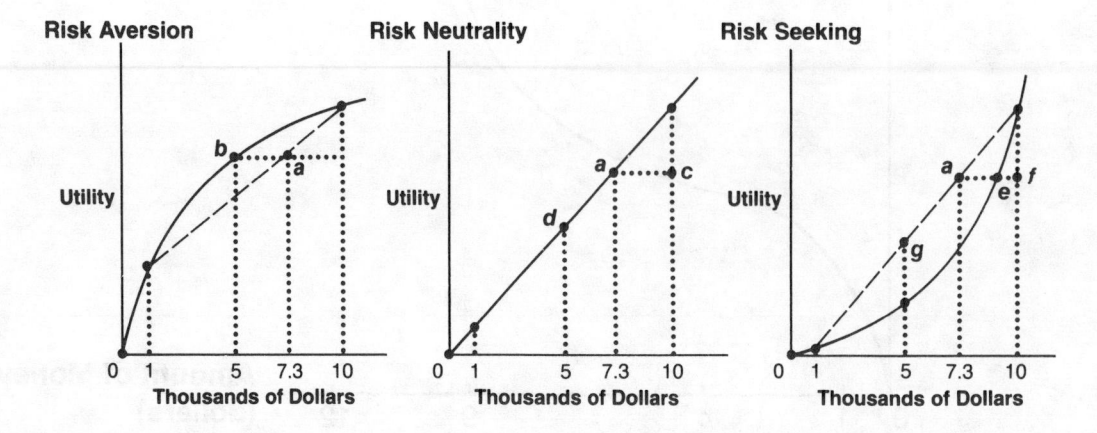

Regardless of the person's attitude toward risk, the expected monetary value of the initial gamble equals $1,000(.3) + $10,000(.7) = $7,300. This provides utility as shown by point *a* in each graph. A risk-averse person might pay as much as $5,000, because utility at *b*, derived from the certain $10,000 minus this premium, equals the utility without insurance at *a*. A risk-neutral person would at most pay $ac = $2,700; the utility corresponding to $10,000 minus the $5,000 premium at *d* being less than that without insurance at *a*. A risk-seeking person would at most pay $ef < $2,700, because utility at *e*, derived from a certain $10,000 minus a < $2,700 premium, equals that without insurance at *a*, while utility at *g*, corresponding to the $5,000 premium, is far below the no-insurance utility at *a*.)

6. F (It's fair when its *EMV* is *zero*.)
7. T
8. T
9. T
10. T

Problems

1. a. The expected monetary value (in millions of dollars) of $A_1 = 1(.1) + 10(.9) = 9.1$; that of $A_2 = 4(.1) + 4(.9) = 4$.
   b. Yes. Instead of getting $4 million with certainty from A_2, the firm would gamble between getting $1 or $10 million from A_1.
   c. No. It is more than fair, because the expected monetary value of the gamble is $9.1 million and that of certainty is $4 million. Thus abandoning certainty for the gamble produces a gain in expected monetary value: $9.1 million − $4 million = $5.1 million. The expected monetary value of taking the plunge into gambling is not zero, but positive.
   d. Any one of these is possible. Risk-seeking and risk-neutral decision makers always accept more-than-fair gambles; risk-averse ones *sometimes* do. See text Table 10.1. "Risk: Attitudes and Behavior."

2. a. The marginal utility declines.
   b. The marginal utility is measured by the *slope* of the total utility function. Note how equal increases in money (horizontal axis) lead to ever-lower marginal utility or increases in total utility ($a > b > c > d > e > f$).

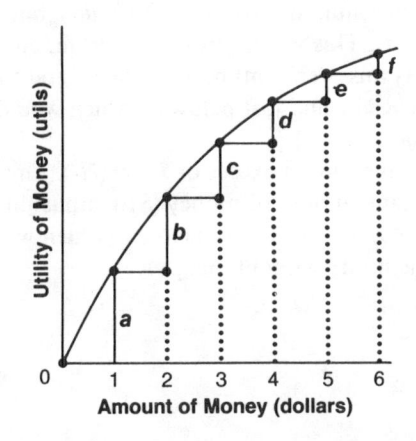

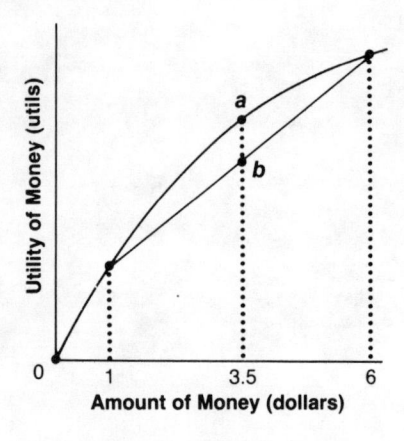

c. The person is risk-averse. The person gets greater utility (such as *a*) from any certain amount (such as $3.50) than from a gamble (such as an equally probable $1 or $6 that has the same expected monetary value (but utility of *b*).

3. a. $10(.9) + 0(.1) = $9
   b. $30(.5) + $50(.5) = $40
   c. $100,000(.01) − $90,000(.01) + 0(.98) = $100

4. The expected monetary value of the gamble would have to equal $1,000. Calling the probabilities x and y, we can write $10(x) + $10,000(y) = $1,000. Because the probabilities must add to 1, we can also write $x + y = 1$. Thus we can solve the equations:

$$x = 1 - y$$
$$\$10(1 - y) + \$10,000(y) = \$1,000$$
$$\$10 - \$10y + \$10,000y = \$1,000$$
$$\$9,990y = \$990$$
$$y = .099099$$
$$x = .900901$$

The probability of getting $10 would have to be .9 (rounded), that of getting $10,000 would have to be .1 (rounded).

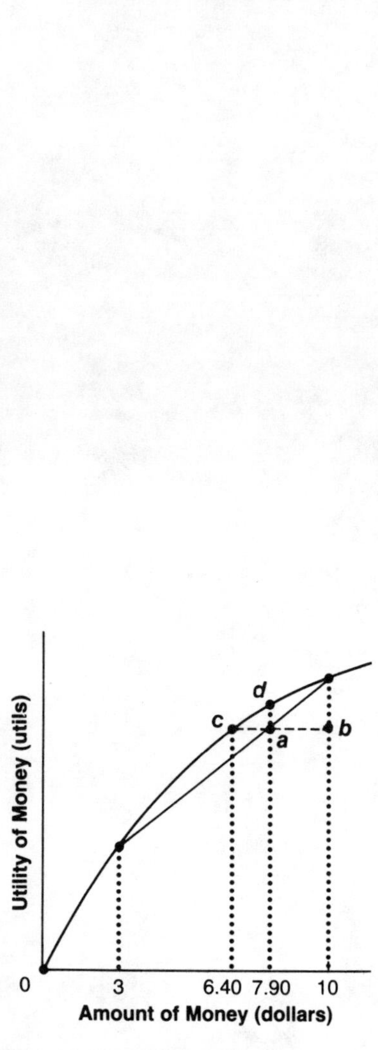

5. The expected monetary value of this person's gamble equals $10(.7) + $3(.3) = $7.90. The associated utility is shown in the graph by *a*, which is the utility the person would get by *not* insuring.
   a. Charge a premium below $2.10 (less than *ab*). This would provide a certain amount of money ($10 minus this premium) in excess of $7.90, hence a utility to the right of *d* (which would clearly be preferred to *a*).
   b. Charge premium of $2.10 (equal to *ab*). This would provide a certain amount of money ($10 minus the premium) equal to $7.90, hence a utility of *d* (which would clearly be preferred to *a*).
   c. Charge a premium in excess of $2.10 (*ab*), but at most equal to $3.60 (*cb*). This would provide a certain amount of money ($10 minus this premium) between $6.40 and $7.89, hence a utility between *c* and just below *d* (which would equal or exceed that at *a*).
   d. Charge a premium in excess of $3.60 (*cb*). This would provide a certain amount of money ($10 minus this premium) below $6.40, hence a utility below *c* (which would be less than *a*, the utility without insurance).

6. The expected monetary value of this person's gamble equals $10(.3) + $3(.7) = $5.10. The associated utility is shown in the graph by *a*, which is the utility the person would get by *not* insuring. Yes, even this risk-seeking person would pay an insurance premium, provided the insurance was more-than-fair. In this example, the person would at most pay $3.75 (*bc*). This would provide a certain amount of money ($10 minus the premium) of $6.25 or more, hence a utility at or to the right of *b* (which would be as good as or better than the utility at *a* without insurance).

7. **a.** The expected monetary value would equal $1(.5) + $4(.5) = $2.50. Certainty of $2.50 would be preferred to the gamble, because *a* > *b*.
   b. The expected monetary value would equal $1(.9) + $9(.1) = $1.80. Certainty of the $1.80 would be preferred to the gamble, because *c* > *d*.
   c. The expected monetary value would equal $1(.1) + $9(.9) = $8.20). The gamble would be preferred to the certainty of $8.20, because *e* > *f*.
   d. The expected monetary value would equal $9(.4) + $12(.6) = $10.80. Certainty of $10.80 would be preferred to the gamble, because *g* > *h*.

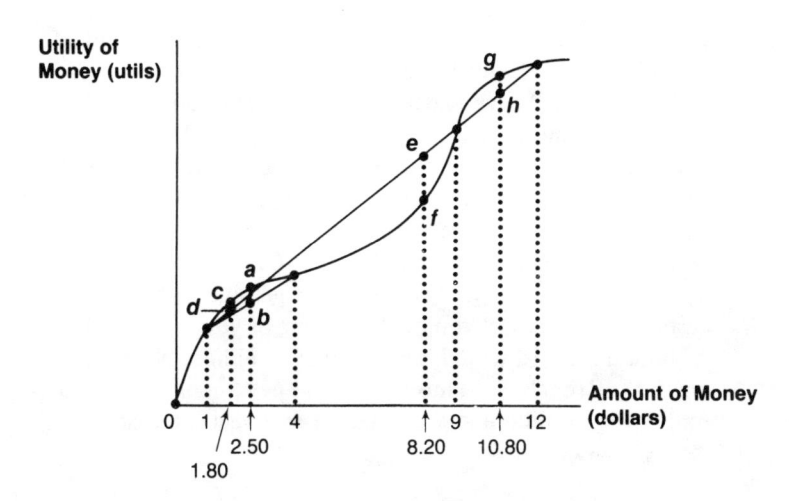

8. **a.** You could protect yourself by means of a selling or short hedge; that is, you could sell a 5,000-unit orange-juice futures contract at $1.30 per unit.
   b. In a month, you could sell 5,000 units of orange juice in the spot market at $.80 (rather than $1), losing 20 cents per unit. You could also buy a 5,000-unit orange-juice futures contract for $1.10 to offset your $1.30 sale, gaining 20 cents per unit. Net result: neither gain nor loss. Effective sale of orange juice at $1 per unit.
   c. In a month, you could sell 5,000 units of orange juice in the spot market at $1.60 (rather than $1), gaining 60 cents per unit. You would also have to buy a 5,000-unit orange-juice futures contract for $1.90 to offset your $1.30 sale, losing 60 cents per unit. Net result: neither gain nor loss. Effective sale of orange juice at $1 per unit.

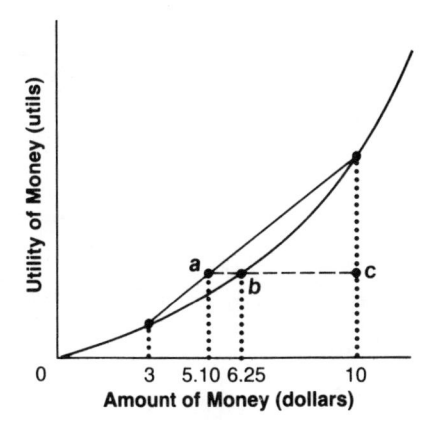

    d. In case (b), you would have sold orange juice at 80 cents instead of $1 per unit, losing from price speculation. In case (c), you would have sold orange juice at $1.60 instead of $1 per unit, gaining from price speculation.

9. **a.** You could protect yourself by means of a buying or long hedge; that is, you could buy a 5,000-ton orange futures contract at $2,200 per ton.

    b. In 6 months, you could buy 5,000 tons of oranges in the spot market at $1,500 per ton (rather than $2,000 per ton), gaining $500 per ton. You could also sell a 5,000-ton orange futures contract for $1,700 per ton to offset your $2,200 purchase, losing $500 per ton. Net result: neither gain nor loss. Effective purchase of oranges at $2,000 per ton.

    c. In 6 months, you could buy 5,000 tons of oranges in the spot market at $2,300 per ton (rather than $2,000 per ton), losing $300 per ton. You could also sell a 5,000-ton orange futures contract for $2,500 per ton to offset your $2,200 purchase, gaining $300 per ton. Net result: neither gain nor loss. Effective purchase of oranges at $2,000 per ton.

    d. In case of (b), you would have bought oranges at $1,500 instead of $2,000 per ton, gaining from price speculation. In case (c), you would have bought oranges at $2,300 instead of $2,000 per ton, losing from price speculation.

10. Let us assign the same subjective probabilities as in Problem 1 but this time to the likelihood of getting a given forecast. You would certainly heavily advertise if you *knew* there would be no earthquake, and your income would then be $10 million. You would certainly not advertise if you knew that there would be an earthquake, and your income would then be $4 million. Thus the expected monetary value of your actions with a perfect forecast would equal 10 million(.9) + 4 million(.1) = $9.4 million. Let us assume that you would have taken the gamble without the forecast. Then the expected monetary value of your action (A_1) without a forecast would have been $9.1 million. Note the answer to 1(c) above. Thus you would pay a maximum of $9.4 million − 9.1 million = $300,000 for a perfect forecast.

BIOGRAPHY 10.1

Joseph A. Schumpeter

Joseph Alois Schumpeter (1883–1950) was born in Triesch, Moravia. His study of law and economics at the University of Vienna was followed by a varied career as banker, jurist, Austrian Minister of Finance (1919–20), and professor at the Universities of Czernowitz, Graz, and Bonn. When Hitler came to power, he emigrated and became a professor at Harvard University. Just before his death, he was honored with the presidency of the American Economic Association.

    Schumpeter was well versed not only in economics, but in history, linguistics, mathematics, philosophy, and sociology. His writings have a broad scope, which is rare in this century. This scope is most evident in his last book, *History of Economic Analysis* (1954), a truly monumental scholarly achievement. Schumpeter is best known, however, for his views on the entrepreneur (noted in this chapter in the text) and for his grand vision of capitalism. In *The Theory of Economic Development* (1912) and in the two-volume *Business Cycles: A Theoretical, Historical, and Statistical Analysis of the Capitalist Process* (1939) he assigns a key role to risk-taking entrepreneurs. Their innovations, along with imitations, Schumpeter argues, produce bursts of investment at irregular intervals, and hence business cycles. They also produce, in the long run, an ever-growing stream of new products at decreasing cost and thus a progressive increase in the economic welfare of the masses. Says Schumpeter:

> The capitalist engine is first and last an engine of mass production which unavoidably means also production for the masses. . . . Queen Elizabeth owned sik stockings. The capitalist achievement does not typically consist in providing more silk stockings for queens but in bringing them within the reach of factory girls in return for steadily decreasing amounts of effort.[1]

Yet Schumpeter also believed that capitalism harbored within itself the seeds for its own destruction. Unlike Marx, he did not think that it would break down under its failures (and by violent revolution), but rather because of its success (and by peaceful evolution). In his *Capitalism, Socialism and Democracy* (1942), Schumpeter predicts that capitalism would fall victim to the very organizations to which it gave birth: In large corporations, the daring entrepreneur will be replaced by the "organization man" (a bureaucrat) or by the committee, which will dampen the innovative zeal. In addition, rising affluence will support a large class of intellectuals who will not appreciate the system. The argument for capitalism, Schumpeter says, is long run and rational. The system is capable of generating impressive economic growth and pouring out an avalanche of consumption goods. Most people, however, are concerned with the short-run and emotional. They focus not on long-run material progress (from which they themselves benefit so much), but on short-run instabilities and inequalities in income (which are an inevitable by-product of the perennial "gale of creative destruction" brought about by technology-advancing entrepreneurs). Before long, a coalition of anti-business intellectuals, government bureaucrats, and labor unions—intent on taming the business cycle and eliminating income inequalities—will smother the capitalist engine with interventionist policies that will discourage the innovator once and for all.

Additional Readings

Harris, Seymour E., ed. *Schumpeter: Social Scientist.* Cambridge: Harvard University Press, 1951.
   A collection of 20 essays on the life and work of Schumpeter.

Seligman, Ben B. *Main Currents in Modern Economics,* vol. 3. Chicago: Quadrangle, 1962, pp. 646–65 and 694–713.
   Critical reviews of the work of Knight and Schumpeter.

---

[1] Joseph A. Schumpeter, *Capitalism, Socialism, and Democracy,* 3rd ed. (New York: Harper & Row, 1962), p. 67.

CHAPTER 11

Labor and Wages

MULTIPLE-CHOICE QUESTIONS

Circle the letter of the *one* answer that you think is correct or closest to correct.

1. For a firm operating in perfectly competitive markets, the dollar marginal benefit from using a resource equals
   a. the marginal value product of the resource.
   b. the marginal physical product of the resource multiplied by the dollar price of the resource.
   c. both (a) and (b).
   d. neither (a) nor (b) above.

2. A firm, which is operating in perfectly competitive markets, can raise its profit by using more of any input as long as the input's marginal value product
   a. exceeds product price.
   b. exceeds input price.
   c. falls short of product price.
   d. falls short of input price.

3. When a firm, which is operating in perfectly competitive markets, is maximizing profits, while the price of its product is $100 and the marginal value product of one of its inputs is $200,
   a. the firm must be paying $200 for a unit of this input.
   b. this input's marginal physical product must equal .5 units of product.
   c. this input's marginal physical product cannot be calculated.
   d. both (a) and (c) are true.

4. A perfectly competitive firm's demand curve for an input may shift to the right as a result of all of the following *except*
   a. an improvement in technology.
   b. an increase in the quantities of complementary inputs.
   c. an increase in input's price.
   d. an increase in product price.

5. Inputs are termed *independent inputs* when a change in the quantity of any one
   a. has no effect on total product.
   b. has no effect on the marginal physical products of the other inputs.
   c. has no effect on its own marginal physical product.
   d. changes total product even if the amounts of other inputs remain unchanged.

6. When the output effect of an input price change works in the opposite direction from the substitution effect,
   a. a firm is using regressive inputs.
   b. an analogy exists to inferior goods.
   c. both (a) and (b) are true.
   d. neither (a) nor (b) is true.

7. The addition of the separate input demand curves of perfectly competitive firms (in order to derive the input's market demand curve) is complicated by the fact that
   a. each firm is assumed to maximize profit, but all firms together cannot possibly do so at the same time.
   b. the marginal physical product of the input clearly declines the more of it that is used.
   c. the market demand for the input is bound to be flatter than the horizontal summation of given marginal-value-product curves.
   d. the assumption of a given product price is invalid if all firms hire more (or less) of the input and, therefore, produce more (or less) output.

When answering questions 8–10, refer to the accompanying graph.

8. The individual shown is indifferent between
   a. a daily income of $72 and no leisure on the one hand and an income of $0A$ and 16 hours of leisure on the other.
   b. a daily income of zero and 24 hour of leisure on the one hand and an income of $0A$ and 16 hours of leisure on the other.
   c. the choices described in (a) or (b).
   d. none of the above.

9. The individual shown
   a. is indifferent between a daily income of $0A$ and 16 hours of leisure.
   b. receives a wage of $3 per hour.
   c. receives a wage of $48 per day.
   d. is correctly described by both (b) and (c).

10. The individual shown
    a. will choose combination B.
    b. will work 8 hours per day.
    c. will earn $24 per day.
    d. is correctly described by all of the above.

11. When a person's marginal rate of substitution equals $5 per hour of leisure,
    a. the person's utility is maximized only if the prevailing wage equals 20¢ per hour.
    b. the person's marginal utility per hour of leisure must be one fifth of that per dollar of income.
    c. the person can increase overall utility by working less, provided the wage equals $3 per hour.
    d. none of the above is true.

12. Imagine the typical graph depicting an individual's income-leisure choices (with income on the vertical axis and leisure on the horizontal one). In such a graph, a price-consumption line
    a. that is vertical implies a vertical supply curve of labor.
    b. that is vertical implies a horizontal supply curve of labor.
    c. is necessarily convex with respect to the origin.
    d. is necessarily concave with respect to the origin.

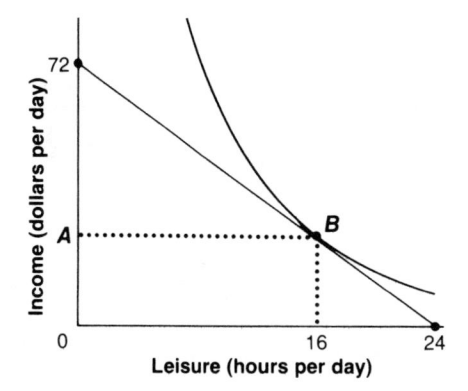

When answering questions 13–15, refer to the following graph depicting an individual's income-leisure choice, initially at *R:*

13. A rise in the wage depicted by budget line *MB* swinging to *MF,* will
    a. lead this consumer to select combination *N.*
    b. lead this consumer to select combination *P.*
    c. raise this consumer's income by *AF.*
    d. lead to (a) and (c).

14. A rise in the wage, depicted by budget line *MB* swinging to *MF,* will produce
    a. an income effect of *AE.*
    b. an income effect of *GH.*
    c. a substitution effect of *LH.*
    d. a substitution effect of *HG.*

15. A rise in the wage, depicted by budget line *MB* swinging to *MF,* will produce
    a. an income effect of *CE.*
    b. a substitution effect of *KD.*
    c. both (a) and (b).
    d. neither (a) nor (b).

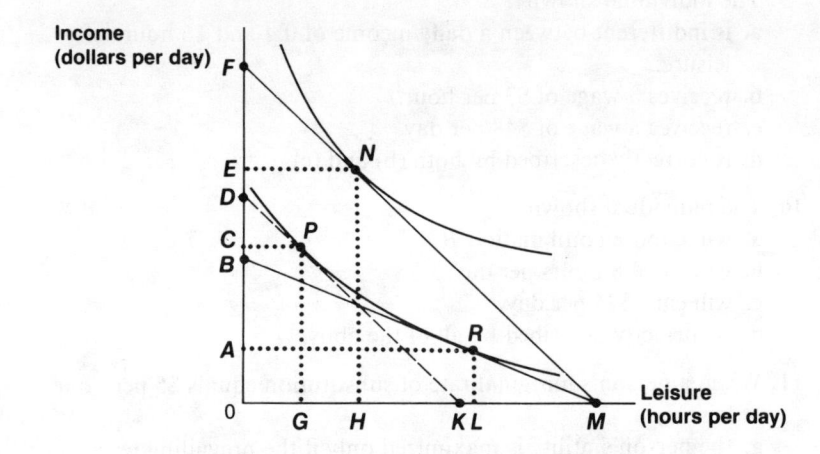

**Income
(dollars per day)**

[Graph with points F, E, D, C, B, A on the vertical axis, N, P, R on curves, and G, H, K, L, M on the horizontal axis labeled Leisure (hours per day)]

16. When the wage rises,
    a. a person tends to consume more leisure (work less) because of the income effect.
    b. a person tends to consume less leisure (work more) because of the substitution effect.
    c. both (a) and (b) are true.
    d. neither (a) nor (b) is true.

17. The income effect of a price change invariably reinforces the substitution effect
    a. for consumption goods that are normal goods.
    b. for consumption goods that are inferior goods.
    c. in the case of leisure.
    d. in the case of (a) and (c).

18. The possibility of the income effect of a price change overpowering the substitution effect
    a. is high for ordinary consumption goods.
    b. is fairly high in the case of leisure.
    c. is zero in the presence of Giffen's paradox.
    d. is not correctly described by any of the above.

19. Under the negative-income-tax system, a reduction of the government grant as earned income rises is expected to produce
    a. a substitution effect: more work.
    b. an income effect: more work.
    c. an income effect: less work.
    d. both (a) and (c).

20. A firm's marginal benefit of using labor could be
    a. the wage of labor.
    b. the marginal revenue product of labor.
    c. the marginal value product of labor.
    d. either (b) or (c).

21. The marginal revenue product of labor equals
    a. its marginal physical product multiplied by marginal revenue.
    b. its average physical product multiplied by marginal revenue.
    c. its total physical product multiplied by marginal revenue.
    d. its marginal physical product multiplied by product price.

22. Assume the following: The use of 1 worker yields output of 500 units; the use of 2 workers yields output of 600 units (all else being equal). An output of 500 units can be sold at $1 per unit; an output of 600 units can be sold at only 90¢ per unit. Under the circumstances, the firm's
    a. marginal revenue equals 40¢.
    b. marginal revenue equals $40.
    c. marginal value product of labor equals $90.
    d. marginal revenue product of labor equals $100.

23. The marginal physical product of labor equals
    a. labor's marginal value product divided by product price.
    b. labor's marginal value product divided by labor's price.
    c. labor's marginal revenue product divided by marginal revenue.
    d. either (a) or (c).

When answering questions 24–26, refer to the graph at the top of page 174.

24. The firm pictured here clearly has
    a. monopoly power in the product market.
    b. monopoly power in the labor market.
    c. monopsony power in the product market.
    d. none of the above.

25. To maximize profit, the firm pictured here would hire
    a. $0G$ workers.
    b. $0E$ workers.
    c. any number of workers who apply (because of the given wage).
    d. a number of workers that cannot be determined without additional information.

26. The workers in this firm are subject to
    a. monopolistic exploitation measured by $ABCD$.
    b. monopolistic exploitation measured by $ABCF$.
    c. monopsonistic exploitation measured by CDF.
    d. monopsonistic exploitation measured by $ECFG$.

Dollars per Worker per Week

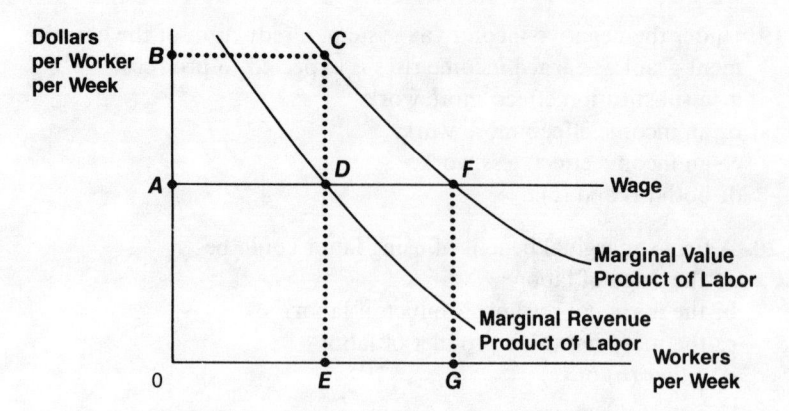

[Graph: B····C at top; horizontal line through A, D, F labeled "Wage"; curves labeled "Marginal Value Product of Labor" and "Marginal Revenue Product of Labor"; vertical dashed lines from D to E and F to G on the horizontal axis labeled "Workers per Week"]

27. Labor is said to be monopolistically exploited when
    a. its marginal revenue product falls short of product price.
    b. its marginal revenue product exceeds product price.
    c. its marginal value product exceeds its wage because marginal revenue product falls short of marginal value product.
    d. its marginal value product exceeds its wage because marginal revenue product exceeds marginal value product.

28. A labor market monopsony may well maximize profit by equating
    a. labor's marginal value product with its wage.
    b. labor's marginal revenue product with its wage.
    c. labor's marginal value product with the marginal outlay on labor.
    d. none of the above.

29. If a monopsony could hire 300 workers at a wage of $5 per hour but could hire 600 workers only at a wage of $8 per hour, its marginal outlay per hour would equal
    a. $3.
    b. $33.
    c. $3,300.
    d. none of the above.

When answering questions 30–33, refer to the graph on page 175.

30. The firm pictured here clearly has
    a. monopoly power in the product market.
    b. monopoly power in the labor market.
    c. monopsony power in the labor market.
    d. both (a) and (c).

31. To maximize profit, the firm pictured here would hire
    a. $0E$ workers.
    b. $0G$ workers.
    c. whatever number of workers applied at the market-given wage.
    d. a number of workers that cannot be determined without additional information.

32. The workers in this firm are subject to
    a. monopolistic exploitation measured by $ECFG$.
    b. monopsonistic exploitation measured by $ABCD$.
    c. both (a) and (b).
    d. neither (a) nor (b).

33. The firm's marginal cost of acquiring labor
    a. varies with the quantity of labor hired and can be read on the supply line.
    b. varies with the quantity of labor hired and can be read on the marginal outlay line.
    c. equals $0A$ regardless of the quantity of labor hired.
    d. equals $0B$ regardless of the quantity of labor hired.

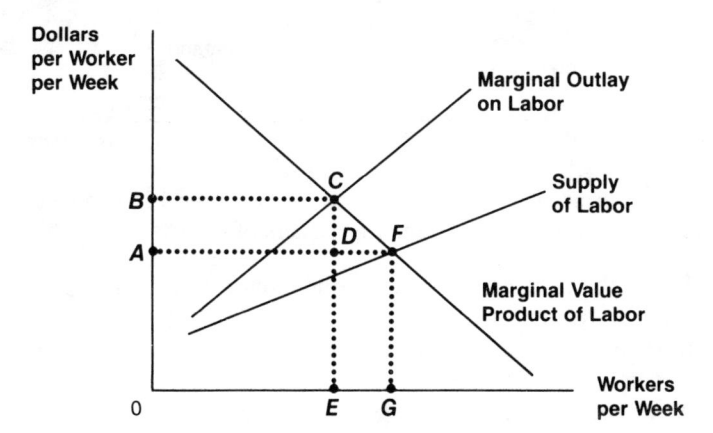

When answering questions 34–37, refer to the graph of a perfectly competitive labor market at the top of page 176.

34. If workers formed a union and pushed the wage from its competitive equilibrium level $0B$ to $0C$, the result would be
    a. labor unemployment of DH.
    b. labor unemployment of DE.
    c. an end to monopolistic exploitation.
    d. both (a) and (c).

35. If the workers, realizing that F was the demand line's midpoint, formed a union and sought to maximize the total income going to workers, they would have to
    a. push the wage above $0B$.
    b. keep the wage at $0B$.
    c. accept a wage below $0B$.
    d. restrict the supply of labor below $0G$.

36. If the workers formed a union and sought to maximize the economic rent going to workers, they would have to
    a. push the wage above $0B$.
    b. keep the wage at $0B$.
    c. restrict the supply of labor below $0G$.
    d. do both (a) and (c).

37. Under perfectly competitive conditions, the workers' economic rent would equal
    a. $0AFG$.
    b. $0BFG$.
    c. ABF.
    d. zero.

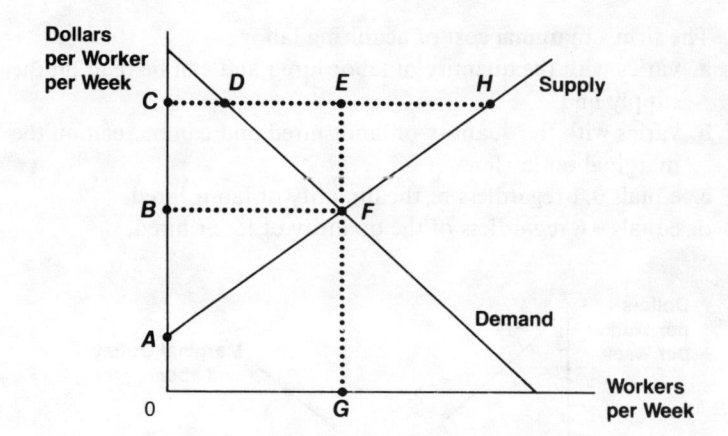

When answering questions 38–41, refer to the graph below about a labor monopsony facing competitive sellers of labor.

38. In the absence of a labor union, the monopsony would
 a. set a wage of $0C$.
 b. set a wage of $0A$.
 c. employ $0E$ workers.
 d. do both (a) and (c).

39. If the monopsony and a labor union now negotiated a wage of $0B$,
 a. employment would rise.
 b. unemployment would appear.
 c. the entire portion of the marginal-outlay-on-labor curve drawn in the graph would become invalid.
 d. all of the above would occur.

40. Furthermore, if the monopsony and a labor union negotiated a wage of $0B$,
 a. monopsonistic exploitation would continue but would decrease.
 b. labor's marginal value product would be reduced along the line shown.
 c. unemployment would actually fall.
 d. all of the above would occur.

41. If the monopsony and a labor union negotiated a wage of $0C$,
 a. employment would rise.
 b. unemployment would be avoided.
 c. only the portion to the left of D of the marginal-outlay-on-labor curve would become invalid.
 d. none of the above would occur.

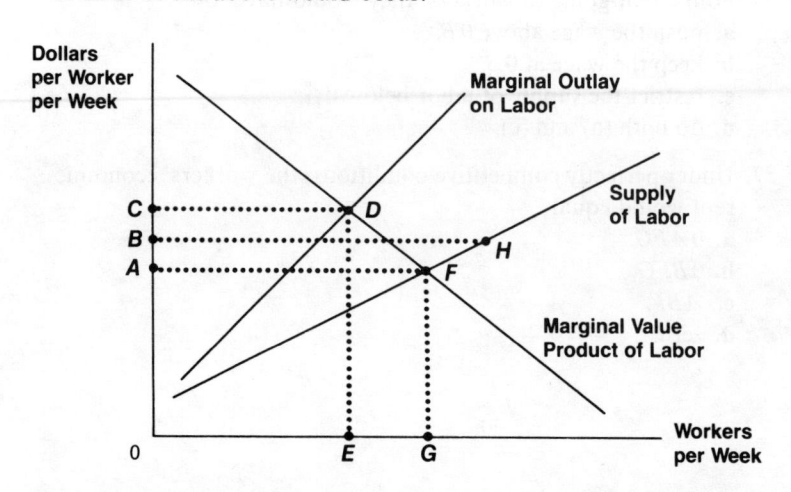

When answering questions 42–43, refer to the following graph depicting a bilateral monopoly:

42. Which of the following is *false*?
  a. Curve *A* is the monopsony's marginal outlay curve.
  b. Curve *B* is the monopoly's marginal cost curve.
  c. Curve *C* is the monopsony's marginal value product curve.
  d. Curve *D* is the monopsony's supply curve.

43. Bargaining between monopoly and monopsony will establish
  a. a price between 0*a* and 0*d*.
  b. a price between 0*b* and 0*c*.
  c. a quantity between 0*e* and 0.
  d. a quantity between 0*f* and 0*g*.

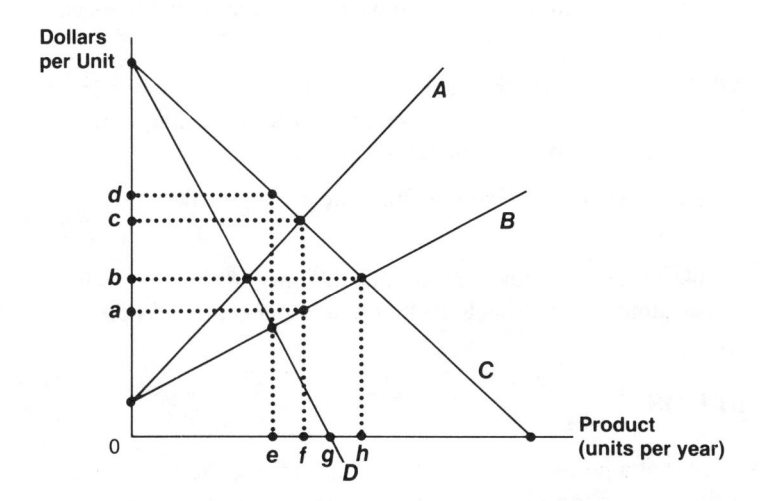

TRUE-FALSE QUESTIONS

In each space below, write a *T* if the statement is true and an *F* if the statement is false.

_____ **1.** For a firm operating in perfectly competitive markets, the hourly wage paid to a worker is in effect the firm's marginal cost of producing a unit of its product.

_____ **2.** A firm operating in perfectly competitive markets can raise its profit by using less of any input whenever the input's marginal physical product falls short of the ratio of input price to product price.

_____ **3.** If a lower price of one input causes a firm to hire additional amounts not only of this input, but of complementary inputs as well, the firm's demand curve for the first input is steeper than any given marginal-value-product curve.

_____ **4.** The substitution effect always makes a firm buy less of an input with a higher price.

_____ **5.** An individual's income-leisure choice is optimal when the wage rate equals that person's marginal rate of substitution between income and leisure.

_____ **6.** The income effect of a change in the price of leisure always works counter to the substitution effect.

_____ 7. Equalizing wage differentials are necessarily smaller than compensating wage differentials.

_____ 8. A profit-maximizing firm may well equate labor's money wage with the ratio of labor's marginal revenue product to marginal revenue.

_____ 9. When labor's marginal value product exceeds its wage because marginal revenue product falls short of marginal value product, labor is said to be monopolistically exploited.

_____ 10. In everyday language, a labor market monopsony is simply referred to as a labor union.

_____ 11. Union shops are firms in which only union members can be hired.

_____ 12. To maximize the income going to workers, a labor union must set a wage so as to reduce to zero the marginal revenue from selling labor.

_____ 13. A labor union cannot eliminate monopolistic exploitation.

_____ 14. When a monopsony can buy 32 units at $2 each or 40 units at $2.50 each, its marginal outlay equals $4.50.

PROBLEMS

Fertilizer (pounds per year)	Strawberries (pounds per year)
0	100
1	150
2	200
3	230
4	240
5	244

1. Suppose that a perfectly competitive firm finds the indicated relationship between inputs of fertilizer and its output of strawberries, all else being equal.
   a. Calculate the firm's demand curve for fertilizer if strawberries sell at 25 cents per pound.
   b. Plot these demand curve data in the following graph, using midpoints of relevant input ranges and connecting the data plots with a smooth curve.

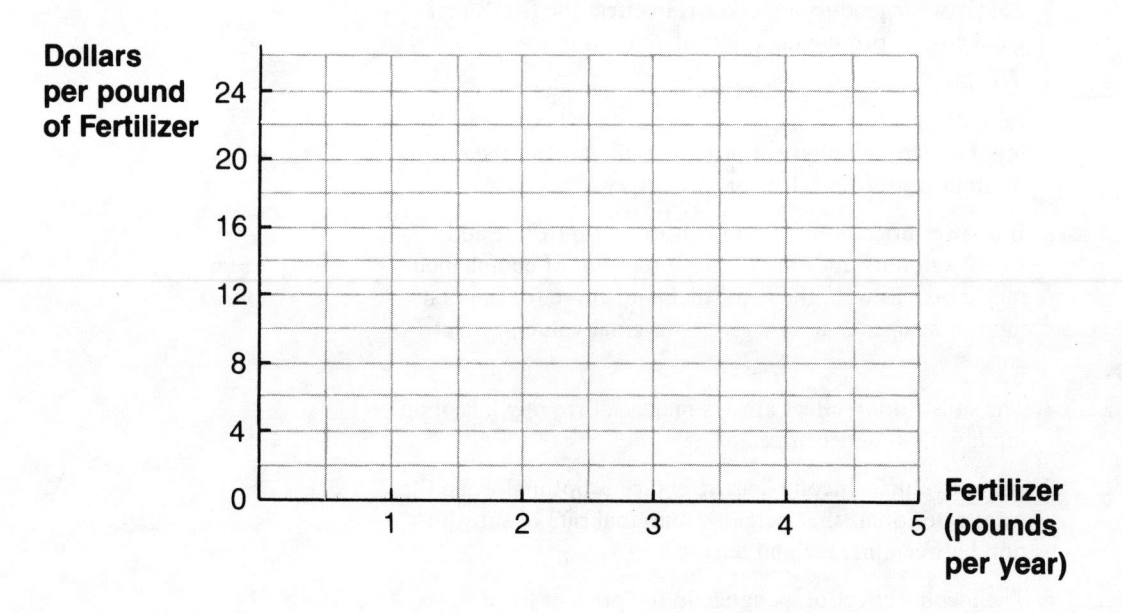

c. How much fertilizer would this profit-maximizing firm buy if the price of fertilizer were $7.50 per pound? Show your answer in the graph.
d. How would you change your answer to (c) if the price of strawberries now doubled? Show this in the graph as well.

2. a. Assume there are 10,000 firms like the one discussed in 1(b) above. In the following graph, plot their market demand for fertilizer, assuming that the given strawberry price never varies from 25 cents per pound.

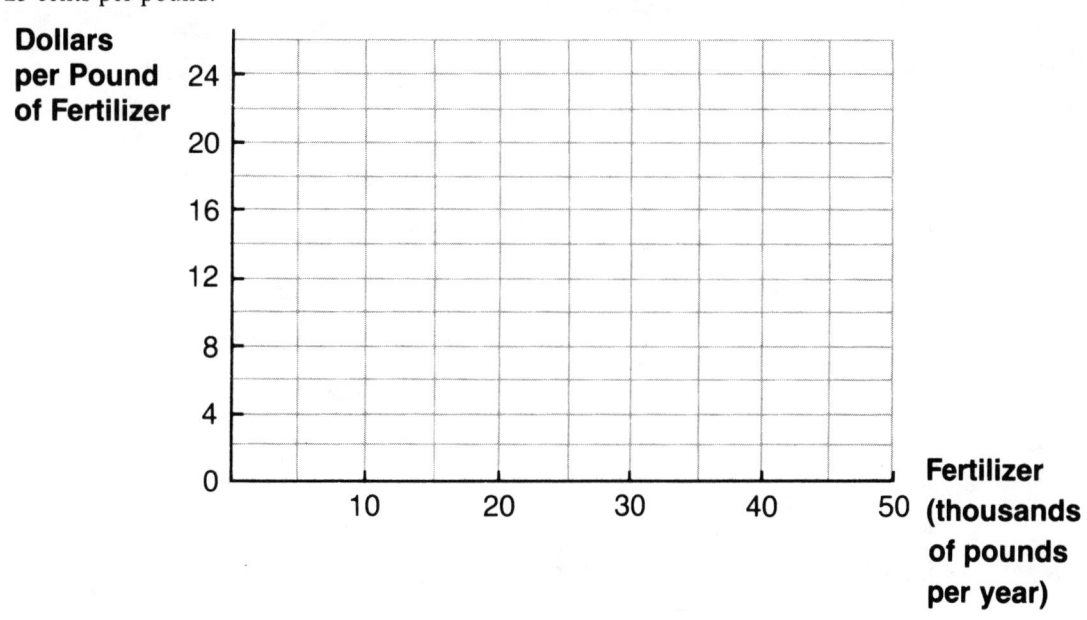

Dollars per Pound of Fertilizer (vertical axis: 0, 4, 8, 12, 16, 20, 24)

Fertilizer (thousands of pounds per year) (horizontal axis: 10, 20, 30, 40, 50)

b. How much fertilizer would these firms buy if its price were $2.50 per pound? If it rose to $10 per pound?
c. How would you correct your answer, if you were told that the higher price of fertilizer so reduced its use and, therefore, the output of strawberries, that the price of strawberries doubled?

	Income (dollars per day)	Leisure (hours per day)
(A)	100	8
(B)	70	10
(C)	42	14
(D)	24	20
(E)	20	24

3. Imagine a person who was indifferent about the indicated combinations of income and leisure.
   a. Does this person have a diminishing marginal rate of substitution? Explain.
   b. Plot the data on the following graph and connect the points with a smooth curve.

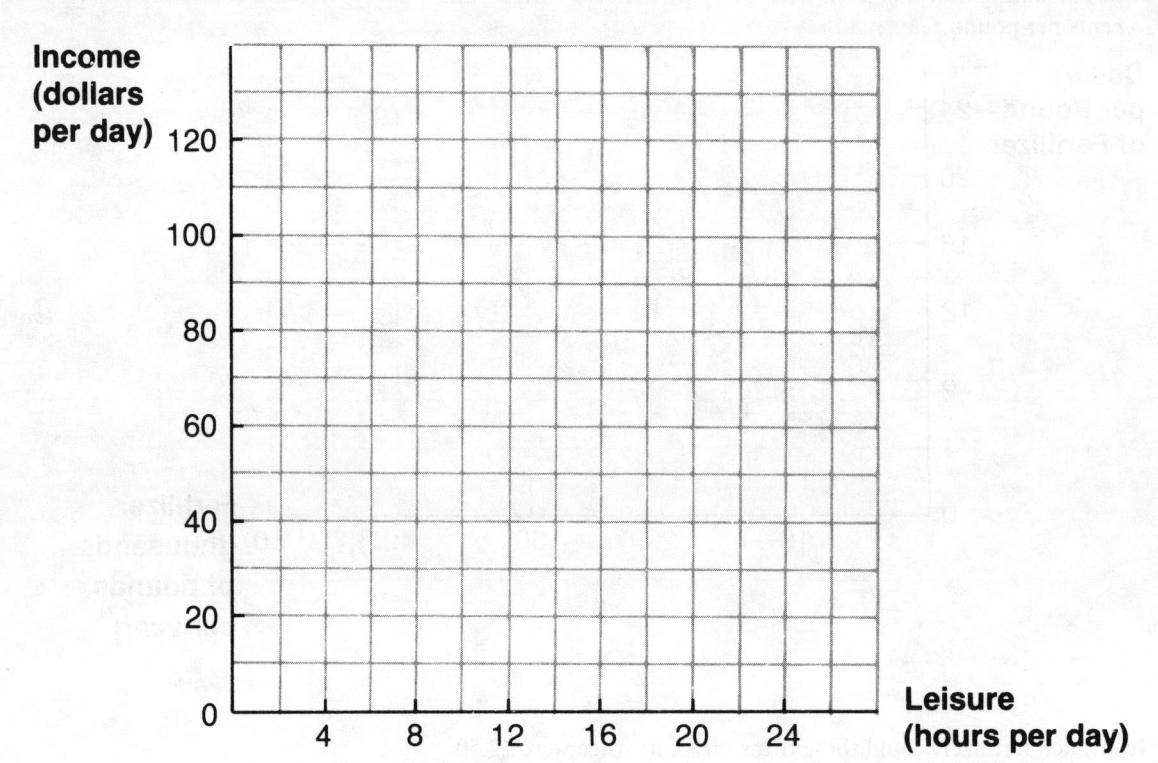

   c. If this person were offered a wage of $4.17 per hour, how many hours would the person work? What would the person's marginal rate of substitution between income and leisure be? Obtain your answers graphically.
   d. How would your answers change, if the wage fell to $2 per hour?
   e. How would your answers change, if the person (still facing a wage of $4.17 per hour) were also offered an unconditional grant of $20 per day?

4. In the following graph, illustrate an individual's income-leisure choice. Then illustrate the new choice following an increase in the wage. Identify the substitution and income effect of the wage hike.

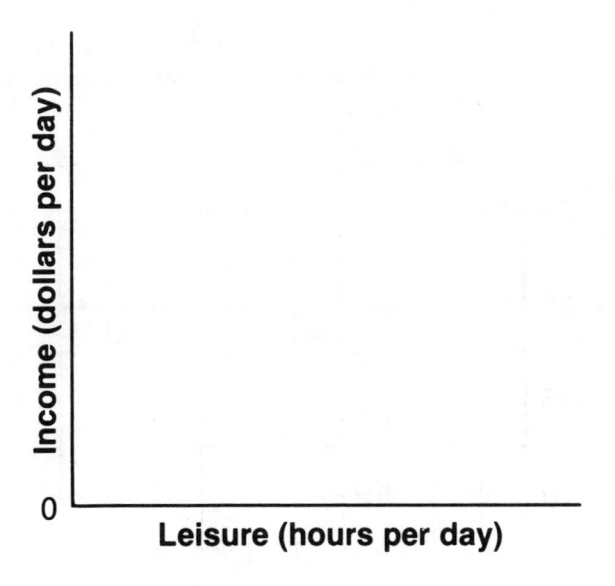

5. Assume that a firm's product-demand curve is described by the equation $Q = 2{,}000 - 200P$, while its marginal revenue curve is $Q = 1{,}000 - 100MR$. Fill in the missing numbers in the following table, always calculating price and marginal revenue with respect to the midpoints between columns (1) and (2) entries, because the entries in columns (3)–(5) also refer to such midpoints.

Workers Employed (1)	Total Physical Product of Labor (units per week) (2)	Marginal Physical Product of Labor (units per worker per week) (3)	Marginal Value Product of Labor (dollars per worker per week) (4)	Marginal Revenue Product of Labor (dollars per worker per week) (5)
1	500			
		400	2,600	1,200
2	900			
		300	1,425	−150
3	1,200			
		200	700	−600
4	1,400			
		100	275	−450
5	1,500			

6. **a.** Plot the data in your columns (4) and (5) with respect to column (1) midpoints in the graph. Connect your entries to form two smooth lines.

Hundreds of Dollars per Worker per Week

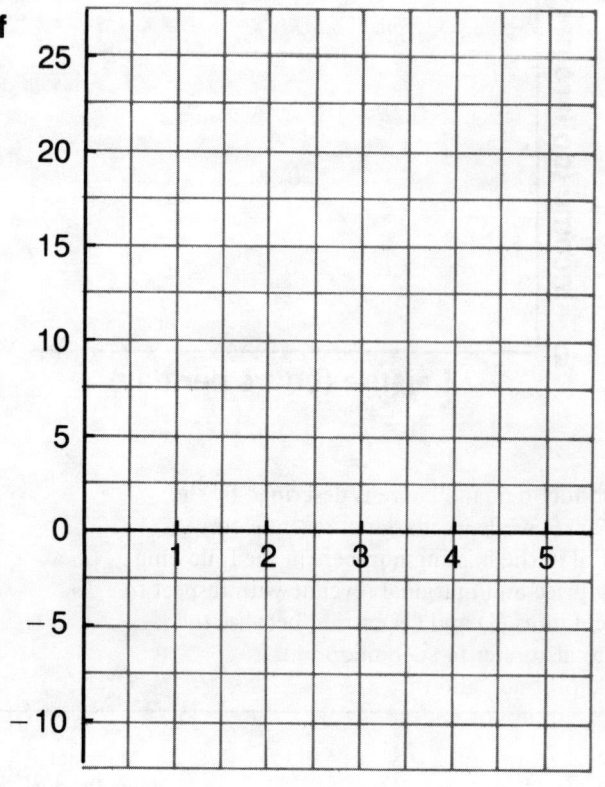

Workers per Week

b. Graphically determine the number of workers hired at a weekly wage of $1,200 per worker, and the extent of monopsonistic and monopolistic exploitation (if any).

7. Fill in the missing numbers in the following table:

Workers employed	Wage	Marginal Outlay
(dollars per worker per week)		
(1)	(2)	(3)
10	50	
20	60	
30	70	
40	80	
50	90	

8. **a.** Plot your column (3) data with respect to the midpoints of the column (1) data in the graph and connect your plots with a smooth line.

[Graph: Marginal Value Product of Labor, Dollars per Worker per Week vs Workers per Week]

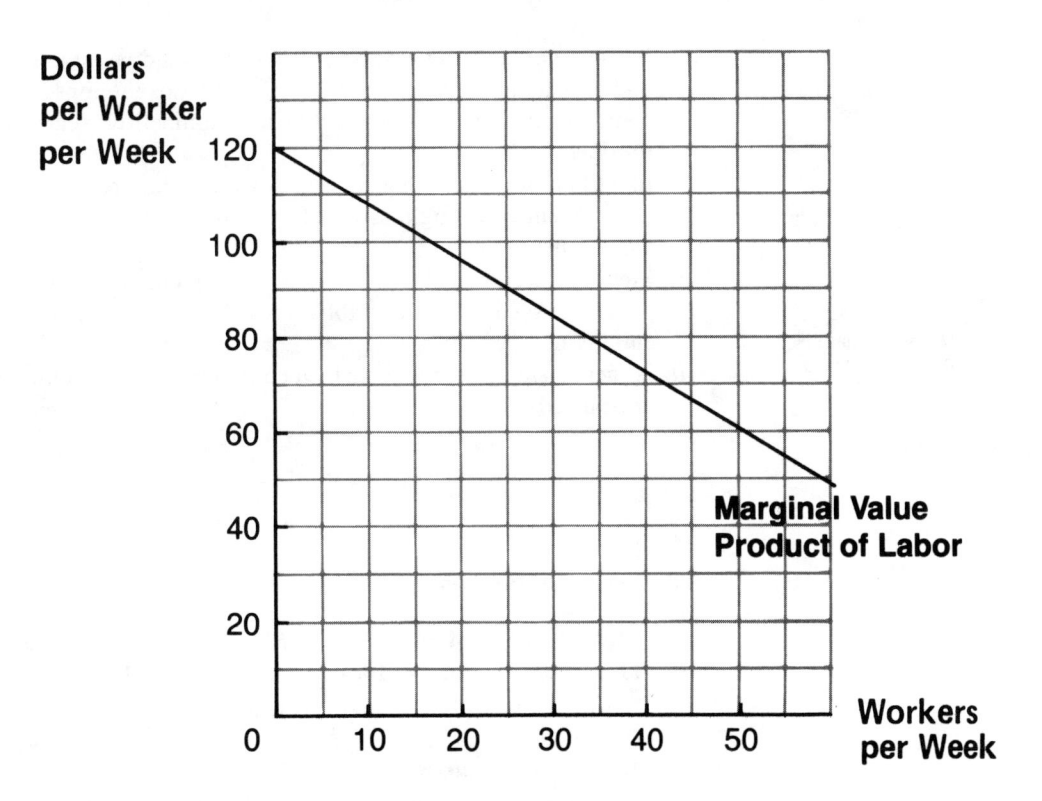

b. Graphically determine the number of workers hired, their wage, the value of their total product, and the extent of monopsonistic and monopolistic exploitation (if any).

9. a. In the following graph, identify the competitive equilibrium wage, as well as the two wage levels that would maximize the workers' economic rent and their total wage income, respectively.

b. Would there be unemployment at any of the three wage levels? If so, how much?

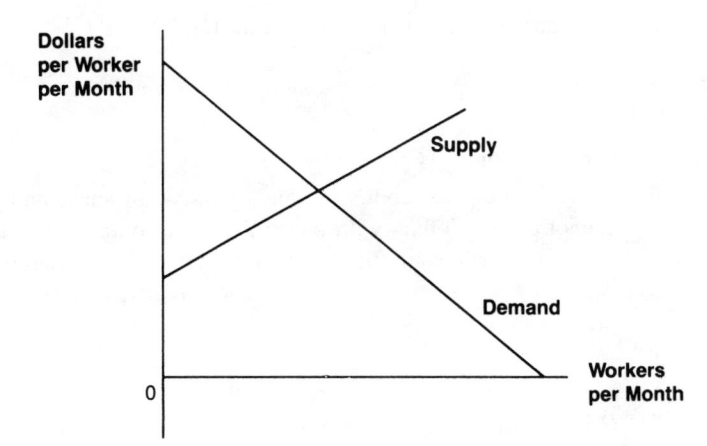

[Graph: Supply and Demand, Dollars per Worker per Month vs Workers per Month]

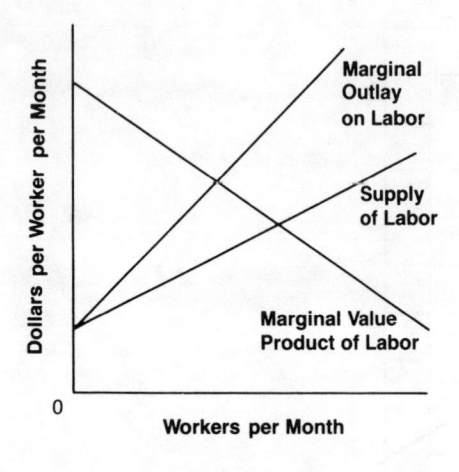

10. **a.** In the accompanying graph, which wage would be established and how many workers would be hired by this monopsony facing competitive sellers of labor and being a competitive seller of its product?
   b. Identify that range of higher wages (which a labor union or government edict might establish) which would actually increase employment.
   c. Identify that range of higher wages (which a labor union or government edict might establish) which would decrease employment.
   d. Is there a similar range for which employment would remain unchanged?

ANSWERS

Multiple-Choice Questions

1. a	2. b	3. a	4. c	5. b	6. c	7. d	8. d	9. b
10. d	11. c	12. a	13. a	14. b	15. d	16. c	17. a	18. b
19. b	20. d	21. a	22. a	23. d	24. a	25. b	26. a	27. c
28. c	29. d	30. c	31. a	32. d	33. b	34. a	35. b	36. d
37. c	38. c	39. d	40. b	41. d	42. d	43. a		

True-False Questions

1. F (The hourly wage is the marginal cost of producing the worker's marginal physical product—which may not equal 1 unit of product. The marginal cost per unit of product, therefore, equals the hourly wage divided by the worker's marginal physical product.)
2. T
3. F (The curve is *flatter* than any given *MVP* curve.)
4. T
5. T
6. T
7. F (The two concepts are identical.)
8. F (That ratio equals labor's marginal physical product, and one cannot equate dollars with physical units of output. A firm can, however, equate the money wage with labor's marginal value product or its marginal revenue product, both being expressed in dollars.)
9. T
10. F (Monopsony means "single buyer." In contrast, a labor union is a single seller and thus a labor market monopoly.)
11. F (Anyone can be hired, but nonunion members must join the union within 30 days to remain employed.)
12. T
13. T
14. T

Problems

1. **a.** The demand for fertilizer is shown by columns (1) and (3): the maximum price the firm would pay for any given quantity of fertilizer equals the associated marginal value product.

Fertilizer (pounds per year) (1)	Marginal Physical Product (pounds of strawberries per year) (2)	Marginal Value Product (dollars' worth of strawberries per year) (3) = (2) · 25¢/pound
0		
	50	$12.50
1		
	50	12.50
2		
	30	7.50
3		
	10	2.50
4		
	4	1.00
5		

Note that the *average* value product, not shown in the table, exceeds the marginal value product in all instances here. For example, when 1 lb of fertilizer is used, the total product is 150 lb. of strawberries; thus, at 25¢ per pound, the value of the total product is $37.50 and so is the average value product of 1 lb. of fertilizer input. Can you see why the average value product equals $25 at 2 lbs. of fertilizer input? $19.17 at 3 lbs.?

b. The curve labeled "Original Demand" corresponds to the calculations made in (a) above.

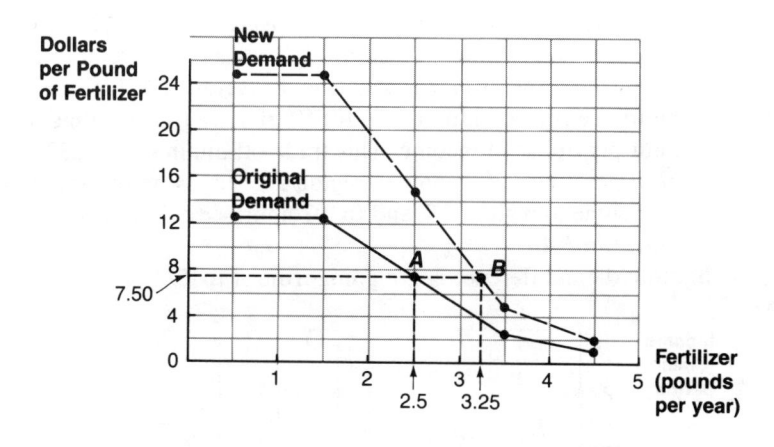

c. 2.5 pounds per year (point *A*).

d. All the entries in column (3) in (a) above would double. Thus a new demand curve (dashed) would emerge, corresponding to the higher price of strawberries. The firm would buy 3.25 pounds per year (point *B*).

2. a. Except for the labeling of the horizontal axis, this market demand line looks just like the individual firm's original demand in the graph for question 1. It is the sum of the individual firm's marginal-value-product curves.

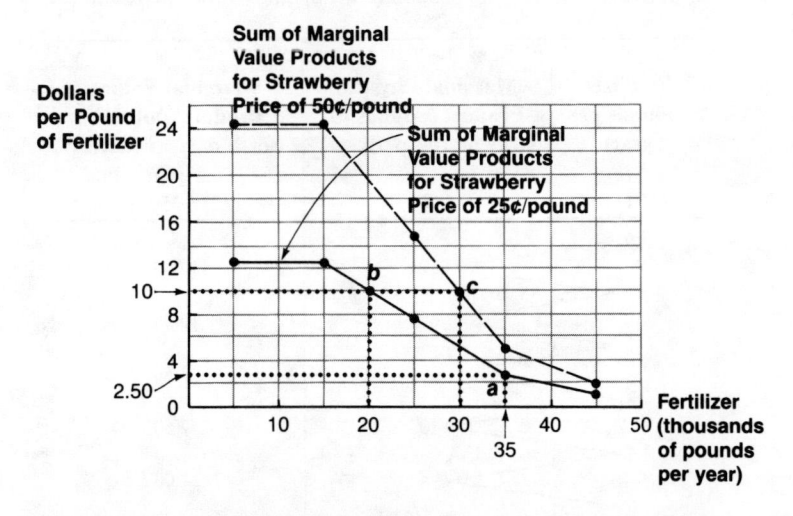

b. 35,000 pounds per year at $2.50 per pound (point *a*). 20,000 pounds per year at $10 per pound (point *b*).

c. At the higher strawberry price, a higher sum-of-marginal-value-product curve would become relevant. Hence the answer is found on this curve: 30,000 pounds per year at $10 per pound (point *c*). The market demand line for fertilizer now does not coincide with either one of the lines shown in the graph for question 2 but is a steeper line (connecting *a* and *c*).

3. a. Yes. In region *AB*, $30 of income per day are traded indifferently for 2 extra hours of leisure; that is, $15 per leisure hour. This trade-off diminishes to $7 per leisure hour in region *BC*, to $3 per leisure hour in region *CD*, to $1 per leisure hour in region *DE*. The reverse is also true. In region *ED*, 4 leisure hours are traded indifferently for $4 of extra income; that is, 1 hour per dollar of income. This trade-off diminishes to .33 hour per dollar of income in region *DC*, to .14 hour per dollar of income in region *CB*, and to .07 hour per dollar of income in region *BA*.

b. Note the indifference curve going from *A* to *E*.

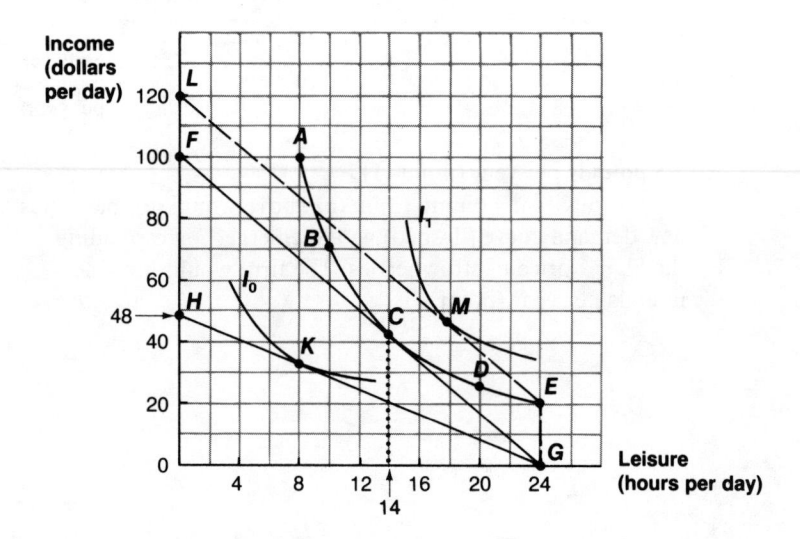

c. The person would have a budget line of *GF* because at $4.17 per hour 24 hours could be traded for $100 per day. The person's optimum would then be at *C*, corresponding to 14 leisure hours, hence 10 work hours per day. The person's marginal rate of substitution would equal the slope of the indifference curve at *C*, hence the slope of budget line *GF*, which is tangent at *C*. Thus the *MRS* would equal $100 per day for 24 hours per day, or $4.17 of income per hour of leisure (a rate that equals the wage).

d. If the wage fell to $2 per hour, the relevant budget line would be *GH*, because at $2 per hour, 24 hours could be traded for only $48 per day. The person would find a new optimum (such as *K*) where this lower budget line was tangent to a lower indifference curve (such as I_0). Depending on where the tangency occurred, leisure would decrease and work increase (as implied by *K*) or leisure would increase and work decrease (not shown). In either case, the person's marginal rate of substitution at the new optimum would equal the slope of the relevant indifference curve (as at *K*) and would thus equal the slope of the new budget line: $2 of income per leisure hour.

e. If a grant of $20 per day were received, $20 of income would be associated even with 24 hours of leisure (point *E*) and $120 of income with zero leisure (point *L*). The new budget line *GEL* (dashed) would parallel the old one (*GF*) to the left of *E*. The person could find a new optimum (such as *M*) where this higher budget line was tangent to a higher indifference curve (such as I_1). Depending on where the tangency occurred, leisure would increase and work decrease (as implied by *M*), or leisure would decrease and work increase (not shown). In either case, the person's marginal rate of substitution at the new optimum would equal the slope of the relevant indifference curve (as at *M*) and would thus equal the slope of the new budget line between *E* and *L*: $4.17 of income per leisure hour.

4. Given original budget line *HA* and indifference curve I_0, an original optimum corresponds to *L*. An increase in the wage swings the budget line to *HC* and yields a new optimum at *M* on higher indifference curve I_1. The substitution effect equals *ED* (a higher price of leisure makes people buy less of it). This is found by drawing hypothetical budget line *GB* parallel to the new budget line *HC* and tangent to old indifference curve I_0. The income effect equals *DF* (a rise in income makes people buy more of normal goods, leisure included). This is found by imagining hypothetical *GB* to shift to actual *HC*. Note: The income effect need not necessarily overpower the substitution effect. Consider panel (a) of text Figure 8.7, "Substitution and Income Effects."

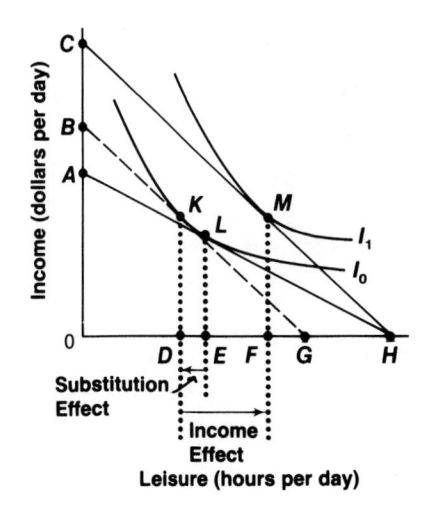

5.

Workers Employed (1)	Marginal Physical Product of Labor (3)	Product Price P	Marginal Value Product of Labor (4) = (3) · P	Marginal Revenue MR	Marginal Revenue Product of Labor (5) = (3) · MR
1.5	400	6.50	2,600	3.00	1,200
2.5	300	4.75	1,425	−.50	−150
3.5	200	3.50	700	−3.00	−600
4.5	100	2.75	275	−4.50	−450

a. *Calculation of product price (and marginal value product):* Assuming the output of 1.5 workers to be 700 (the average output of 1 and 2 workers), product price can be calculated from the equation $Q = 700 = 2,000 - 200P$ as $P = 6.50$. The remaining entries in the product price column can be similarly calculated (from $Q = 1,050 = 2,000 - 200P$ or $Q = 1,300 = 2,000 - 200P$ or $Q = 1,450 = 2,000 - 200P$). Marginal value product is, of course, simply the marginal physical product multiplied by the associated product price.

b. *Calculation of marginal revenue (and marginal revenue product):* Assuming the output of 1.5 workers to be 700, marginal revenue can be calculated from the equation $Q = 700 = 1,000 - 100MR$ as 3.00. The remaining entries in the marginal-revenue column can be similarly calculated (from $Q = 1,050 = 1,000 - 100MR$ or $Q = 1,300 = 1,000 - 100MR$ or $Q = 1,450 = 1,000 - 100MR$). Marginal revenue product is, of course, simply the marginal physical product multiplied by the associated marginal revenue.

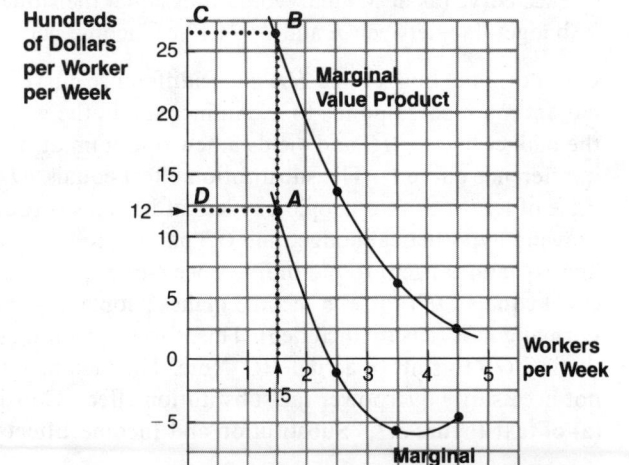

6. a.

b. The solution corresponds to point A: 1.5 workers. There is no monopsonistic exploitation (marginal outlay does not exceed the wage); monopolistic exploitation equals $ABCD$ (because marginal value product exceeds marginal revenue product).

7. When 10 workers earn $50 per week each, the firm's weekly wage bill equals $500. When 20 workers earn $60 per week each, the firm's weekly wage bill equals $1,200. Thus an extra $700 must be spent to get an extra 10 workers, and the marginal outlay (for the 10–20 worker range, to be plotted at the 15-worker midpoint) is $700 divided by 10, or $70 per worker. This is the first missing entry. The remaining ones can be similarly calculated as $90, 110, and 130 per worker.

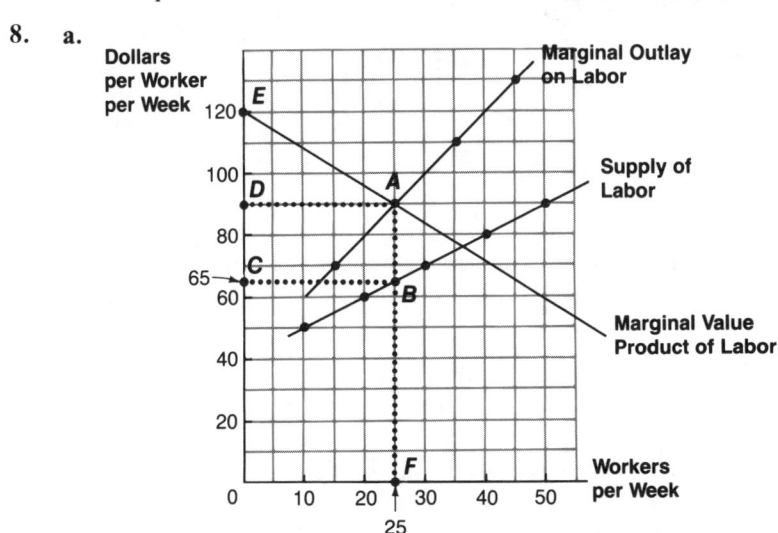

8. **a.**

   b. This requires plotting also the supply curve of labor, as given in columns (1) and (2) of the problem 3 table. The equality of marginal outlay and marginal value product at A determines the profit-maximizing labor quantity: 25 workers. The corresponding wage can be read off on the supply curve at B and equals $65 per worker per week. The value of the workers' total product is the sum of marginal value products up to the labor quantity employed, or area $0EAF$. Monopsonistic exploitation equals $ABCD$ because marginal outlay exceeds the wage. There is no monopolistic exploitation because marginal value product is not shown to diverge from marginal revenue product.

9. **a.** The competitive equilibrium wage corresponds to the supply-and-demand intersection A. It equals $0F$. The wage level that would maximize the workers' economic rent (assuming workers would act as a monopoly seller of labor) equals $0E$, which matches point B, corresponding to the intersection at C of the workers' marginal revenue from selling labor and their marginal cost of doing so (embodied in their supply line). The maximum rent equals area $BCDE$. The wage level that would maximize the workers' total wage income equals $0G$, which matches point H, the midpoint on the straight demand line at which the price elasticity of demand equals $|1|$ and the marginal revenue from selling labor is zero (point I). The maximum income equals area $0GHI$.

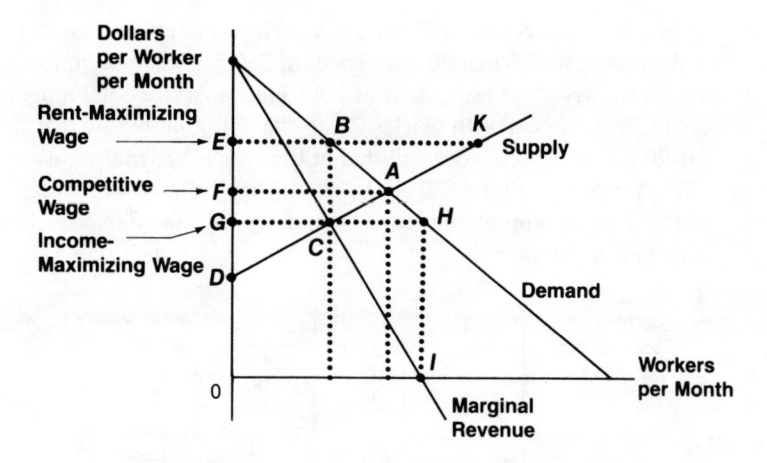

b. At the rent-maximizing wage, there would be unemployment of *BK*. At the competitive wage, there would be none. At the income-maximizing wage, there would be a labor shortage of *CH*.

10. a. The firm would establish a wage of $0A$ and hire $0E$ workers, corresponding to the marginal outlay and marginal value product intersection at *C*. (Caution: The supply-of-labor curve tells us at which wage $0E$ workers would work.)
    b. All such wages above $0A$ and just below $0B$ would increase employment.
    c. All wages above $0B$ would decrease employment.
    d. Not a range, but a single wage: $0B$ itself. *Note:* All this is explained in detail in text Figure 11.16, "Labor Union vs. Labor Monopsony."

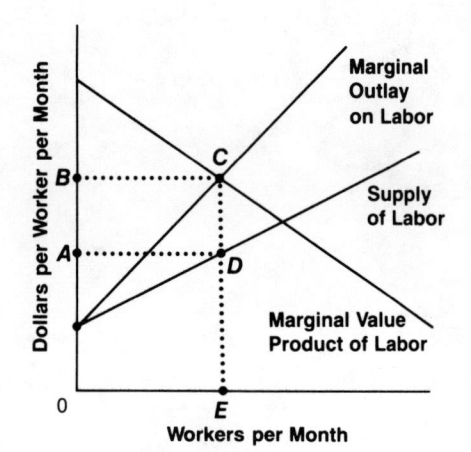

BIOGRAPHY 11.1

Joan V. Robinson

Joan Violet Robinson (1903–1983) was born in England, was educated at London and Cambridge, and has taught at Cambridge University since 1931. As Antoine Cournot (see Biography 7.1) before her and Edward Chamberlin (see Biography 8.1) in her own time, Joan Robinson was a pioneer in the study of the behavior of firms in imperfectly competitive markets. "It is customary," she writes in *The Economics of Imperfect Competition* (1933), "in setting out the principles of economic theory, to open with the analysis of a perfectly competitive world and to treat monopoly as a special case. . . . This process can with advantage be reversed. . . . It is more proper to set out the analysis of monopoly, treating perfect competition as a special case." The book covered some of the same ground as Chamberlin's but also other matters, from price discrimination to monopolistic and monopsonistic exploitation.

"The fundamental cause of exploitation," says Robinson, "will be found to be the lack of perfect elasticity in the supply of labor or in the demand for commodities. . . . Thus the function of a trade union or a minimum wage law in removing exploitation lies not so much in the fact that it improves the bargaining strength of the workers as in the fact that by means of a 'common rule' it reproduces artificially the condition of perfect elasticity of supply of labor to individual employers"[1] [Note the horizontal, union-imposed wage lines in the three parts of Figure 11.4 as examples of this common rule.] Robinson also showed that *monopolistic* exploitation could exist even in the absence of monopsony (see Figure 11.13), and that unions and minimum wages could not remove it.

The Economics of Imperfect Competition established for its author a worldwide reputation at a young age; this renown has been maintained through a lifelong penchant for travel, debate, and social criticism. (Chamberlin, in contrast, abstained from social criticism, except in his last book, which painted a frightful picture of industry domination by labor unions.) Early on, Joan Robinson became a vehement Marxian critic of the market economy, denouncing the system of private property on which it rests and the great evils of unemployment and injustice that she attributed to it. These "heretical" ideas have found expression in such books as *Private Enterprise and Public Control* (1945), *An Essay on Marxian Economics* (1942), *Economic Philosophy* (1962), *Freedom and Necessity* (1970), and *Aspects of Development and Underdevelopment* (1979).

While promoting her own revolution in microeconomics, Robinson worked tirelessly to expound that of her colleague, John M. Keynes, in macroeconomics and to extend his theory to long-run questions of capital accumulation and growth. Thus she came to write *Introduction to the Theory of Employment* (1937), *The Rate of Interest and Other Essays* (1952), *The Accumulation of Capital* (1956), *Essays in the Theory of Economic Growth* (1962), and (with John Eatwell) *An Introduction to Modern Economics* (1973). Her *Collected Economic Papers,* which appeared in five volumes between 1951 and 1979, show how thoroughly the land of economcis is crisscrossed with the footprints of this prolific scholar. Of Chamberlin's long quest (noted in Chapter 8) to have his product recognized as different from hers, she once said, "I'm sorry I ruined his life."

Additional Readings

Eichner, Alfred S. "Joan Robinson's Legacy," *Challenge,* May–June 1984, pp. 42–46.

Gram, Harvey, and Vivian Walsh. "Joan Robinson's Economics in Retrospect," *The Journal of Economic Literature,* June 1983, pp. 518–50.

---

[1] Joan V. Robinson, *The Economics of Imperfect Competition* (London: Macmillan, 1933), p. 281–82.

CHAPTER 12

Capital and Interest

MULTIPLE-CHOICE QUESTIONS

Circle the letter of the *one* answer that you think is correct or closest to correct.

1. A household that saves
   a. is trading current consumption for future consumption.
   b. must become a supplier in the loanable-funds market.
   c. may do so by selling IOU's or ownership claims.
   d. is correctly described by (a) and (c).

2. Böhm-Bawerk thought that people typically had a positive time preference because
   a. they expected a more ample provision of goods in the future than in the present.
   b. they systematically overestimated their future wants.
   c. in their imaginations future scarcity was usually more intense than present scarcity.
   d. of both (b) and (c).

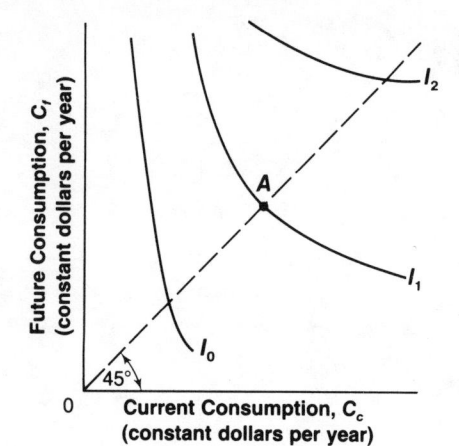

When answering questions 3–5, refer to the accompanying graph and assume equal units are used on both axes.

3. The slope of indifference curve I_0 has an absolute value of greater than one at all points. This slope denotes
   a. a lower marginal utility of future than of current consumption goods.
   b. that future scarcity is imagined to be less intense than present scarcity.
   c. a positive time preference.
   d. all of the above.

4. The slope of indifference curve I_2 has an absolute value of less than one at all points. This slope denotes
   a. a higher marginal utility of future than of current consumption goods.
   b. a lower marginal utility of future than of current consumption goods.
   c. that future scarcity is imagined to be less intense than present scarcity.
   d. both (b) and (c).

5. The slope of indifference curve I_1 has an absolute value of greater than one to the left of A and an absolute value of less than one to the right of A. This slope denotes
   a. the neutral time preference of a typical miser.
   b. positive time preference to the left of A.
   c. negative time preference to the right of A.
   d. both (b) and (c).

When answering questions 6–7, refer to the graph.

6. A consumer with the market opportunities pictured here
   a. has a positive time preference.
   b. faces an effective real interest rate of 20 percent per year.
   c. enjoys a higher marginal utility from future than from current consumption goods.
   d. is correctly described by all of the above.

7. A consumer with the market opportunities pictured here
   a. may well face a nominal rate of interest of 100 percent a year.
   b. may well face rapid inflation.
   c. is correctly described by both (a) and (b).
   d. is correctly described by neither (a) nor (b).

When answering questions 8–9, refer to the graph.

8. The consumer whose optimum is depicted at A
   a. is indifferent between x units of future consumption and 80 units of current consumption.
   b. is lending 20 units of current consumption goods for 40 units of future consumption goods.
   c. is earning an effective real interest rate of 200 percent per year.
   d. is correctly described by all of the above.

9. The consumer whose optimum is depicted at A
   a. has an income of 100 units of current consumption goods.
   b. has a marginal rate of time preference of 4:1.
   c. is consuming $x = 20$ units of future consumption goods.
   d. is correctly described by all of the above.

10. The income effect of a changed real interest rate
    a. works counter to the substitution effect.
    b. reinforces the substitution effect.
    c. induces more future as well as more current consumption.
    d. increases saving and lending.

When answering questions 11–12, consider the accompanying graph.

11. The consumer's initial optimum at A implies
    a. an upward-sloping supply curve of loanable funds.
    b. a real interest rate of 50 percent per year.
    c. a marginal rate of time preference of 2:1.
    d. all of the above.

12. A rise in the real interest rate from 100 to 300 percent per year
    a. produces a substitution effect that is stronger than the income effect.
    b. produces an income effect that reinforces the substitution effect.
    c. is associated with a backward-sloping supply curve of loanable funds (assuming all saving is lent).
    d. has effects that cannot possibly be determined from this graph.

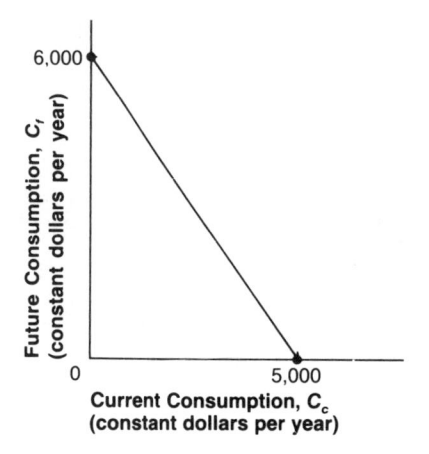

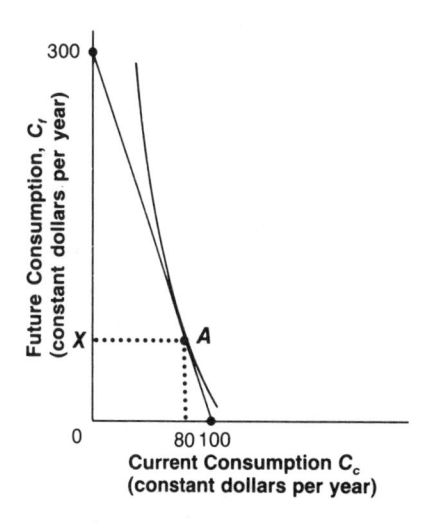

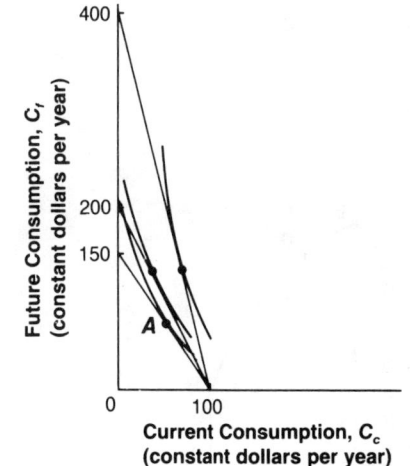

13. Capital budgeting involves
    a. identifying available investment opportunities.
    b. selecting investment projects to be carried out.
    c. arranging for the financing of chosen investment projects.
    d. all of the above.

14. An investment project consisting of annual cash flows (exclusive of financing costs) of $-100, +50, +50, +50$
    a. is worthwhile if the investors already have the money and need not borrow it, because the project then produces an overall profit of 50.
    b. is worthwhile if the investors can borrow the money at 10 percent interest per year, because the project then produces an overall profit of 20.
    c. is correctly described by both (a) and (b).
    d. is correctly described by neither (a) nor (b).

15. If the net present value of an investment project is calculated as $50,000 with the help of a discount rate of 17 percent per year, then
    a. the project is worthwhile to undertake if the prevailing interest rate is 5 percent per year.
    b. the investors will earn $50,000 more (in terms of present dollars) than they could earn when lending out their own money at 17 percent interest per year instead of sinking it into the project.
    c. the investors will earn $50,000 (in terms of present dollars) even if they borrow all the funds to finance the project and do so at 17 percent interest per year.
    d. all of the above are correct.

16. A profit-maximizing firm, which can borrow all it wants in a perfect market for loanable funds, will include in its capital budget
    a. all possible investment projects with zero net present values.
    b. all possible investment projects with positive net present values.
    c. all possible investment projects with positive internal rates of return.
    d. both (b) and (c).

When answering questions 17–18, refer to this graph of a society's intertemporal production-possibilities frontier.

17. By using given resources and technology in the best possible way, the people of this society could produce
    a. 3 units of current consumption goods as well as 3.5 units of capital goods.
    b. 2 units of current consumption goods as well as 6 units of future consumption goods.
    c. 1 unit of current consumption goods as well as .5 units of future consumption goods.
    d. 4 units of current consumption goods as well as 7.5 units of future consumption goods.

18. Potential investment projects available to the people of this society promise
    a. a 400 percent return starting at A.
    b. a 200 percent return starting at B.
    c. a negative return starting at D.
    d. all of the above.

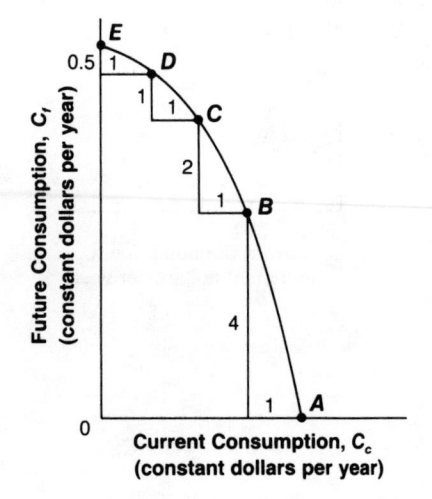

19. An asset that produces with certainty a net income stream of $500 for 5 years has a capitalized value
   a. of $2,500.
   b. of more than $2,500.
   c. of less than $2,500.
   d. that cannot be determined, even approximately, given the above information only.

20. An increase in the rate of interest
   a. raises the capitalized value of an income stream gradually as time goes by.
   b. raises the capitalized value of an income stream instantly.
   c. lowers the capitalized value of an income stream gradually as time goes by.
   d. lowers the capitalized value of an income stream instantly.

21. It is quite possible for some investment projects
   a. to have no meaningful net present value.
   b. to have no meaningful internal rate of return.
   c. to have several (positive or negative) net present values.
   d. to be characterized by all of the above.

22. Which one of the following is reliable under any and all circumstances?
   a. The internal-rate-of-return criterion.
   b. The payback method.
   c. The net-present-value criterion.
   d. The coefficient of relative effectiveness.

23. In spite of ever-increasing rates of consumption, for many natural resources
   a. prices have been steadily falling over time.
   b. the search for substitutes has decelerated over time.
   c. the ratio of known global reserves to consumption has remained constant over time.
   d. all of the above are true.

24. Suppose you own a bauxite deposit and can lend and borrow all the money you like at an interest rate of 10 percent a year. Assuming 1. you want to maximize profit, 2. it costs you $10 to mine a ton this year and $20 to mine it next year, and 3. you can sell a ton of bauxite now for $60, you will
   a. increase your current rate of mining if you expect a ton of bauxite to sell for $70 in a year.
   b. increase your current rate of mining if you expect bauxite to sell for $100 in a year.
   c. not change your current rate of mining if you expect bauxite to sell for $60 in a year.
   d. do all of the above.

25. According to the *separation theorem,*
   a. an isolated individual, such as Robinson Crusoe, inevitably must consume the identical combination of current and future consumption goods as that which is being produced.
   b. as long as loanable-funds markets are perfectly competitive, and the same rate of interest is applicable to borrowing and lending, an individual's productive optimum is entirely independent of the individual's intertemporal preferences.

c. a corporate manager who maximizes the wealth of the corporation cannot possibly maximize the separate wealth increments that go to individual stockholders.

d. under certain assumptions, the price of nonrenewable natural resources rises over time at a rate equal to the pure rate of interest.

TRUE-FALSE QUESTIONS

In each space below, write a *T* if the statement is true and an *F* if the statement is false.

_____ 1. *Capital market* and *loanable-funds market* are synonymous terms.

_____ 2. When consumers subjectively value present goods more highly than future goods of like kind and number, they are said to have a *present time preference*.

_____ 3. People who believe that present circumstances of abundance are likely to give way to intense future scarcity may well exhibit a negative time preference.

_____ 4. The percentage by which the purchasing power returned to a lender exceeds the purchasing power lent is referred to as the *real rate of interest*.

_____ 5. When a supply curve of loanable funds is backward-sloping, the income effect of higher real interest is overpowering the substitution effect.

_____ 6. The compound interest formula says $FV_t = PV_0(1 + r)^t$.

_____ 7. According to the discounting formula, $57 due in 13 years is now worth $\frac{\$57}{1.13^{10}}$, provided the interest rate equals 10 percent per year.

_____ 8. Whenever the net-present-value and internal-rate-of-return investment criteria conflict, a profit-maximizing firm is wise to ignore the former criterion.

_____ 9. Under perfectly competitive markets and certainty, an increase in the pure rate of interest will lead to a faster increase over time in the net price of nonrenewable natural resources.

_____ 10. There are many similarities between investing in physical goods and investing in people.

PROBLEMS

1. Consider a consumer with a neutral time preference and a present income sufficient to buy 500 units of consumption goods per year. In the following graph, indicate how this consumer would divide consumption between the present and the future, if the effective real rate of interest were 20 percent per year.

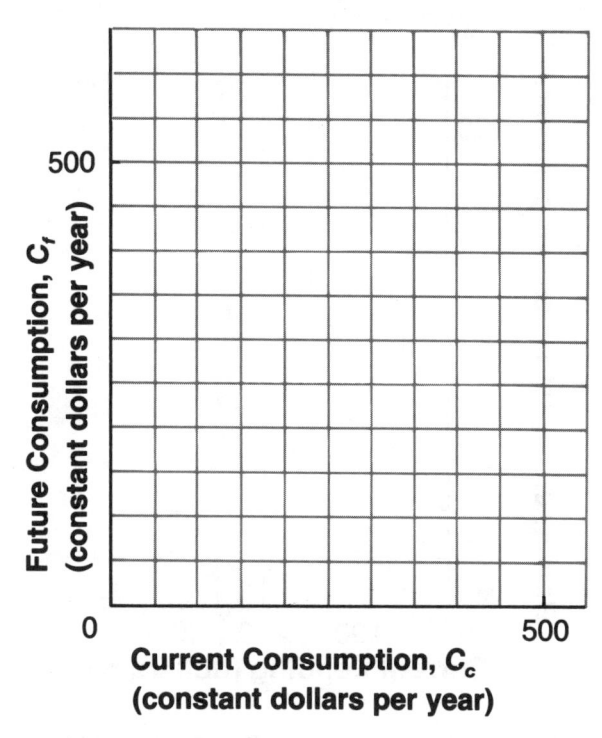

2. Consider graph (a), which depicts a consumer with a positive time preference.

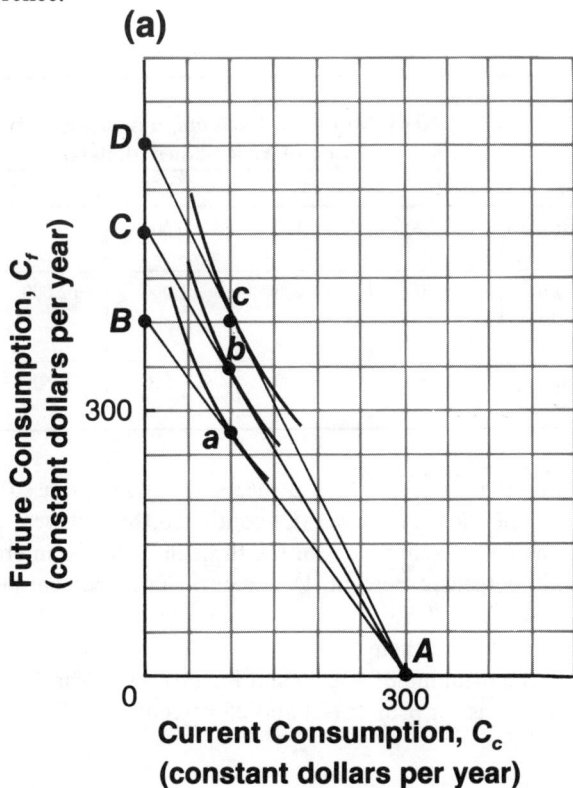

a. How does one know the consumer has a positive time preference?
b. What is the meaning of the swing in the budget line from AB to AC to AD?
c. In graph (b), plot this consumer's supply of loanable funds insofar as you have the relevant data.

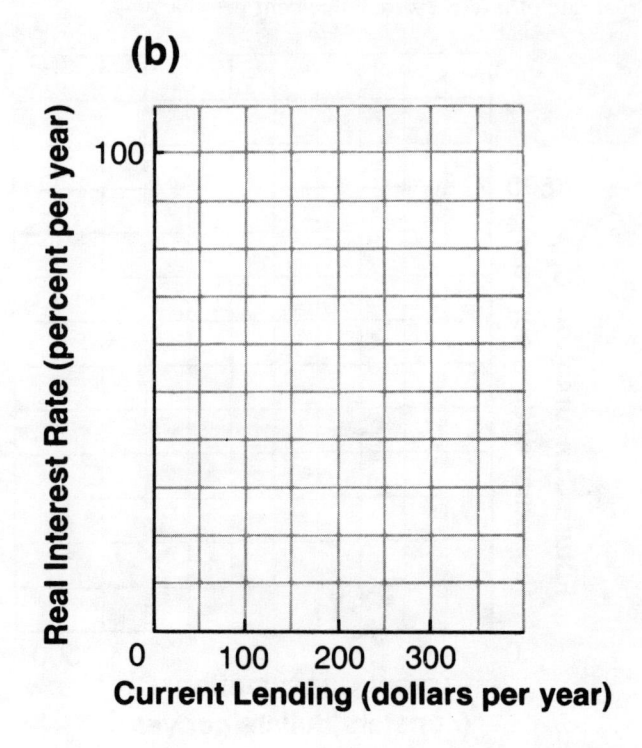

(b)

[Graph: Real Interest Rate (percent per year) on vertical axis from 0 to 100; Current Lending (dollars per year) on horizontal axis from 0 to 300]

d. How much is the consumer going to consume in the future?

3. Suppose a household faced the following possible investment projects:

Project	Net Cash Flows, Excluding Financing Cost, at End of Year Shown (dollars)			
	0	1	2	3
(A) Home-study course	−1,000	+100	+500	+700
(B) Surgery to improve health	−5,000	+2,000	+2,000	+2,000
(C) Migration for better job	−2,000	+500	+500	+500 → year 5

a. Assuming that the benefits of projects (A) and (B) ceased after year 3, while those of project (C) continued through year 5, calculate the net present values of the three projects for interest rates of 0 percent, 5 percent, 15 percent, and 25 percent per year. (Note: Text Table 12.2 "Compound Interest and Discount Factors" may be of help.)
b. Explain the meaning of your result for project (A) and 5 percent, as well as for project (C) and 25 percent.

c. In the following graph, draw this household's demand for loanable funds, assuming a perfect market for funds and certainty.

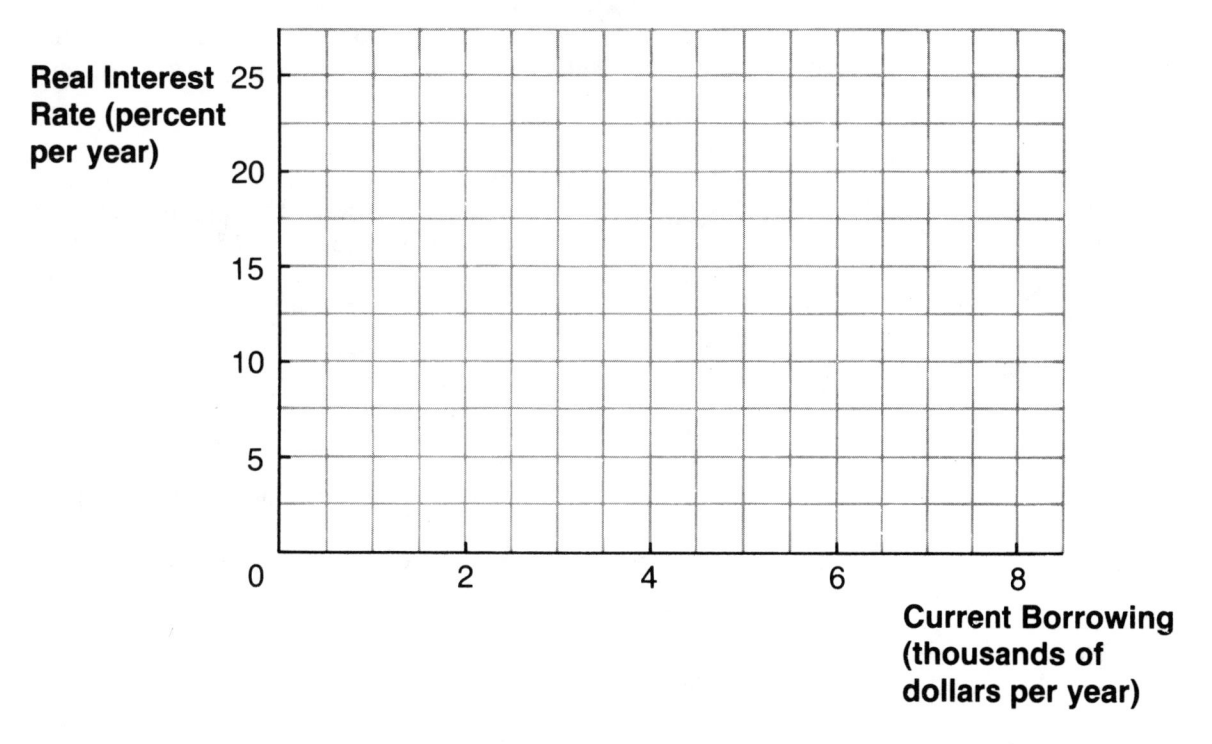

Real Interest Rate (percent per year)

Current Borrowing (thousands of dollars per year)

d. If the household used the payback method of evaluating household investment projects and established a maximum payback period of 2 years, what would be the verdict on projects (A)–(C)? What if the maximum period were set at 3 years?

e. Certain investments in human capital (such as taking home-study courses, going to college, or migrating for better job opportunities) are more readily undertaken by 20-year-olds than by 50-year-olds. Why do you think this is so?

4. In years past, when making choices among alternative investment projects, such as hydroelectric vs. fossil-fueled power plants, the Soviets consistently chose the former. Can you explain it? (*Hint:* Consider the fact that the former type of plants have much higher initial outlays, but lower annual operating costs, than the latter and that the Soviets ignored interest.)

5. Not so long ago, the Chase Manhattan Bank advertised: "Surprise! Only the Chase Savings Center will pay your savings interest in advance. Deposit $4,000, for example, in a 4-year savings plan. Today. The Chase Savings Center will pay you $976.81 in interest. Today." Ignoring the complication introduced by inflation, was this a good deal?

6. A life-insurance company made this offer to a student (age 20): "You pay us $150 now and every year for 19 more years or until you die, if you die first. We pay you $5,000 at age 40 or at death, if you die first. Because $150 times 20 equals $3,000, you are certain to get back every penny you pay and then some." What was the net present value of the policy to the insurance company? (Assume a relevant interest rate of 8 percent per year and that the student will in fact survive to age 40.)

7. The Exxon Travel Club recently offered a choice of any one of the following to a sweepstakes winner:
   a. a $79,000 solar-heated home or that amount in cash.
   b. an annual birthday gift of $16,000 for 5 years.
   c. an annual birthday gift of $5,000 for life up to a maximum of $250,000.
   Which was best?

8. You own a factory abroad that yields annual net receipts of $16,000. The foreign government wants to expropriate you, but one of its officials tells you that a proper bribe could delay the expropriation for 5 years. Assuming you have no moral qualms about bribes, what is the maximum bribe worth paying?

9. "Woodcutter, spare the tree." Can you give a reason for such a statement (apart from a sentimental one)?

10. The American Solar Heat Corporation of Danbury, Connecticut advertised a solar hot water heater ("good for 2 billion years") at $1,260. It was said to reduce hot water bills by 70 percent. If your annual hot water bill were $120 and you could borrow the money for 15 percent, would it be worth your while to buy the solar heater?

ANSWERS

Multiple-Choice Questions

1. a 2. a 3. d 4. a 5. d 6. b 7. c 8. c 9. a
10. a 11. b 12. c 13. d 14. d 15. d 16. b 17. b 18. c
19. c 20. d 21. b 22. c 23. c 24. a 25. b

True-False Questions

1. F (Capital markets and *asset* markets are synonymous.)
2. F (They are said to have a *positive* time preference.)
3. T
4. T
5. T
6. T
7. F (It is now worth $\$57/1.10^{13}$.)
8. F (The firm is wise to ignore the *latter* criterion.)
9. T
10. T

Problems

1. Such a consumer would face budget line AB, because 500 units of current consumption could be sacrificed for 600 units of future consumption at the indicated rate of interest. Such a consumer would also have indifference curves between future and current consumption, such as I, with this characteristic: The absolute value of the slope of such curves would equal 1 at a point such as a (where a 45-degree line from the origin intersects such curves). The slope would be $>|1|$ to the left of a (denoting positive time preference) and $<|1|$ to the right of a (denoting negative time preference). Given the positive interest rate embodied in budget line AB, the consumer's optimum must lie to the left of a, at a point such as b. As the graph is drawn, the consumer would now consume 130 units and later 444 units. This makes sense, because 130 units consumed now imply $500 - 130 = 370$ units saved, and they turn into 444 units at 20 percent interest per year.

2. a. The indifference curves that are tangent to the budget lines at a, b, and c, have a slope of $>|1|$ at all points. This means the consumer always insists on getting more than 1 unit of future consumption for the sacrifice of 1 unit of current consumption.
    b. The real interest is rising from 33.33 percent to 66.66 percent to 100 percent per year. Note: $(0B/0A)100 = (400/300)100 = 133.33$, implying a 33.33 percent rate of interest. And so on.
    c. This consumer consumes the same amount ($100 per year) at all of these interest rates, as is visible in graph (a) when noting the values of points a, b, and c, on the horizontal axis. The consumer, therefore, saves $300 - $100 = $200 per year at all of these interest rates, as is shown in graph (b) here.
    d. This can be seen by reading off the values of points a, b, and c on the vertical axis (or by calculating how much $200 of saving and lending turns into at the three alternative interest rates): $266.66; $333.33; $400.

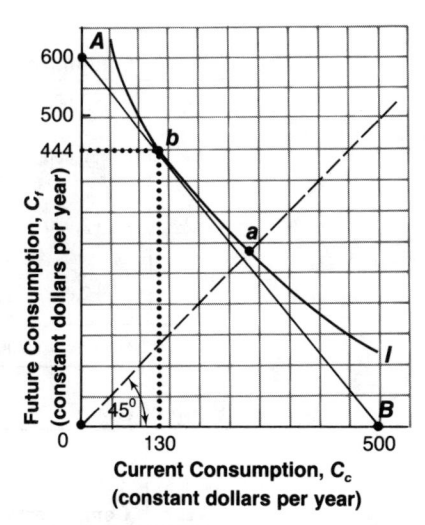

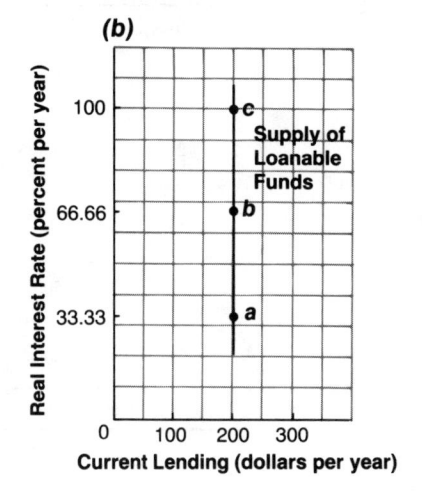

3. a.

    Project A:

		\multicolumn{4}{c}{Present values of components at annual interest rate of}
Year	Components	0%	5%	15%	25%
0	−1,000	−1,000	−1,000.00	−1,000.00	−1,000.00
1	+100	+100	$\frac{100}{1.05} = 95.24$	$\frac{100}{1.15} = 86.96$	$\frac{100}{1.25} = 80.00$
2	+500	+500	$\frac{500}{1.1025} = 453.51$	$\frac{500}{1.3225} = 378.07$	$\frac{500}{1.5625} = 320.00$
3	+700	+700	$\frac{700}{1.1576} = 604.70$	$\frac{700}{1.5209} = 460.25$	$\frac{700}{1.9531} = 358.40$
	Net present value:	+300	+153.45	−74.72	−241.60

Project B:

Year	Components	\multicolumn{4}{c}{Present values of components at annual interest rate of}
		0%	5%	15%	25%
0	−5,000	−5,000	−5,000.00	−5,000.00	−5,000.00
1	+2,000	+2,000	$\frac{2000}{1.05} = 1,904.76$	$\frac{2000}{1.15} = 1,739.13$	$\frac{2000}{1.25} = 1,600.00$
2	+2,000	+2,000	$\frac{2000}{1.1025} = 1,814.06$	$\frac{2000}{1.3225} = 1,512.29$	$\frac{2000}{1.5625} = 1,280.00$
3	+2,000	+2,000	$\frac{2000}{1.1576} = 1,727.71$	$\frac{2000}{1.5209} = 1,315.01$	$\frac{2000}{1.9531} = 1,024.01$
	Net present value:	+1,000	+446.53	−433.57	−1,095.99

Project C:

Year	Components	\multicolumn{4}{c}{Present values of components at annual interest rate of}
		0%	5%	15%	25%
0	−2,000	−2,000	−2,000.00	−2,000.00	−2,000.00
1	+500	+500	$\frac{500}{1.05} = 476.19$	$\frac{500}{1.15} = 434.78$	$\frac{500}{1.25} = 400.00$
2	+500	+500	$\frac{500}{1.1025} = 453.51$	$\frac{500}{1.3225} = 378.07$	$\frac{500}{1.5625} = 320.00$
3	+500	+500	$\frac{500}{1.1576} = 431.93$	$\frac{500}{1.5209} = 328.75$	$\frac{500}{1.9531} = 256.00$
4	+500	+500	$\frac{500}{1.2155} = 411.35$	$\frac{500}{1.749} = 285.88$	$\frac{500}{2.4414} = 204.80$
5	+500	+500	$\frac{500}{1.2763} = 391.76$	$\frac{500}{2.0114} = 248.58$	$\frac{500}{3.0518} = 163.84$
	Net present value:	+500	+164.74	−323.94	−655.36

b. *Project A and 5 percent:* The project is equivalent to spending $1,000 now for the privilege of receiving $1,153.45 instantly. The latter sum, invested at 5 percent, would turn into $1,211.12 at the end of year 1. Taking out $100 would leave $1,111.12, which would grow into $1,166.68 at the end of year 2. Taking out $500 would leave $666.68, which would grow into the $700 final benefit of the project by the end of year 3.
Project C and 25 percent: The project is equivalent to spending $2,000 now for the privilege of receiving instantly $655.36 less, or $1,344.64. The latter sum, invested at 25 percent, would turn into $1,680.80 at the end of year 1. Taking out $500 would leave $1,180.80, which would grow into $1,476 at the end of year 2. Taking out $500 would leave $976, which would grow into $1,220 at the end of year 3. Taking out $500 would leave $720, which would grow into $900 at the end of year 4. Taking out $500 would leave $400, which would grow into the $500 final benefit of the project by the end of year 5.

c. At 0 percent and 5 percent interest, this household would be wise to carry out all three projects, because their net present values would be positive. Thus $8,000 would be demanded (points *A* and *B*). At 15 percent and 25 percent interest, none of the projects would be worthwhile. Thus the demand for loanable funds would be zero (points *C* and *D*). The above calculations do not tell us what the quantity demanded will be at interest rates *between* 5 percent and 15 percent per year, but, clearly, one project after another will become unprofitable as the rate rises above 5 percent per year.

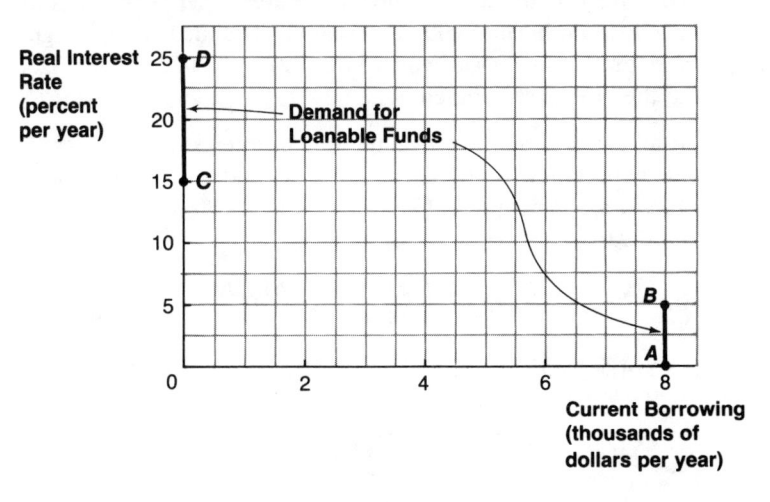

d. *For 2 years:* No project would be acceptable, because payback occurs after 2.57 years for project *A*, after 2.5 years for project *B*, after 4 years for project *C*. *For 3 years:* Projects *A* and *B* would be judged acceptable.

e. There are more years of benefit for a 20-year-old, thus the likelihood of a positive net present value is greater.

4. Imagine two plants producing the same electric-power benefits but having costs as shown in the table. The Soviets would compare the sum of costs: $100 + 40(10) = 500$ vs. $30 + 40(20) = 830$, and they would choose the former. Yet the pure interest rate in the Soviet Union was in fact not zero but high (reflecting great impatience by consumers to consume now as well as a great time productivity of postponing present consumption to produce capital and future consumption). At a high interest rate (which would heavily discount future costs), the fossil-fuel alternative was in fact preferable. (You may wish to calculate the present values for the above two streams of cost at, say, 20 percent interest per year.)

	Year 0	Year 1–40
hydroelectric plant:	100	10 each
fossil-fuel plant:	30	20 each

5. A person taking up the offer would be exposed to the cash flow shown in Table A. With the help of text Table 12.2, "Compound Interest and Discount Factors," we can calculate the present value of $4,000 payable in 4 years as shown in Table B. The offer was clearly a good deal if the saver's alternative interest-earning opportunities were 5 percent per year or less. (At 5 percent, for example, the deal was equivalent to spending $3,023.19 now and getting $3,290.83 back instantly.) The offer was a bad deal if the saver could earn elsewhere 10 percent per year or more. (At 10 percent, for example, the deal was equivalent to spending $3,023.19 now and getting $2,732.05 back instantly.) The precise dividing line between good and bad deals lay between 7 percent and 8 percent.

TABLE A

At the End of Year	Dollars
0	−3,023.19
1	0
2	0
3	0
4	+4,000

TABLE B

At Interest Rate of	Present Value
0%	$4,000
5%	$\dfrac{4,000}{1.2155} = \$3,290.83$
10%	$\dfrac{4,000}{1.4641} = \$2,732.05$
15%	$\dfrac{4,000}{1.749} = \$2,287.02$
20%	$\dfrac{4,000}{2.0736} = \$1,929.01$
25%	$\dfrac{4,000}{2.4414} = \$1,638.40$

6. The net present value of the stream of $150 annual receipts equals

$$\$150 + \frac{150}{1.08} + \frac{150}{1.08^2} + \ldots \frac{150}{1.08^{19}} = \$1,605.21.$$

The net present value of the $5,000 payment in 20 years equals

$$\frac{-5,000}{1.08^{20}} = \frac{-5,000}{4.661} = -\$1,072.73.$$

(By investing this amount now at 8 percent per year, the company would have $5,000 in 20 years.) Thus the student's signature on the policy was worth $1,605.21 − $1,072.73 = $532.48 to the insurance company.

7. It all depends on the prevailing rate of interest and the person's life expectancy. At 10 percent per year and a life expectancy of 10 years, the net present values work out as follows:
   a. $79,000
   b. $\dfrac{\$16,000}{1.1} + \dfrac{\$16,000}{1.21} + \dfrac{\$16,000}{1.331} + \dfrac{\$16,000}{1.4641} + \dfrac{\$16,000}{1.6105} =$

   $14,545.45 + 13,223.14 + 12,021.04 + 10,928.22 + 9,934.80 = \$60,652.65$

   c. $\dfrac{\$5,000}{1.1} + \dfrac{\$5,000}{1.1^2} + \ldots \dfrac{\$5,000}{1.1^{10}} = \$30,722.87$

8. It depends on the prevailing rate of interest. At 10 percent per year, the net present value of the 5-year stream of $16,000 equals $60,652.65—see 7(b) above. This would be the maximum worth paying. If you paid it (and could trust the promised outcome), you could be indifferent about expropriation now and in 5 years. Note: An initial $60,652.65, invested at 10 percent per year, would enable you to take out precisely $16,000 per year for 5 years.

9. Cutting down the tree now might yield $1,000 in net receipts—equivalent to $1,100 in a year, if an interest rate of 10 percent per year was available. Letting the tree stand might result in 15 percent natural growth (and net receipts of $1,150, if it was felled in a year).

10. No. Spending $1,260 under the circumstances would mean forgoing $189 in annual interest income forever. The annual savings would only come to 70 percent of $120, or $84.

BIOGRAPHY 12.1

Eugen von Böhm-Bawerk

Eugen von Böhm-Bawerk (1851–1914) was born in Brünn, Moravia. His father was vice-governor of Moravia—the descendant of a long line of civil servants. Although Eugen was more interested in physical science, family tradition won out. He studied law at the University of Vienna, then political science at Heidelberg, Leipzig, Jena.

At the age of 30, he was made a professor at the University of Innsbruck. Along with Carl Menger and Friedrich von Wieser, he became the founder of a special brand of economic theory, called the

Austrian School. On three occasions (in 1895, 1897, and from 1900–1904), Böhm-Bawerk served the Austro-Hungarian Empire as Minister of Finance. During the last decade of his life, while a professor of political economy at the University of Vienna, he gave a series of famous seminars and became the inspiring teacher of many, including Joseph A. Schumpeter (see Biography 10.1). In those days, Böhm-Bawerk displayed a formidable talent as a debater; he was always quick to concede his opponents' good points but ever-ready to destroy their errors with irrefutable logic.

Böhm-Bawerk, the theorist, focused on the role of *time* in economic life and, thus, on the role of interest and capital. Indeed, his major work is *Capital and Interest,* which appeared in two volumes: The *History and Critique of Interest Theories* (1884) provided a painstaking review of explanations for interest that had been proffered since ancient times. The *Positive Theory of Capital* (1889) presented Böhm-Bawerk's own exposition. (*Further Essays on Capital and Interest* appeared as an appendix to this volume in 1909.)

Schumpeter called his teacher the "bourgeois Marx," for like Marx (see Biography 14.3) who wrote *Capital,* Böhm-Bawerk held a grand vision of the economic process in which interest and capital played crucial roles. Yet Böhm-Bawerk thoroughly disagreed with Marx and even wrote *Karl Marx and the Close of His System* (1896). Marx, for example, viewed interest as a form of exploitation that arose in the *process of production.* According to Marx, exploitation resulted when a small number of capitalists had a monopoly in the ownership of physical capital goods badly needed by workers. Hence capitalists could force workers to hand over part of the "workers" output. Böhm-Bawerk instead viewed interest as arising in the *process of exchange,* as a phenomenon linked with barter across time of present for future goods. Karl Marx, the revolutionary, saw capital as a historical concept, as physical goods about which class conflict arose. Böhm-Bawerk, the scientist, viewed capital as a theoretical concept, as the present value of future income streams. Unlike Marx, Böhm-Bawerk viewed interest and capital as concepts that would manifest themselves in any economic system, regardless of time and place—even in a socialist system.

BIOGRAPHY 12.2

Irving Fisher

Irving Fisher (1867–1947) was born at Saugerties, New York, the son of a Congregational minister. As did his father, Fisher studied at Yale. Mathematics was his favorite subject. He won first prize in a math contest even as a freshman; his doctoral dissertation, *Mathematical Investigations in the Theory of Value and Prices* (1892) was a landmark in the development of mathematical economics. It won immediate praise from no lesser figures than Francis Y. Edgeworth (see Biography 2.3) and Vilfredo Pareto (see Biography 14.1). Some 55 years later, Ragnar Frisch (eventual winner of the 1969 Nobel Prize in Economic Science) would say about Fisher: "He has been anywhere from a decade to two generations ahead of his time . . . it will be hard to find any single work that has been more influential than Fisher's dissertation."[1] No wonder that Fisher was a full professor of political economy at Yale within seven years of graduation. He stayed there during his entire career.

Fisher's main contributions lay in the theory of utility and consumer choice, the theory of interest and capital, and the theory of statistics (index numbers, distributed lags). These contributions are reflected in such works as *The Nature of Capital and Income* (1906) and *The Theory of Interest* (1930), a revision of a 1907 book. "Dedicated to the memory of John Rae and of Eugen von Böhm-Bawerk who laid the foundations," *The Theory of Interest* carries the subtitle *As Determined by Impatience to Spend Income and Opportunity to Invest It*. Other major works include *The Purchasing Power of Money* (1911), a great pioneering venture in econometrics, and *The Making of Index Numbers* (1922). These works established Fisher's reputation as the country's greatest scientific economist. As such, he served, in 1918, as president of the American Economic Association and was a founder and the first president of the Econometric Society. He played a major role in the establishemnt of the Cowles Foundation (now at Yale) as a means to nurture mathematical and quantitative research in economics.

But there was also another side to Fisher. By becoming a passionate crusader in many causes that he believed essential to human welfare, he managed to dim his reputation as an important contributor to the scientific foundation of economics. Fisher's father had died of tuberculosis; he himself had a three-year bout with the disease. Subsequently, he became compulsive about fresh air and promoting the health of his family, his country, and then the world. Fisher campaigned for "biologic living," diet and exercise programs, eugenics research, the conservation of racial vigor, and abstention from alcohol and tobacco.

Fisher was also an inventor, taking out patents on the internal mechanisms of the piano, a tent for tuberculosis patients, and a visible-card index system. The index system was very successful and eventually gave rise to Remington Rand, bringing him a fortune. Fisher then campaigned for "the compensated dollar" (the gold content of which would be changed in accordance with the price index) and "100 percent money" (which would be based on a 100 percent reserve requirement). He pictured depressions as merely "dances of the dollar," easily avoidable by sound money. He spent more than $100,000 of his own funds seeking to develop support for his monetary proposals; his ill-timed belief in the 1920s that prosperity would stay forever cost him between $8 million and $10 million during the Great Depression. Before long, he was ridiculed as a health faddist and monetary crank. Yet Fisher's scientific work has stood the test of time. The theory of interest and capital discussed in this chapter is a major part of his work.

---

[1] Ragnar Frisch, "Irving Fisher at Eighty," *Econometrica,* April 1947, pp. 71–72.

BIOGRAPHY 12.3

Theodore W. Schultz

Theodore William Schultz (1902–   ) was born among German settlers on a farm near Arlington, South Dakota. He studied agricultural economics first at South Dakota State College and then at the University of Wisconsin. He taught the same subject at Iowa State, then moved on to a distinguished career at the University of Chicago, where he has remained. In 1960, he served as president of the American Economic Association; in 1972, he received its prestigious Walker Medal.

Throughout his career, Schultz has focused on the role of the agricultural sector in the economy. He has written a number of books on U.S. agricultural policy and others on the process of economic development in poor countries, including *The Economic Organization of Agriculture* (1953), *Economic Growth and Agriculture* (1968), and *Distortions of Agricultural Incentives* (1978).

His interest in economic growth led Schultz to discover the crucial role of human capital, not only in the poor countries he studied, but also in rich ones. Having served in post-World War II Germany, with its physical capital in ruins, Schultz was struck by the extraordinary speed of recovery. He attributed this rate of recovery to human capital, which, like a ghost, was still there. Because people had the skills to rebuild their physical capital stock and had the knowledge to operate it, they quickly regained their former levels of output. Indeed, their invisible human capital had gone on growing even while bridges, factories, and houses were being destroyed. Prewar output levels were soon surpassed with the help of a larger stock of human capital. In his *Transforming Traditional Agriculture* (1976), Schultz drew parallels between the experience of war-torn Europe and that of poor countries, such as India, where investments in health care and education produced extraordinary increases in productivity. But he also pointed out how misguided governmental policies that fix prices below equilibrium levels can easily counteract the potential benefits derivable from human capital accumulation. (In the mid-1960s, when the Indian government held down the prices of wheat and rice to aid industrialization, the government destroyed farmers' incentives and produced stagnating output levels, just when new knowledge might have achieved the opposite.)

Schultz's work on human capital, including *The Economic Value of Education* (1963) and *Investment in Education: The Equity-Efficiency Quandary* (1972), has, however, inspired applications far beyond the ones cited here. These applications have already been noted in Chapter 10 and are explored in a selection of articles edited by Schultz on the *Economics of the Family: Marriage, Children, and Human Capital* (1974). In 1979, Schultz was awarded the Nobel Memorial Prize in Economic Science (jointly with Sir Arthur Lewis) for his work on human capital.

Additional Readings

Fellner, William, et al. *Ten Economic Studies in the Tradition of Irving Fisher.* New York: Wiley, 1967.

>   Essays honoring the memory of Fisher, written at the centennial of his birth.

Fisher, Irving Norton. *My Father Irving Fisher.* New York: Comet Press, 1956.

> The story of Fisher's life and work.

Kuenne, Robert E. *Eugen von Böhm-Bawerk.* New York: Columbia University Press, 1971.

> The story of Böhm-Bawerk's life and work.

Schumpeter, Joseph A. *Ten Great Economists: From Marx to Keynes.* New York: Oxford University Press, 1951.

> Chapters 6 and 8 discuss the lives and works of Böhm-Bawerk and Fisher.

Seligman, Ben B. *Main Currents in Modern Economics.* Chicago: Quadrangle, 1962, vol. 2, pp. 294–310, and vol. 3, pp. 637–46.

> A critical review of the work of Böhm-Bawerk and Fisher.

APPENDIX 12A MARKETS FOR BONDS AND STOCKS

In order to focus on the essential nature of interest (as a phenomenon arising from time preference and time productivity), the discussion in text Chapter 12 was based on a simplified view of reality. Households were treated only as savers and suppliers of funds. Firms, in turn, were looked upon only as investors and demanders of funds. The two groups were viewed as trading directly with each other and as trading household funds for IOU's issued by firms under conditions of certainty.

Now consider this: Households can also borrow funds, and they frequently do, notably for the purchase of houses, durable consumer goods, education, and health care. Firms can also be lenders of funds; a multitude of financial intermediaries (ranging from commercial, investment, and savings banks to life insurance and finance companies, and to mutual funds and pension funds) can and often do mediate these exchanges. In addition, certainty is often absent. Funds are often exchanged not for risk-free IOU's, but for risky ones (and for risky ownership claims). We cannot and need not discuss all of these complications, but we can lend a greater degree of realism to our previous analysis by considering some of them, in particular the role of different types of securities and of the organized securities exchanges.

Major Types of Securities

The supply of funds in capital markets can also be viewed as a demand for securities; the demand for funds can similarly be viewed as a supply of securities.

A **security** is simply a written legal instrument (a certificate of indebtedness or an ownership claim) that bestows upon its holder certain rights to future funds. Bonds, stocks, and various hybrids are the most important types of securities.

A **bond** is a certificate of indebtedness, a promissory note, an IOU. Bonds are issued by governments and corporations. Federal government bonds carry no risk of default (because the federal government can create money); the interest rate on long-term federal bonds, after adjustment for inflation, is usually regarded as the best indicator of the pure rate of interest. The yields on other types of securities are higher because they reflect, among other things, different degrees of risk. Consider corporate bonds. They are usually issued in denominations of $1,000 or more. They promise to pay a fixed and stated dollar amount of **coupon interest** at regular intervals and the bond's face value at a specified maturity date. The bonds can be resold by their original purchasers to other people prior to the maturity date. Legally, the issuing firm must pay the interest and principal as promised, regardless of its earnings. But unlike the federal government, a firm can go bankrupt. Thus corporate bonds are more risky than those of the federal government (and bonds of Fly-by-Night Airlines are likely to be more risky than those of TWA). In the case of bankruptcy, bondholders (being creditors) have first claim on a firm's assets.

A share of **common stock** is an ownership claim on a corporation. It entitles the holder to a voice in the firm's management and to a share of profit or loss. Positive profit may be divided among stockholders and paid out in the form of a **dividend,** if the board of directors decides to declare one, or it may be retained and received by the individual stockholder as a **capital gain;** that is, an increase in the market price of stock. Losses, similarly, will be reflected in a decrease in the stock's price or the occurrence of a **capital loss.** (Note: Dividends are usually paid in cash. On occasion, however, stockholders receive a **stock dividend,** a dividend that is in the form of extra shares of stock. A 3 percent stock dividend, for example, gives 3 new shares to anyone already holding 100 shares. This is not to be confused with a **stock split,** the substitution of a larger number of new shares for old ones. A 2 for 1 split would replace 100 old shares with 200 new ones, resulting in a halving of the price per share. This lower price might attract more investors, which is the purpose of the split.) A share of stock has no maturity date but can be freely traded to other people. In the case of bankruptcy, stockholders (being owners) have last claim on the firm's assets.

A large number of hybrids exist between the two basic types of securities just discussed. **Preferred stock** pays dividends before any are paid to holders of common stock but never pays more than a stated percentage of the face value of the stock. Holders of preferred stock also lose the right to run company affairs but have a claim on assets prior to holders of common stock. **Cumulative preferred stock** is promised dividends at regular intervals or the payment of accumulated missed dividends before any dividends are paid to common stockholders. **Callable preferred stock** can be bought back by the firm at a previously stated price per share. **Convertible preferred stock** as well as a **convertible bond,** can be exchanged, at the holder's will, for common stock at a stipulated ratio. An **income bond** pays interest only if the firm's earnings are large enough.

A **warrant** is a pledge by a firm to sell to its present stockholders a share of new common stock at a stated price until a stated date. The holder can let the warrant expire, use it to buy the new stock, or sell it to someone else. (If the warrant specified a price of $75 per share, but the going market price of stock was $100, the warrant could be sold for as much as $25.)

A **put option** is a contract giving the holder the right to sell a fixed number of shares at a stated price until a stated date. A **call option** is a contract giving the holder the right to buy a fixed number of shares at a stated price until a stated date. (Such "puts" and "calls" are expensive, costing as much as 10 percent of the applicable stock's value. Consider how a call option might be used: The stock of Fly-by-Night Airlines might sell at $10 per share. Someone could buy 500 shares for $5,000, or instead, a call option for 500 shares at $10 each, paying, say, 10 percent of their value, or $500. If the price falls to $5 per share, the buyer can let the option lapse and has lost $500—but would have lost $2,500 had the shares been bought outright. If the price rises to $15 per share, the buyer can exercise the option and buy 500 shares for $5,000, then sell them for $7,500, and make a gain of $2,500 − $500 = $2,000. The buyer would have gained $2,500 had the shares been bought outright. Note: Outright buying of shares involves absolutely larger losses and gains than does the option route, but smaller losses and gains per dollar invested.)

Major Securities Exchanges

Many millions of securities, such as those discussed in the previous section, are traded on every business day, and trading goes on in many places. When a corporation first issues securities, an investment bank usually *underwrites* the issue: that is, guarantees its full sale. The issuing firm must also prepare a comprehensive document, or *prospectus,* that provides a variety of information about the firm—information that must be cleared by the Securities and Exchange Commission for accuracy and completeness. If the corporation is a new firm, its stocks and bonds subsequently are likely to be traded in the OTC or **over-the-counter market,** a vast informal network for the exchange of securities, held together by stockbrokers interconnected by telephone and teletype. Eventually, a new firm's securities may also be traded on one of 14 regional exchanges, such as the Boston, Cincinnati, Detroit, Midwest, Pittsburgh, Philadelphia, or Pacific Stock Exchanges. Yet the above markets account for only about 15 percent of the daily trading volume. The remaining 85 percent of trades are made at New York City's New York and American Stock Exchanges. These trading places deal only with stocks and bonds issued by firms that have grown beyond local importance and that meet certain minimum requirements—concerning, for instance, the number of stockholders.

The *New York Stock Exchange,* also known as the "Big Board," was founded in 1792 and then operated under a tree on Wall Street. Nowadays, it is the world's largest central auction market for securities. In this market, millions of buyers and sellers have their orders executed within minutes and at a common price. The trading is facilitated by 1,375 member brokers, each of whom must buy a "seat" on the exchange (so called because traders used to be seated in the auction room).

The *American Stock Exchange,* also known as the "Curb," was founded in 1849 and also began in the street. Brokers stood on the curb, trading securities and giving hand signals to clerks who hung out of windows and recorded the transactions. The trading moved indoors only in 1921. There are 650 members today.

In 1980, a new exchange was being born in New York Citys Wall Street area. Like the old Curb Exchange (now the American Stock Exchange), it operated out on the street, but its favored object of trade was not stocks and bonds, but marijuana. On a paved-over vacant lot, Harlem and Brooklyn youngsters posted themselves at regular locations around the edge of the lot. They dealt in small lots because small-quantity sales carried small penalties (a $10 fine), but "we stash the rest all over the place and go get more when we sell out," one dealer said. He was making several hundred dollars on a good day, but only $15 if he was busted during lunch when most customers were around. (Dealers bought a pound of grass for $500, rolled 150 joints from an ounce, and sold a joint for $1. There was almost $2,000 of profit in a pound.) As a New York police officer put it: "Where else can you get such good hours and high pay—and with no taxes to pay on what you make? It's the best business you could be in." (See Dave Lindorff, "Fair Weather Brings Market Surge," *Forbes,* March 17, 1980, p. 166.)

Note: Most of the securities traded daily on the New York Stock Exchange and the American Stock Exchange are not new issues providing new funds to firms but old issues that are being exchanged among seucrity holders. Nevertheless, this secondary market is crucial for firms because they are enabled to sell stock without a maturity date and bonds with far-off maturity dates, precisely because every buyer knows that there is a vast market in which shares can be traded to other people, and their initial financial investment can be liquidated, at a moment's notice, though with capital gain or loss.

The Random Walk

Stock prices are formed as a result of these trading activities. They reflect the composite (and often rapidly changing) judgment of outsiders of the past performance and future prospects of each firm. Every weekday, during trading hours, these prices are flashed to hundreds of U.S. cities second by second via a communication network tens of thousands of miles long. And with but little delay, through the facilities of Western Union, Radio Suisse, and Reuters Economic Service, the same news is reported in major cities throughout the world, from Montreal to Paris and London, from Zürich to Frankfurt and Tokyo.

Eventually, a variety of market-price indicators are published, including the Dow-Jones Averages (of 65 stocks), Standard and Poor's Average (of 500 stocks), and the New York Stock Exchange Comprehensive Average. Experts, of course, always try to outguess the market. Despite the occasional claims of self-styled experts, however, individual stock prices change in a rather unpredictable fashion. Their movement has been likened to a **random walk,** akin to the *Brownian motion* one sees in a microscope due to the chance impacts of invisible and unpredictable atoms on just-visible huge molecules or colloidal particles. Table 15A.1 illustrates how the typical newspaper reports securities trading.

Section (A) of Table 12A.1 refers to the stock market page. All stock prices are stated in dollar amounts per share; thus $67\frac{3}{4}$ means $67.75 per share, and $49\frac{1}{8}$ means $49.125 per share. These figures refer, respectively, to the highest and lowest price a particular stock reached in the past 52 weeks. (The quoting of prices by eighths is a throwback to the old pirate days and the Spanish gold "pieces of eight.") The abbreviation of the firm's name (Exxon) is followed by the current annual dividend per share ($4.80), then the **yield,** or the percentage return on the current stock price represented by the latest annual dividend (here 8.4 percent per year, which is $4.80 divided by $57).

Then comes the **price-earnings ratio** (equal to 6 here), which is the number of times by which the firm's latest 12-month earnings per share must be multiplied to obtain the current price per share. The relevant earnings per share equal the after-tax profit available to common shareholders divided by outstanding shares of common stock.

TABLE 12A.1

The Stock- and Bond-Market Page

(A) Stocks

52-week High	52-week Low	Stock	Div.	Yld. %	PE Ratio	Sales 100s	High	Low	Last	Chg.
67¾	49⅛	Exxon	4.80	8.4	6	2930	57¼	55⅜	57	+1⅝

(B) Bonds

Bond		Yld.	Sales $1,000	High	Low	Last	Chg.
Exxon	6½ 98	11.1	40	59	58⅛	58½	+⅜

(C) Stock Options

Option	Price	April Vol.	April Last	July Vol.	July Last	October Vol.	October Last	N.Y. Close
Exxon	50	127	7⅜	10	9½	19	10⅝	57

This table shows excerpts from newspaper reports on the March 28, 1980, trading of stocks, bonds, and stock options. Sections (A) and (B) refer to the combined trading on major and regional exchanges and the O.T.C. market; section (C) refers to Chicago trading only.

Sales refers to the volume of stock traded on this day (293,000 shares). The *High* and *Low* in the next columns refer to the particular day's trading prices ($57.25 and $55.375 per share). Then follows the day's last reported price ($57 per share) and the change between the day's last reported price and the previous closing price (+$1.625 per share).

Note: The prices shown here always refer to the **market value** of stock; that is, the price per share established by supply and demand. This should not be confused with the stock's **book value,** or the ratio of the firm's net worth to the number of shares outstanding. Nor should it be confused with the **par value,** which is the minimum price per share considered acceptable on the day of issue that is printed on the certificate and becomes meaningless thereafter.

Section (B) of Table 12A.1 refers to the bond market page. The name of the firm (Exxon) is followed by the coupon interest rate (6½ percent per year) and the last two digits of the year of maturity (here 1998). Yield now represents the percentage return on the current bond price of the coupon interest (here 11.1 percent, calculated from the ratio of $6.50 to $58.50). Sales volume was $40,000 for the day.

As in the case of stocks, the data in the columns labeled *High, Low,* and *Last* refer to the bond's price during the day ($59, $58.125, and $58.50, respectively). *Change* represents the difference between the day's last reported price and the previous closing price (here +$.375).

Section (C) refers to the stock option page. The firm's name (Exxon) is followed by the price per share specified in the option ($50). Then come various dates for the exercise of the option and the New York closing price per share ($57), already noted in section (A). For April options, for example, trading volume (in 100s) was 12,700 options, and the last quoted price of an option was $7.375.

Note: Over-the-counter quotations (not shown separately in Table 12A.1) are similarly reported in newspapers. They list **bid prices** (which potential buyers are ready to pay) and **ask prices** (which potential sellers are ready to accept). Actual prices can be assumed to form within the range indicated by the bid and ask prices.

Key Terms

ask prices	**income bond**
bid prices	**market value of stock**
bond	**over-the-counter market *(OTC)***
book value of stock	**par value of stock**
callable preferred stock	**preferred stock**
call option	**price-earnings ratio**
capital gain	**put option**
capital loss	**random walk**
common stock	**security**
convertible bond	**stock dividend**
convertible preferred stock	**stock split**
coupon interest	**warrant**
cumulative preferred stock	**yield**
dividend	

Additional Readings

Friend, Irwin. "The Economic Consequences of the Stock Market." *The American Economic Review,* May 1972, p. 212–19.

> A discussion of the efficiency of the stock market and related institutions in allocating investment funds.

Malkiel, Burton G. *A Random Walk Down Wall Street.* New York: Norton, 1973.

New York Stock Exchange, *Annual Report.*

"Smith, Adam," *The Money Game.* (New York: Random House, 1967).

> An amusing, but well-informed account of the stock market.

CHAPTER 13

General Equilibrium

MULTIPLE-CHOICE QUESTIONS

Circle the letter of the *one* answer that you think is correct or closest to correct.

1. The partial-equilibrium approach
   a. rejects the *ceteris paribus* clause.
   b. typically yields inaccurate predictions because it ignores interrelationships.
   c. is correctly described by both (a) and (b).
   d. is correctly described by neither (a) nor (b).

2. A general equilibrium is
   a. the end result of a mathematical process that solves partial equilibria in sequence.
   b. a state of the economy in which the optimizing decisions by multitudes of decision makers are compatible with each other because all input and output markets are in equilibrium at the same time.
   c. correctly described by both (a) and (b).
   d. correctly described by neither (a) nor (b).

3. When a change in the demand for pork leads to a change in the demand for beef, this changed demand for beef is called
   a. a spillout effect.
   b. an impact effect.
   c. a feedback effect.
   d. a substitution effect.

4. If a given consumer in a perfectly competitive market for pork were to cut purchases at any given price,
   a. this cut would be referred to as a fall in the consumer's quantity demanded.
   b. this cut would be referred to as a fall in the consumer's demand.
   c. a surplus in the market would develop, and the price of pork would fall.
   d. both (b) and (c) are true.

5. If all the buyers of oranges reduced their demand in a perfectly competitive market that was in long-run equilibrium, it would not be surprising if there soon occurred
   a. an increase in demand (due to the substitution effect).
   b. an increase in quantity supplied.
   c. an increase in quantity demanded.
   d. all of the above.

6. If all the buyers of oranges reduced their demand in a perfectly competitive market that was in long-run equilibrium, one expects
   a. firms first to make temporary losses.
   b. firms to make zero economic profits eventually.
   c. a fall in quantity supplied at first, a fall in supply eventually.
   d. all of the above.

7. It would not be surprising if a rise in the demand for cheese in a competitive market economy produced such spillout effects as
   a. an increase in the demand for milk.
   b. an increase in the quantity of cheese supplied in the short run.
   c. an increase in the supply of cheese in the long run.
   d. all of the above.

8. It would not be surprising if a rise in the demand for meat in a competitive market economy produced such feedback effects as
   a. a rise in the price of meat.
   b. a further rise in the demand for meat.
   c. a fall in the quantity of meat demanded.
   d. any of the above.

9. If the demand for shoes fell in a constant-cost, perfectly competitive industry, one could predict
   a. that the price of shoes would fall at first, then rise again.
   b. that the industry's output would fall at first, then rise again.
   c. that firms would make zero economic profits throughout the adjustment process.
   d. all of the above.

10. Long-run general equilibrium exists when the Walrasian system of equations yields
    a. a simultaneous equilibrium in all markets.
    b. zero economic profits for all firms.
    c. zero quasi rent for reproducible resources.
    d. all of the above.

11. The Walrasian theory of *tâtonnement* is best viewed as
    a. a short-hand description of how auctioneers find equilibrium prices in real-world markets.
    b. a short-hand description of how equilibrium quantities are established in real-world markets.
    c. a poetic vision of how a market economy could solve the Walrasian equations.
    d. all of the above.

12. Critics of Walras have argued
    a. that an equality of the numbers of independent equations and unknowns in the Walrasian system was neither a sufficient nor a necessary condition for the existence of a general equilibrium.
    b. that the Walrasian equations cannot have a solution in the realm of real numbers (the only realm that has any economic meaning) but that the solution is bound to involve imaginary and negative numbers (that grossly violate economic reality).
    c. both (a) and (b).
    d. neither (a) nor (b).

13. Intermediate goods are goods produced by domestic producers during a period and then
    a. used up by the same or other domestic producers during the same period in the making of other goods.
    b. used up by the same domestic producers during the same period in the making of other goods.
    c. used up by other domestic producers during the same period in the making of other goods.
    d. used up by (domestic or foreign) producers or consumers during the next period.

14. Which of the following would be part of the 1985 set of final goods?
    a. Milk produced in January and instantly fed to sheep by domestic producers of wool.
    b. Grapes produced in September and turned into wine in October by domestic producers.
    c. Milk produced in January and instantly fed to human babies.
    d. None of the above.

15. A country's set of final goods could include
    a. electric power sold to domestic firms.
    b. electric power sold to foreign firms.
    c. both (a) and (b).
    d. neither (a) nor (b).

When answering questions 16–20, refer to the following input-output table. Assume that it lists annual flows of milk in gallons, annual flows of cheese and bread in pounds, and annual flows of labor in hours. Assume that there are no other activities in this economy.

Suppliers \ Recipients	of Intermediate Goods — Milk Producers	Cheese Producers	Bread Producers	of Final Goods — Households
Milk producers	100	500	50	1,000
Cheese producers	0	10	10	2,000
Bread producers	0	0	30	3,000
Households (supplying labor)	500	200	900	300

16. The first column of a technical-coefficients table derived from this input-output table would show
    a. milk, cheese, bread, and labor inputs required per unit of milk produced.
    b. milk, cheese, bread, and labor inputs required per unit of milk delivered to households.
    c. milk, cheese, bread, and labor inputs required per unit of milk used by milk producers.
    d. none of the above.

17. The second column of a technical-coefficients table derived from this input-output table would be stated in terms of
    a. pounds of cheese.
    b. pounds of cheese per different units of various inputs.
    c. different units of various inputs per pound of cheese.
    d. none of the above.

18. The third column of a technical-coefficients table derived from this input-output table would read (in respective units)
    a. .05, .005, .01, 3.
    b. 2/33, 1/202, 1/101, 9/19.
    c. 0, 0, .01, .1.
    d. like none of the above.

19. The fourth column of a technical-coefficients table derived from this input-output table would
    a. read 1, 2, 3, .3.
    b. read .5, .1, .3, 1.
    c. equal the first column.
    d. not exist.

20. The first row of a technical-coefficients table derived from this input-output table would read
    a. 1/6, 50/71, 5/99 (all stated in gallons of milk per unit of respective outputs).
    b. 1/6, 50/71, 5/99 (all stated in different units).
    c. 2/33, 25/101, 5/303.
    d. like none of the above.

21. The Leontief inverse matrix is simply another name for
    a. the reproduction schema of Karl Marx.
    b. the equations of Walras.
    c. the tableau of Quesnay.
    d. none of the above.

When answering questions 22–23, use the Leontief inverse below:

	1 ton of wheat	1 barrel of milk	1 ton of bread
Wheat (tons)	1.01	0.05	0.28
Milk (barrels)	0.12	1.36	0.41
Bread (tons)	0.05	0.05	1.56

22. According to this Leontief inverse, the provision of one extra ton of bread to starving people abroad would require the extra production of (among other things)
    a. 2.25 tons of bread.
    b. 1.66 tons of bread.
    c. 1.56 tons of bread.
    d. 1 ton of bread.

23. According to this Leontief inverse, the provision of one extra ton of wheat plus one extra barrel of milk plus one extra ton of bread to starving people at home would require the extra production of
    a. 1.01 tons of wheat, 1.36 barrels of milk, and 1.56 tons of bread.
    b. the amounts shown in (a) plus additional amounts that cannot be determined from the given information.
    c. 1.18 tons of wheat, 1.46 barrels of milk, and 2.25 tons of bread.
    d. 1.34 tons of wheat, 1.89 barrels of milk, and 1.66 tons of bread.

24. When every cell in an economy's input-output table contains a nonzero entry, the economy is said to
    a. show a random pattern of interindustry transactions.
    b. show a hierarchical pattern of interindustry transactions.
    c. be subject to triangulation.
    d. be completely interdependent.

25. When some industries deliver their entire output to themselves or to final demanders but absorb inputs from all sectors, while other industries deliver output to all sectors but use as inputs only their own output and primary resource services, the economy is called
    a. completely interdependent.
    b. hierarchical.
    c. diagonal.
    d. none of the above.

26. Goods that are produced by domestic producers during a period, and then used up by the same or other domestic producers during the same period in the making of other goods, are called
    a. final goods.
    b. intermediate goods.
    c. producer goods.
    d. consumption goods.

27. Electric power is
    a. an intermediate good.
    b. a consumption good.
    c. a final good.
    d. possibly any one of the above.

28. *Capitalism* is an economic system wherein
    a. most resources are privately owned.
    b. capital resources are more important than human or natural ones.
    c. most capital is human capital.
    d. most goods are produced with financial capital.

29. In the circular-flow diagram of the capitalist market economy, the monetary counterflow to the flow of resource services may well include
    a. wages.
    b. rentals.
    c. profits.
    d. all of the above.

30. Modern economic systems are extremely complex because
    a. everywhere and everyday people have to wrestle with scarcity.
    b. people are participants in a vast scheme of specialization.
    c. the people of different societies have chosen vastly different arrangements for making the painful choices about allocating resources and apportioning goods that scarcity forces upon them.
    d. of all of the above.

31. In the United States of the mid-1980s, economic choices are being made by
    a. about 88 million households.
    b. about 20 million firms.
    c. about 80,000 governments (local, state, and federal).
    d. all of the above.

32. In a society organized on the basis of specialization and exchange, everything that one person does
    a. comes to hang together with the actions of all others in an endless web.
    b. has to be approved of by governmental planners.
    c. requires, directly and indirectly, appropriate complementary actions by thousands of other people who must be guided by government accordingly.
    d. is correctly described by all of the above.

33. The separate economic activities of people engaged in a division of labor can be coordinated
    a. spontaneously by a human manager.
    b. deliberately by a human manager.
    c. deliberately by the market.
    d. in any one of the above ways.

34. Which of the following is *not* equivalent to the Visible Hand?
    a. Market coordination.
    b. Managerial coordination.
    c. Deliberate coordination.
    d. Central planning.

TRUE-FALSE QUESTIONS

In each space below, write a *T* if the statement is true and an *F* if the statement is false.

_____ 1. A situation in one part of the economy that contains no innate tendency to change is referred to as *partial equilibrium*.

_____ 2. A general equilibrium can take the place of either a managerial or a market coordination of individual actions.

_____ 3. *Partial equilibrium* is another term for *short-run equilibrium*, just as *general equilibrium* is another term for *long-run equilibrium*.

_____ 4. In a competitive market economy, an increase in the demand for wine is likely to produce spillout effects, such as an increased demand for grapes.

_____ 5. It is quite possible for one equation with two unknowns to have a solution in the realm of real numbers.

_____ 6. Grapes produced in September, exported in October, and turned into wine by November, would be part of the exporting country's set of intermediate goods for that year.

_____ 7. An input-output table shows the web of interrelationships in an economy by listing the stocks of various commodities and services in the hands of their suppliers.

_____ 8. Leontief's input-output analysis assumes the existence of constant returns to scale.

_____ 9. All entries in a given row of a technical-coefficients table are stated in units of an identical input (per unit of different outputs).

_____ 10. Each entry in a given column of a Leontief inverse matrix is stated in identical units.

_____ 11. When production is organized on the basis of specialization and exchange, people tend to be richer than they would be under self-sufficiency.

_____ 12. The division of labor provides great benefits but it also burdens people by necessitating central economic planning to assure that the specialized activities of people are well coordinated.

_____ 13. Economic order is a state of affairs in which the specialized activities of all the people engaged in the division of labor are well coordinated by a human manager.

_____ 14. Managerial coordination, deliberate coordination, and Visible Hand are synonymous.

_____ 15. Spontaneous coordination, in contrast to deliberate coordination, is costless.

PROBLEMS

1. Calculate a table of technical coefficients from the following input-output table:

Suppliers \ Recipients	Of Intermediate Goods and Primary Resources			Of Final Goods and Primary Resources				Total
	Truck producers (1)	Fuel-oil producers (2)	Corn producers (3)	Domestic households (4)	Domestic producers (5)	Domestic government (6)	Foreigners (7)	(8)
(A) Truck producers (millions of trucks)	1	2	3	0	7	2	5	20
(B) Fuel-oil producers (millions of barrels of oil)	2	5	6	400	200	100	−663	50
(C) Corn producers (millions of tons of corn)	1	0	30	20	−261	10	500	300
(D) Labor owners (millions of labor hours)	200	1	30	100	✗	300	69	700
(E) Landowners (millions of acre hours)	800	550	930	100	✗	500	20	2,900
(F) Capital owners (millions of machine hours)	400	50	150	200	✗	20	80	900

2. The Leontief inverse applicable to the above input-output table is given as the small table below. Use it to calculate a new input-output table in which final demands are drastically different, as shown by the encircled numbers. (Make your calculations to four decimal places).

	Total Output Required if Delivery to Final Users Is to Equal		
	1 truck (1)	1 barrel of oil (2)	1 ton of corn (3)
(A) Trucks (number)	1.0583	0.0470	0.0128
(B) Fuel oil (barrels)	0.1189	1.1164	0.0261
(C) Corn (tons)	0.0588	0.0026	1.1118

Recipients \ Suppliers	Of Intermediate Goods and Primary Resources			Of Final Goods and Primary Resources				Total[a]
	Truck producers (1)	Fuel-oil producers (2)	Corn producers (3)	Domestic households (4)	Domestic producers (5)	Domestic government (6)	Foreigners (7)	(8)
(A) Truck producers (millions of trucks)				0	1	20	1	
(B) Fuel-oil producers (millions of barrels of oil)				100	2	1	−100	
(C) Corn producers (millions of tons of corn)				20	100	10	10	
(D) Labor owners (millions of labor hours)				100	✕	300	0	
(E) Landowners (millions of acre hours)				100	✕	500	0	
(F) Capital owners (millions of machine hours)				200	✕	20	0	

[a] Row entries may not add up to total because of rounding.

ANSWERS

Multiple-Choice Questions

1. d	2. b	3. a	4. b	5. c	6. d	7. a	8. b	9. a
10. d	11. c	12. a	13. a	14. c	15. b	16. a	17. c	18. d
19. d	20. c	21. d	22. c	23. d	24. d	25. b	26. b	27. d
28. a	29. d	30. b	31. d	32. a	33. b	34. a		

True-False Questions

1. T
2. F (A general equilibrium is a conceivable end result of either type of coordination.)
3. F (Either equilibrium can refer to the short run or the long run. E.g.: If every firm in a perfectly competitive industry had equated marginal cost and output price, a *short-run* (profit-maximizing) equilibrium would exist. Yet if every firm also earned positive economic profit, further changes would occur, and *long-run* industry equilibrium would be reached only when economic profits were zero. Similarly, if all input and output markets were in short run equilibrium at the same time, a *short-run* general equilibrium would exist. Yet there would not yet exist a *long-run* general equilibrium, if some reproducible resources still earned positive quasi rent, if some firm earned positive economic profit.)
4. T
5. T
6. F (These grapes would be part of the country's set of *final* goods for that year.)

7. F (All entries in the table are *flows*. They do not refer to a given moment of time, as would be true of stocks. They refer to time periods, e.g., per week, per month, per year.)
8. T
9. T
10. F (Entries are stated in the units given at the head of each row and are referenced per unit of whatever is listed at the head of the column.)
11. T
12. F (Central economic planning or the deliberate order is one method of assuring coordination, but the market economy or the spontaneous order is another; each of these methods has its costs, as the text shows.)
13. F (Not necessarily by a human manager.)
14. T
15. F (Both methods have costs, as the text explains.)

Problems

1.

	Inputs Required by Average Producer to Make		
	1 truck (1)	**1 barrel of oil** (2)	**1 ton of corn** (3)
(A) Trucks (number)	$\frac{1}{20} = .05$	$\frac{2}{50} = .04$	$\frac{3}{300} = .01$
(B) Fuel oil (barrels)	$\frac{2}{20} = .1$	$\frac{5}{50} = .1$	$\frac{6}{300} = .02$
(C) Corn (tons)	$\frac{1}{20} = .05$	$\frac{0}{50} = 0$	$\frac{30}{300} = .1$
(D) Labor (labor hours)	$\frac{200}{20} = 10$	$\frac{1}{50} = .02$	$\frac{30}{300} = .1$
(E) Land (acre hours)	$\frac{800}{20} = 40$	$\frac{550}{50} = 11$	$\frac{930}{300} = 3.1$
(F) Capital (machine hours)	$\frac{400}{20} = 20$	$\frac{50}{50} = 1$	$\frac{150}{300} = .5$

2.

Recipients \ Suppliers	Of Intermediate Goods and Primary Resources			Of Final Goods and Primary Resources				Total[a]
	Truck producers (1)	Fuel-oil producers (2)	Corn producers (3)	Domestic households (4)	Domestic producers (5)	Domestic government (6)	Foreigners (7)	(8)
(A) Truck producers (millions of trucks)	1.2608	0.3848	1.5695	0	1	20	1	23.2826 0.1410 1.7920 25.2156
(B) Fuel-oil producers (millions of barrels of oil)	2.5216	0.9619	3.1391	100	2	1	−100	2.6158 3.3492 3.6540 9.6190
(C) Corn producers (millions of tons of corn)	1.2608	0	15.6953	20	100	10	10	1.2936 0.0078 155.6520 156.9534
(D) Labor owners (millions of labor hours)	252.1560	0.1924	15.6953	100	✗	300	0	668.0437
(E) Landowners (millions of acre hours)	1,008.6240	105.8090	486.5555	100	✗	500	0	2,200.9885
(F) Capital owners (millions of machine hours)	504.3120	9.6190	78.4767	200	✗	20	0	812.4077

[a] Row entries may not add up to total because of rounding.

BIOGRAPHY 13.1

Adam Smith

Adam Smith (1723–1790) was born in Kirkaldy, Scotland. He studied at Glasgow, then at Oxford, only to return to Glasgow as a teacher of moral philosophy (economics had not yet been invented as a separate discipline). Smith wrote only two books; both brought him instant fame. His first book, *The Theory of Moral Sentiments,* was published in 1759; his second, *An Inquiry into the Nature and Causes of the Wealth of Nations,* in 1776. The latter book has been called the fountainhead of economic science; it earned Smith the title "father of economics."

The single most important source of the wealth of nations, Smith argues in his later book, is the division of labor. "This division of labor, from which so many advantages are derived," he says, "is not originally the effect of any human wisdom, which foresees and intends that general opulence to which it gives occasion. It is the necessary, though very slow and gradual, consequence of a certain propensity in human nature which has in view no such extensive utility; the propensity to truck, barter, and exchange one thing for another."[1] Through-

---

[1] Adam Smith, *An Inquiry into the Nature and Causes of the Wealth of Nations* (Homewood, Ill.: Richard D. Irwin, 1963), Book 1, Chap. 1, p. 11.

out his book, Smith emphasizes the importance of economic liberty; that is, of free competition among individuals pursuing their self-interest as they choose to define it. The free, spontaneous interaction of people in the marketplace—all persons having only their own narrow, but not necessarily selfish, ends in mind—would bring about, argues Smith, the general benefit of humanity that nobody intended. In contrast, governmental attempts to guide or regulate the market would end up doing more harm than good.

Consider his famous story of beavers and deer. In an attempt to explain the relative prices of beavers and deer, Smith wrote:

> In that early and rude state of society which precedes both the accumulation of stock and the appropriation of land, the proportion between the quantities of labor necessary for acquiring different objects seems to be the only circumstance which can afford any rule for exchanging them for one another. If, among a nation of hunters, for example, it usually costs twice the labor to kill a beaver which it does to kill a deer, one beaver should naturally exchange for or be worth two deer. It is natural that what is usually the produce of two days' or two hours' labor, should be worth double of what is usually the produce of one day's or one hour's labor.[2]

Let us examine how Smith reached his conclusion. He pictured a primitive nation of hunters all of whom, apparently, had equal access to the resources provided by nature and all of whom, apparently, relied exclusively on their own labor and not on fancy capital ("stock" as he called it) to catch their prey. He assumed that a hunter would usually need two days to kill a beaver but only one day to kill a deer. Under these circumstances, anyone going into the forest and interacting only with nature could in fact exchange one beaver for two deer: By using two days for hunting deer, one would gain two deer (the benefit) but sacrifice one beaver (the opportunity cost). Conversely, by using two days for hunting beaver, one would catch one beaver (the benefit) but lose two deer (the opportunity cost).

Therefore, argued Smith, one beaver would also have to exchange for two deer when people met in their villages and interacted *with each other*. Smith imagined what would happen if one beaver were to be traded in the market for *one* deer: Hunters of beaver, instead of hunting for two days to get a beaver, could then hunt for one day to get a deer and trade it for a beaver in the market. Thus they could have an extra day of leisure without any sacrifice of goods available to them. Or they could hunt for two days to get two deer, trading in one (or both) at the marketplace. Thus they could perform the same amount of work, while receiving more goods (two beavers instead of one or one beaver and one deer instead of one beaver only).

Surely, concluded Smith, most people would interpret more leisure (without an added sacrifice of goods) or more goods (without an added sacrifice of leisure) as an increase in their welfare. Therefore, many people would begin to act accordingly: The hunting of beaver would drop or even cease altogether, while the hunting of deer would increase. If the forest continued to yield a deer for every day's hunt in spite of all the extra hunters (and that Smith assumed), the supply of deer in the market would rise and rise, but the supply of beaver would all but vanish. People seeking to buy beavers with deer would have a

---

[2]*Ibid.*, Chap. 6, p. 38.

hard time finding sellers; but any remaining hunters of beaver would find it easy to raise their price and demand more than one deer. Only when the price of beaver in the market had risen to two deer (and become equal to the "price" in the forest), concluded Smith, would the shortage of beaver in the marketplace disappear—along with the surplus of deer.

Thus Smith tried to explain the structure of relative prices by the relative *labor* costs of production. Under his assumptions (natural resources being abundant, capital resources being of negligible importance, human resources being freely mobile between various activities, and people acting so as to promote their welfare as much as possible), he was right.

As long as nature yielded one beaver or two deer for the same amount of effort (and all people were free to hunt what they liked), a beaver's price in the market ultimately would have to be twice that of a deer, reflecting the relative opportunity costs in the process of production. Any well-intentioned government that tried to help consumers of beavers by decreeing, let us say, a halving of their price (as measured in deer) would simply cause the supply of beavers to dry up. That would leave the very people government wanted to help without beavers altogether!

Thus Smith proved to be a brilliant microeconomist when he noted that the structure of relative prices, in a society wherein resources were mobile, was far from arbitrary and could not be altered at will (away from the structure of opportunity costs) without undesirable consequences. While it did not matter whether the absolute price of beaver was £1 or £100, under the assumptions underlying Smith's example, any price of a deer that was not half that of a beaver in the marketplace would cause nothing but trouble.

BIOGRAPHY 13.2

Marie Esprit Léon Walras

Marie Esprit Léon Walras (1834–1910) was born in Évreux, France. At a young age, he studied classics, literature, and science but then failed twice to be admitted to the prestigious École Polytechnique in Paris because of his poor math. He studied briefly at the École des Mines and then began to drift, writing an unsuccessful novel, engaging in free-lance journalism, lecturing on social reform, and managing a bank for cooperatives (which failed). His father, Auguste, who was an economist (and classmate of Cournot), urged him to study economics. Walras did study economics—not formally, but on his own. By 1860, when he published a polemic against Proudhon *(L'Économie Politique et La Justice: Examen Critique et Réfutation des Doctrines Économiques de M. P.-J. Proudhon),* he knew he wanted to unite mathematics with economic theory. His opportunity to pursue this area of study came after he read a paper at an international congress on taxation at Lausanne (where, ironically, Proudhon's essay won 1st prize). In the audience was a man who later founded a chair of political economy at the University of Lausanne. In 1870, Walras was the first to be appointed to that chair, even though he lacked formal training;

there he stayed till he retired and his student Pareto (see Biography 14.1) replaced him.

Schumpeter (see Biography 10.1) called Walras the greatest of all economists; certainly the Walrasian analysis of general equilibrium (which is introduced in this chapter in the text) is an outstanding landmark on the road economics has traveled to the status of an exact science. In addition, Walras became the third independent discoverer of the principle of diminishing marginal utility, besides Menger and Jevons (see Biography 2.2). Both of these great achievements appeared in *Éléments d'Économie Politique Pure* (1847–77), which was followed by two supplements, *Études d'Économie Sociale* (1896) and *Études d'Économie Politique Appliquée* (1898).

Although Walras's work is now recognized as a great achievement second to none, his efforts at the time were viewed with indifference or hostility by students and most colleagues alike, with Enrico Barone, Irving Fisher (Biography 12.2), and Vilfredo Pareto being the most notable exceptions. In addition, Walras alienated many by attaching as much importance to his questionable ideas about social justice, land nationalization schemes, sound monetary management, and the like as to his superb achievement in pure theory. Thus Walras found himself isolated, but he was a prolific letter writer. The copies he kept of his own correspondence and the letters he received show how Walras walked a solitary path, with little encouragement other than what he found in himself. In one letter he says to a friend: "If one wants to harvest quickly, one must plant carrots and salads; if one has the ambition to plant oaks, one must have the sense to tell oneself: my grandchildren will owe me this shade."[1]

Walras writes in his *Notice Autobiographique* (1904):

> On the afternoon of June 23, 1903, I met again at the door of my office the young professor, Henry L. Moore, of Columbia University of New York, who, after having explained to me the difficulties he himself had encountered in America, said: "You must recognize, my dear M. Walras, that for a scientific revolution such as you wish to make in economics, it requires 50 years."
>
> "That is the exact period," I responded.

There is a monument to Walras now at the University of Lausanne. Rightly, it bears no other inscription but *"Équilibre Économique."*

BIOGRAPHY 13.3

Wassily W. Leontief

Wassily W. Leontief (1906–   ) was born in Petrograd, Russia, the son of an economist. He studied at the Universities of Leningrad and Berlin, subsequently did research at the University of Kiel, and advised the Chinese government in Nanking. In 1931, he joined the faculty of Harvard. He stayed at Harvard until his retirement and subsequent move to New York University. While at Harvard, in 1970, he served as president of the American Economic Association. In 1973, "for his input-output methods of quantifying interdependencies in an economy and using them to predict large-scale trends," he was awarded the Nobel Memorial Prize in Economic Science.

---

[1] Cited in Joseph A. Schumpeter, *History of Economic Analysis* (New York: Oxford University Press, 1954), p. 829.

Even though Leontief had experimented with primitive chessboard balances of the Soviet economy in the 1920s while at Leningrad, his major work on input-output analysis appeared much later as *The Structure of [the] American Economy: 1919–1939: An Empirical Application of Equilibrium Analysis* (1941). Other important works include *Studies in the Structure of the American Economy: Theoretical and Empirical Explorations in Input-Output Analysis* (1953); *Input-Output Economics* (1966); *Essays in Economics, vol. 1: Theories and Theorizing* (1966), *vol. 2: Theories, Facts, and Policies* (1977); and *The Future of the World Economy: A United Nations Study* (1977).

During World War II, the U.S. government was the first to develop input-output tables; nowadays, such tables are commonly used around the world. But everywhere, analysts have run into the problem of having insufficient data that advancing theoretical knowledge and high-speed computers are ready to use. Said Leontief in his presidential address to the American Economic Association:

> Economics today rides the crest of intellectual respectability and popular acclaim. . . . But I submit that the consistently indifferent performance in practical applications is in fact a symptom of a fundamental imbalance in the present state of our discipline. The weak and all too slowly growing empirical foundation clearly cannot support the proliferating superstructure of pure, or should I say, speculative economic theory. . . . The task of securing a massive flow of primary economic data can be compared to that of providing the high energy physicists with a gigantic accelerator. The scientists have their machines while the economists are still waiting for their data. In our case not only must the society be willing to provide year after year the millions of dollars required for maintenance of a vast statistical machine, but a large number of citizens must be prepared to play, at least, a passive and occasionally even an active part in actual fact-finding operations. It is as if the electrons and protons had to be persuaded to cooperate with the physicist. . . . Economists should be prepared to take a leading role in shaping this major social enterprise. . . . [The] public has amply demonstrated its readiness to back the pursuit of knowledge. It will lend its generous support to our venture, too, if we take the trouble to explain what it is all about.[1]

Additional Readings

Gramm, Warren S. "The Selective Interpretation of Adam Smith," *Journal of Economic Issues,* March 1980, pp. 119–41.

> An argument that economists are not reading Adam Smith correctly.

Hicks, John R. "Léon Walras," *Econometrica,* October 1934, pp. 338–48.

> On the life and work of Walras.

Jaffé, William, ed. *Correspondence of Léon Walras and Related Papers,* 3 vols. Amsterdam: North-Holland, 1965.

Schumpeter, Joseph A. *Ten Great Economists: From Marx to Keynes.* New York: Oxford University Press, 1951, chap. 2.

> On Walras.

Seligman, Ben B. *Main Currents in Modern Economics, vol. 2: The Reaffirmation of Tradition.* Chicago: Quadrangle, 1962, pp. 367–86 and pp. 434–41.

> A discussion of Walras's work and a discussion of Leontief's work.

---

[1] Wassily Leontief, "Theoretical Assumptions and Nonobserved Facts," *The American Economic Review,* March 1971, pp. 1–7.

Smith, Adam. *An Inquiry into the Nature and Causes of the Wealth of Nations.* Homewood, Ill.: Richard D. Irwin, 1963, originally 1776.

    The first systematic treatment of economic science.

Walker, Donald A. "Léon Walras in the Light of His Correspondence and Related Papers," *Journal of Political Economy,* July/August 1970, pp. 685–701.

APPENDIX 13A DERIVING THE LEONTIEF INVERSE

We begin our brief excursion into the mathematics of general equilibrium analysis by defining a **matrix** as any rectangular array of numbers. Thus the set of technical coefficients for intermediate goods, in rows (A) to (C) of text Table 13.3, "Technical Coefficients," can be symbolized by the capital letter T and written in matrix form as follows:

$$T = \begin{bmatrix} .10 & .15 & .04 \\ .20 & .25 & .01 \\ .10 & .10 & .20 \end{bmatrix}$$

This matrix happens to be a square one, and the **main diagonal** of a square matrix is defined as all elements running from the upper left to the lower right corner. A matrix that consists of 1s along the main diagonal, all other elements being zero, is called an **identity matrix.** We shall symbolize it by I, and write

$$I = \begin{bmatrix} 1 & 0 & 0 \\ 0 & 1 & 0 \\ 0 & 0 & 1 \end{bmatrix}$$

One matrix can be subtracted from another if both have identical numbers of rows and columns. Subtraction involves, as the example below shows, taking the difference of corresponding elements.

$$I - T = L$$

$$\begin{bmatrix} 1 & 0 & 0 \\ 0 & 1 & 0 \\ 0 & 0 & 1 \end{bmatrix} - \begin{bmatrix} .10 & .15 & .04 \\ .20 & .25 & .01 \\ .10 & .10 & .20 \end{bmatrix} = \begin{bmatrix} .90 & -.15 & -.04 \\ -.20 & .75 & -.01 \\ -.10 & -.10 & .80 \end{bmatrix}$$

Matrix L, resulting from subtracting our technical-coefficients matrix from an identity matrix, is called a **Leontief matrix.** When inverted, this matrix is called the Leontief inverse.

An **inverse matrix** is a matrix that, multiplied by the original matrix, yields an identity matrix. Labeling the inverse of the Leontief matrix L by L^{-1}, we can say that $L \times L^{-1} = I$. Before we can find L^{-1}, however, we have to introduce additional concepts, namely, the concepts of a *determinant,* a *minor,* and a *cofactor.* At first sight, it is easy to confuse a determinant with a matrix, for a **determinant** is a number of elements arranged in rows and columns to form a square, such as

$$D = \begin{vmatrix} x_{11} & x_{12} \\ x_{21} & x_{22} \end{vmatrix}$$

wherein the subscripts identify rows and columns.

A visual distinction used is that a determinant is enclosed in straight lines, a matrix in square brackets. A more basic distinction is, however, that a determinant *can be evaluated, yielding a single number*. The type of determinant illustrated above always is equal to

$$D = (x_{11} \cdot x_{22}) - (x_{21} \cdot x_{12})$$

The above determinant, having two rows and two columns, is called a second-order determinant. The elements of a higher-order determinant can be expressed as minors and cofactors. Let us look at a third-order determinant, such as

$$D' = \begin{vmatrix} x_{11} & x_{12} & x_{13} \\ x_{21} & x_{22} & x_{23} \\ x_{31} & x_{32} & x_{33} \end{vmatrix}$$

The **minor** of any element of a third-order determinant consists of the second-order determinant remaining when the row and column of the element in question are deleted. Indicating minors by the symbol *m*, we can write the minor of x_{11} as

$$m_{11} = \begin{vmatrix} x_{22} & x_{23} \\ x_{32} & x_{33} \end{vmatrix}$$

Similarly, the minor of x_{23} would be

$$m_{23} = \begin{vmatrix} x_{11} & x_{12} \\ x_{31} & x_{32} \end{vmatrix}$$

The **cofactor** of an element, which we shall label *C*, equals the element's minor with an appropriate sign. If the sum of the subscripts of the element is *even*, as for x_{11}, the cofactor has a *plus* sign, if it is *odd*, as for x_{23}, a *minus* sign. Hence the cofactors of x_{11} and x_{23} are

$$C_{11} = + \begin{vmatrix} x_{22} & x_{23} \\ x_{32} & x_{33} \end{vmatrix} \text{ and } C_{23} = - \begin{vmatrix} x_{11} & x_{12} \\ x_{31} & x_{32} \end{vmatrix}$$

Each of the cofactors can, of course, be evaluated as was determinant *D*, giving us

$$C_{11} = + [(x_{22} \cdot x_{33}) - (x_{32} \cdot x_{23})] \text{ and } C_{23} = - [(x_{11} \cdot x_{32}) - (x_{31} \cdot x_{12})]$$

A third-order determinant will obviously then have nine such cofactors. A third-order determinant is evaluated with the help of the minors of the first row, so that

$$D' = (x_{11} \cdot m_{11}) - (x_{12} \cdot m_{12}) + (x_{13} \cdot m_{13})$$

Now we are ready to proceed with the inversion of our matrix *L*, which consists of six steps:

(1) Writing the matrix as a determinant and evaluating it.

$$D = \begin{vmatrix} .90 & -.15 & -.04 \\ -.20 & .75 & -.01 \\ -.10 & -.10 & .80 \end{vmatrix}$$

$$= .90 \begin{vmatrix} .75 & -.01 \\ -.10 & .80 \end{vmatrix} - (-.15) \begin{vmatrix} -.20 & -.01 \\ -.10 & .80 \end{vmatrix}$$
$$+ (-.04) \begin{vmatrix} -.20 & .75 \\ -.10 & -.10 \end{vmatrix}$$

$$= .90\{.75(.80) - [(-.10)(-.01)]\} + .15\{-.20(.80) - [(-.10)(-.01)]\} - .04\{-.20(-.10) - [(-.10)(.75)]\}$$

$$= .90(.599) + .15(-.161) - .04(.095)$$

$$= .5391 - .02415 - .0038 = .51115$$

(2) Identifying all cofactors of the determinant.

$$C_{11} = + \begin{vmatrix} .75 & -.01 \\ -.10 & .80 \end{vmatrix} = .599$$

$$C_{12} = - \begin{vmatrix} -.20 & -.01 \\ -.10 & .80 \end{vmatrix} = .161$$

$$C_{13} = + \begin{vmatrix} -.20 & .75 \\ -.10 & -.10 \end{vmatrix} = .095$$

$$C_{21} = - \begin{vmatrix} -.15 & -.04 \\ -.10 & .80 \end{vmatrix} = .124$$

$$C_{22} = + \begin{vmatrix} .90 & -.04 \\ -.10 & .80 \end{vmatrix} = .716$$

$$C_{23} = - \begin{vmatrix} .90 & -.15 \\ -.10 & -.10 \end{vmatrix} = .105$$

$$C_{31} = + \begin{vmatrix} -.15 & -.04 \\ .75 & -.01 \end{vmatrix} = .0315$$

$$C_{32} = - \begin{vmatrix} .90 & -.04 \\ -.20 & -.01 \end{vmatrix} = .017$$

$$C_{33} = + \begin{vmatrix} .90 & -.15 \\ -.20 & -.75 \end{vmatrix} = .645$$

(3) Arranging the cofactors in matrix form.

$$C = \begin{bmatrix} .599 & .161 & .095 \\ .124 & .716 & .105 \\ .0315 & .017 & .645 \end{bmatrix}$$

(4) Finding the **adjoint matrix,** which is the matrix of cofactors transposed—that is, with rows and columns interchanged.

$$A = \begin{bmatrix} .599 & .124 & .0315 \\ .161 & .716 & .017 \\ .095 & .105 & .645 \end{bmatrix}$$

(5) Finding the Leontief inverted matrix by dividing each element in the adjoint matrix by the value of the determinant found in step 1.

$$L^{-1} = \begin{bmatrix} 1.1718673 & 0.2425902 & 0.0616257 \\ 0.314976 & 1.4007629 & 0.0332583 \\ 0.1858554 & 0.2054191 & 1.2618605 \end{bmatrix}$$

(6) Checking the result by multiplying the original matrix L by its inverse L^{-1}, which should yield an identity matrix I. Matrices can be multiplied only if they are *conformable,* that is, if the number of columns of the first equals the number of rows of the second. This is, of course, the case here with L and L^{-1}. The product will have as many rows as the first and as many columns as the second. The rule for multiplication can easily be derived from the detail that follows. It involves finding each element of the product matrix by cumulative multiplication of the elements of the first row of the first matrix by those of the first, second, and third column of the second matrix, and so on.

$$L \cdot L^{-1} = I$$

$$\begin{bmatrix} .90 & -.15 & -.04 \\ -.20 & .75 & -.01 \\ -.10 & -.10 & .80 \end{bmatrix} \cdot \begin{bmatrix} 1.1718673 & 0.2425902 & .0616257 \\ 0.314976 & 1.4007629 & .0332583 \\ 0.1858554 & 0.2054191 & 1.2618605 \end{bmatrix}$$

$$= \begin{bmatrix} (\ .90 \times 1.1718673) + (-.15 \times .314976) + (-.04 \times .1858554) \\ (-.20 \times 1.1718673) + (\ .75 \times .314976) + (-.01 \times .1858554) \\ (-.10 \times 1.1718673) + (-.10 \times .314976) + (\ .80 \times .1858554) \end{bmatrix}$$

$$\begin{matrix} (\ .90 \times .2425902) + (-.15 \times 1.4007629) + (-.04 \times .2054191) \\ (-.20 \times .2425902) + (\ .75 \times 1.4007629) + (-.01 \times .2054191) \\ (-.10 \times .2425902) + (-.10 \times 1.4007629) + (\ .80 \times .2054191) \end{matrix}$$

$$\begin{bmatrix} (\ .90 \times .0616257) + (-.15 \times .0332583) + (-.04 \times 1.2618605) \\ (-.20 \times .0616257) + (\ .75 \times .0332583) + (-.01 \times 1.2618605) \\ (-.10 \times .0616257) + (-.10 \times .0332583) + (\ .80 \times 1.2618605) \end{bmatrix}$$

$$= \begin{bmatrix} 1 & 0 & 0 \\ 0 & 1 & 0 \\ 0 & 0 & 1 \end{bmatrix}$$

This proves our calculation of the inverse as correct. The inverse L^{-1} is, of course, presented as text Table 13.4, "The Leontief Inverse Matrix."

CHAPTER 14

Efficiency and Equity

MULTIPLE-CHOICE QUESTIONS

Circle the letter of the *one* answer that you think is correct or closest to correct.

1. Technical efficiency is said to exist when a firm finds it impossible, with given technical knowledge,
   a. to produce a larger output from a given set of inputs.
   b. to produce a given output with less of one or more inputs without increasing the amount of other inputs.
   c. to do (a) or (b).
   d. to make a larger profit from a given set of inputs.

2. Which of the following concepts is *not* equivalent to two of the others?
   a. Economic efficiency.
   b. X-efficiency.
   c. Allocative efficiency.
   d. Static efficiency.

3. *Economic inefficiency* is a situation in which
   a. total economic welfare can be raised with certainty.
   b. total economic welfare cannot be raised further.
   c. any change brings with it a net benefit for some people and a net cost for others, thus leaving the effect on overall welfare uncertain.
   d. it is impossible, through some reallocation of resources or goods, to make some or all people better off without making others worse off.

4. If one producer used 20 units of land (along with other inputs) to produce 80 units of corn in a year, while another used 20 units of land (along with other inputs) to produce 40 units of corn, we would know
   a. that technical inefficiency existed so far as the second producer was concerned.
   b. that economic inefficiency existed in the economy as a whole.
   c. neither (a) nor (b).
   d. both (a) and (b).

5. If one producer's marginal rate of transformation equaled 1 unit of labor for 7 bushels of lettuce, while another's equaled 1 unit of labor for 9 bushels of lettuce,
   a. their combined output would equal 16 bushels of lettuce.
   b. their combined output could be increased without the use of additional inputs.
   c. technical inefficiency would exist.
   d. all of the above would be true.

6. If the marginal rate of transformation equaled 1 unit of labor for 500 units of output in Firm A, but 1 unit of labor for 900 units of output in Firm B, a move in the direction of economic efficiency
   a. would involve moving labor from A to B.
   b. might involve a change in A's *MRT* to 1 for 600.
   c. might involve a change in B's *MRT* to 1 for 600.
   d. would be correctly described by all of the above.

7. If Producer A produced 20 million cars plus 30 million bicycles in a year, while Producer B produced 50 million cars plus 60 million bicycles, it would be clear to any trained economist that
   a. society's output could be increased unambiguously if A specialized in cars and B in bicycles.
   b. society's output could be increased unambiguously if A specialized in bicycles and B in cars.
   c. both firms were using their resources in technically inefficient ways.
   d. all of the above answers are wrong.

8. If a producer's marginal rate of transformation were 1 sweater for 10 pairs of socks and a consumer's marginal rate of substitution were 1 sweater for 20 pairs of socks, one could raise social welfare with certainty by
   a. producing (and consuming) another sweater at the expense of socks.
   b. producing (and consuming) more socks at the expense of a sweater.
   c. doing neither (a) nor (b), because more sweaters would mean fewer socks and more socks would mean fewer sweaters—hence *ambiguity* as to the effect on overall welfare.
   d. producing more sweaters as well as more socks.

9. A consumer who consumed 17 units of food plus 1 unit of clothing in a year instead of 10 units of food plus 20 units of clothing would be better off—provided the consumer's marginal rate of
   a. substitution was 7 units of food for 19 units of clothing.
   b. substitution was 10 units of food for 19 units of clothing.
   c. substitution was 5 units of food for 19 units of clothing.
   d. transformation was 7 units of food for 19 units of clothing.

10. If one consumer consumed 10 units of good a plus 20 units of good b in a year, while a second consumer consumed 200 a plus 50 b, we would know
    a. the allocation of goods among consumers was economically inefficient.
    b. the allocation of goods among consumers was unfair.
    c. both (a) and (b).
    d. neither (a) nor (b).

11. At the Pareto optimum,
    a. any *MRT* equals any corresponding *MRS*.
    b. it is possible to reallocate resources or goods in such a way as to raise total economic welfare with certainty.
    c. social welfare exceeds the sum of individual welfares by the maximum possible amount.
    d. all of the above are true.

12. If, as a result of a reallocation of resources or goods, a million people asserted that they were better off, while a single person claimed to be worse off, Pareto would

a. consider total economic welfare to have increased.
  b. consider total economic welfare to have decreased (provided the lone dissenter's preferences were superior).
  c. claim ignorance about the effect on total economic welfare.
  d. possibly assert any of the above (depending on whether the Pareto optimum had been reached).

13. If perfectly competitive markets existed and the prices of shoes and labor equaled, respectively, $30 per pair and $10 per hour, all profit-maximizing firms would employ labor in shoe production precisely to the degree that yielded
    a. a marginal physical product of $\frac{1}{3}$ pair of shoes per labor hour.
    b. a marginal rate of transformation of 3 labor hours for 1 pair of shoes.
    c. a marginal rate of transformation of 1 labor hour for $\frac{1}{3}$ pair of shoes.
    d. all of the above.

14. Which of the following is *not* a reason why profit maximization under perfect competition leads to an equality of different firms' marginal rates of transformation between any two goods?
    a. Under the circumstances, profit maximization involves producing output quantities that equate the output's rising marginal cost with its price.
    b. All firms face an identical price for any given input.
    c. All firms face an identical price for any given output.
    d. The reciprocal of the marginal cost ratio of two goods is nothing else but Pareto's marginal rate of transformation between them.

15. A marginal cost of $5 for butter and $10 for cheese implies
    a. a marginal rate of transformation of 1 unit of butter for 2 units of cheese.
    b. a marginal rate of transformation of 1 unit of butter for $\frac{1}{2}$ unit of cheese.
    c. that it is more profitable to produce butter than cheese.
    d. both (b) and (c).

16. When the marginal costs of producing TV sets and radios equal, respectively, $300 and $20 in Firm A but $200 and $20 in Firm B, we know
    a. that both firms may well be maximizing profit.
    b. that a Pareto condition of economic efficiency is violated.
    c. both (a) and (b).
    d. neither (a) nor (b).

17. When first measuring the welfare loss implied by economic inefficiency, Harberger
    a. focused on the fact that firms with monopsony power in the input market always choose input levels at which input price falls short of marginal outlay.
    b. focused on the fact that firms with monopoly power in the output market always choose output levels at which output price exceeds marginal cost.
    c. assumed, quite realistically, that the price elasticity of demand was $> |1|$ at the output levels likely to be chosen by imperfectly competitive firms.
    d. did all of the above.

When answering questions 18-19, refer to the following graph, which can be used to illustrate Harberger's method of measuring the welfare loss from imperfectly competitive markets.

18. After identifying an industry's actual price and quantity as $0B$ and $0E$, Harberger identified the welfare loss from economic inefficiency as area
   a. $0ADE$.
   b. $ABCD$.
   c. $ECFG$.
   d. CDF.

19. Harberger also argued that long-run economic profit under perfect competition would equal
   a. zero.
   b. area $ABCF$.
   c. area CDF.
   d. area $ECFG$.

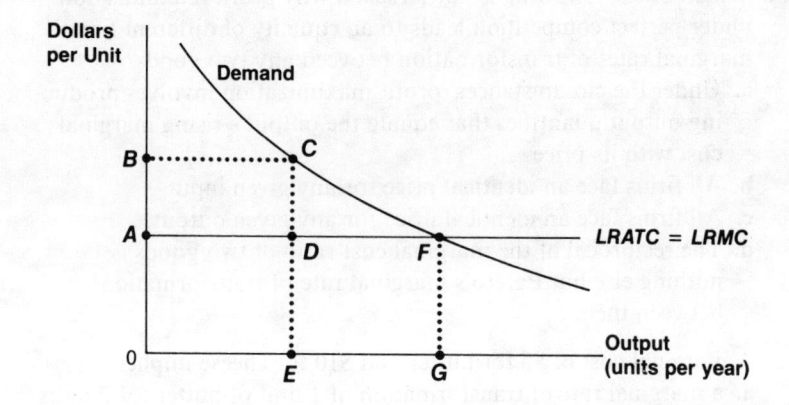

20. Critics of Harberger's trivial estimate of the deadweight welfare loss caused by monopoly power have argued that he understated the monopoly price distortion by
   a. assuming a unitary price elasticity of demand.
   b. exaggerating expenses connected with acquiring and defending monopoly power.
   c. using the average rate of return on invested capital in manufacturing as a proxy for the normal interest return on invested capital that competitive conditions would produce.
   d. using the average rate of return on invested capital in agriculture and services as a proxy for the normal interest return on invested capital that competitive conditions would produce.

When answering questions 21-24, refer to the graph, often used to illustrate the welfare loss from imperfectly competitive markets, on page 237.

21. If imperfect competition resulted in price $0C$ and quantity $0G$, and if a Harberger-type estimate of competitive long-run average total cost (including implicit interest) corresponded to $LRATC_2$, the Harberger measure of deadweight loss would equal
   a. $BCDE$.
   b. DEH.
   c. $BCDH$.
   d. none of the above.

22. If we accepted Leibenstein's argument that monopoly power allows firms to raise not only prices but also costs (as from $LRATC_1$ to $LRATC_2$), the Harberger measure of deadweight loss would become
 a. *ACDF.*
 b. *ACDL.*
 c. *DLF.*
 d. none of the above.

23. Under imperfect competition, Leibenstein would identify pure X-inefficiency equal to
 a. *ABHL.*
 b. *EHLF.*
 c. *HLI.*
 d. *ABEF.*

24. Under perfect competition, Leibenstein would expect
 a. a price of *0A*.
 b. an output of *0K*.
 c. a price of *0B*.
 d. an output of *0G*.

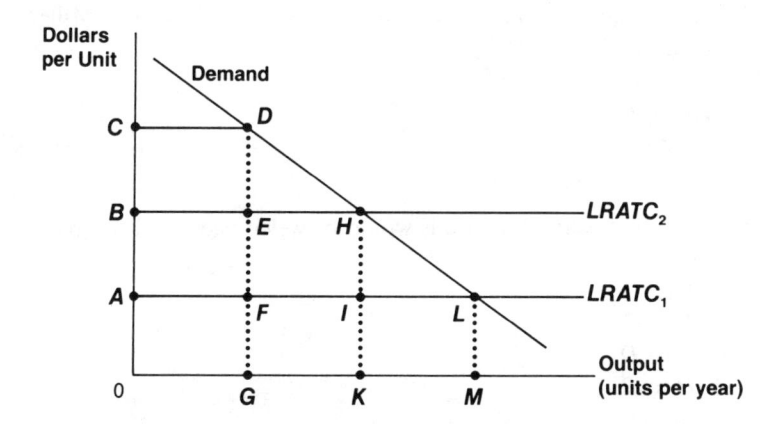

25. Hayek's *knowledge problem* is concerned with
 a. the fleeting knowledge of the particular circumstances of place and time.
 b. scientific knowledge.
 c. knowledge of general applicability.
 d. all of the above.

26. The marginal rate of transformation is
 a. the rate at which a consumer is willing to exchange, as a matter of indifference, a little bit of one variable (say, butter) for a little bit of another variable (say, meat).
 b. the rate at which a consumer is willing to exchange, in order to be better off, a little bit of one variable (say, butter) for a little bit of another variable (say, meat).
 c. the rate at which a producer is technically able to exchange, in the process of production, a little bit of one variable (say, jackets) for a little bit of another variable (say, pants).
 d. both (a) and (c).

27. When the welfare of some people can be increased at the expense of the welfare of others, we know that the initial situation is one of
    a. economic efficiency.
    b. economic inefficiency.
    c. either (a) or (b).
    d. economic inequity.

When answering questions 28–32, refer to the Edgeworth-box diagram below.

28. If bread and meat were allocated between Miller and Smith as shown by point *H*, the allocation would clearly be
    a. economically unfair.
    b. economically inefficient.
    c. equitable.
    d. equitable as well as efficient.

29. All allocations of goods that are on the contract curve are
    a. equally efficient.
    b. equally fair.
    c. neither efficient nor fair.
    d. unfair.

30. Which reallocation of goods would be welcomed by both Miller and Smith?
    a. *B* to *C*.
    b. *H* to *C*.
    c. *G* to *D*.
    d. None of the above.

31. Which reallocation of goods would be welcomed by Miller but resisted by Smith?
    a. *E* to *D*.
    b. *F* to *C*.
    c. *G* to *D*.
    d. *H* to *B*.

32. Which reallocation of goods would be a matter of indifference to either Miller or Smith?
    a. *A* to *E*.
    b. *B* to *D*.
    c. *F* to *C*.
    d. *H* to *B*.

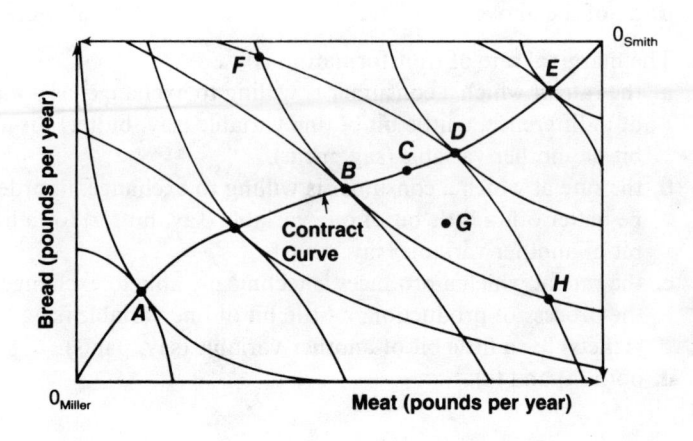

33. In the Edgeworth-box diagram, the tangency of two consumption-indifference curves denotes
    a. economic equity.
    b. economic inequity.
    c. economic efficiency.
    d. none of the above, at least not necessarily.

34. The advocates of distributive economic justice argue that goods should be apportioned among people
    a. in an economically efficient way.
    b. by some authority seeking to act justly.
    c. as depicted along the Edgeworth contract curve.
    d. as noted in (a) and (c).

35. Commutative economic justice clearly exists
    a. all along the Edgeworth conflict curve.
    b. all along the Edgeworth contract curve.
    c. whenever the percentage of all goods going to any one person is determined by some authority with reference to some personal characteristic that establishes the recipient as meritorious.
    d. in none of the above cases.

When answering questions 36–40, refer to the graph, often used to defend an absolutely equal distribution of income among people, on page 240.

36. If a total of $200 per week were to be apportioned between persons A and B, the receipt by A of $120 would
    a. give A a total utility of *qoln*.
    b. give A a total utility of *acmn*.
    c. give A a marginal utility of *ln*.
    d. result in both (a) and (c).

37. If a total of $200 per week were to be apportioned between persons A and B, the receipt by B of $140 would
    a. give A a total utility of *acdf*.
    b. give B a total utility of *fdpq*.
    c. give B a marginal utility of *fd*.
    d. result in all of the above.

38. If a total of $200 per week were to be apportioned between persons A and B so as to maximize utility, this maximum utility would equal
    a. *qoik* plus *acjk*.
    b. *qogh* plus *acgh*.
    c. *qpjk* plus *abik*.
    d. none of the above.

39. An absolutely equal apportionment of money income is usually advocated in the face of uncertainty as to the location of
    a. points *c* and *o*.
    b. points *b* and *p*.
    c. point *g*.
    d. point *k*.

40. Income equality is advocated because
    a. equal dollar deviations from equality raise total welfare as often as they lower it.
    b. area *gij* is smaller than *ilmj*.
    c. area *gij* is smaller than *gde*.
    d. of both (a) and (b).

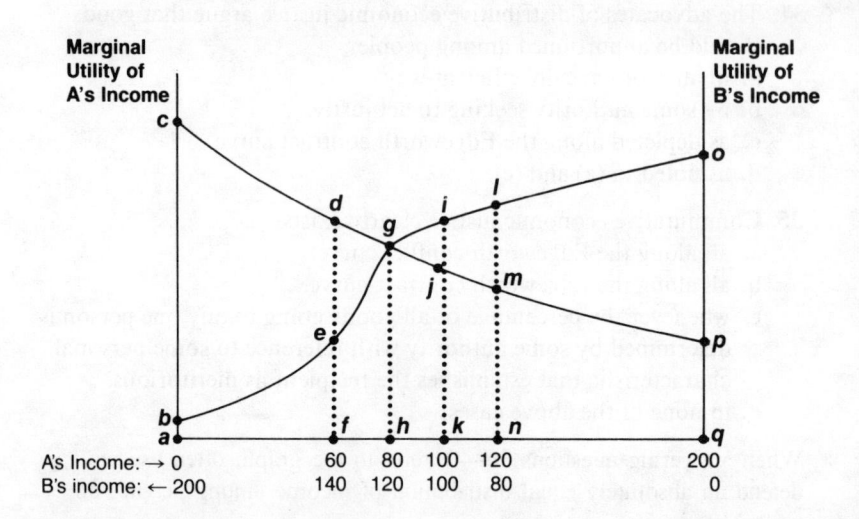

41. Which of the following is most closely linked to the notion of commutative justice?
    a. Equal opportunity.
    b. Equal end result.
    c. Communism.
    d. Socialism.

42. Under a regime of commutative justice,
    a. central planners would define the "true needs" of people.
    b. central planners would make sure that people contributed their fair share of work (and produced the right kinds of goods).
    c. neither (a) nor (b) would occur.
    d. both (a) and (b) would occur.

When answering questions 43–46, consider the accompanying graph.

43. A hypothetial Lorenz curve, which pictures a perfectly equal distribution of wealth among all families, is
    a. right-angled line 0*AB*.
    b. right-angled line 0*CB*.
    c. straight line 0*B*.
    d. not shown in this graph.

44. The line of perfect equity
    a. equals vertical 0*A*.
    b. equals diagonal 0*B*.
    c. equals horizontal 0*C*.
    d. cannot be objectively defined.

45. The Gini coefficient equals
    a. area *a* divided by (*b* + *c*).
    b. area *b* divided by (*b* + *c*).
    c. distance 0*B* along the Lorenz curve divided by distance 0*B* along the diagonal.
    d. distance 0*CB* divided by distance 0*B* along the diagonal.

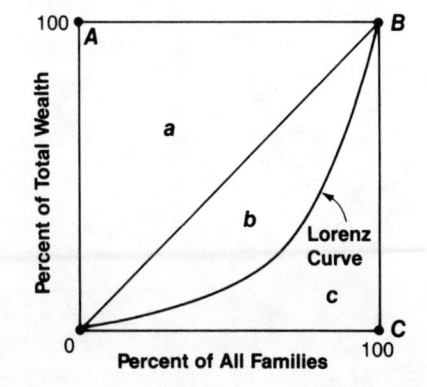

46. Any change in family wealth distribution in the direction of greater equality would
    a. shift the Lorenz curve toward straight line 0B.
    b. shift the Lorenz curve toward right-angled line 0CB.
    c. shift straight line 0B toward the Lorenz curve.
    d. increase the Gini coefficient.

47. Someone who wants to promote commutative justice is most likely to advocate
    a. the elimination of private or governmental practices that differentially restrict people's opportunities to utilize their resources.
    b. taxing the rich and subsidizing the poor.
    c. governmental aid in kind (commodity distributions, medical clinics, public housing).
    d. governmental aid in cash (agricultural subsidies, rent supplements, welfare payments).

48. In the United States at present, about one half of total income is received by
    a. 85 percent of all families (and the other half by the other 15 percent of them).
    b. 75 percent of all families (and the other half by the other 25 percent of them).
    c. 65 percent of all families (and the other half by the other 35 percent of them).
    d. 55 percent of all families (and the other half by the other 45 percent of them).

When answering questions 49–51, refer to the accompanying graph, usually used to illustrate the limits of income redistribution.

49. Starting at A, a redistribution of income from Rich to Poor along line AB denotes
    a. a rise in the Gini coefficient.
    b. no change in the Gini coefficient (because it does not apply to this graph).
    c. the presence of strong disincentive effects.
    d. the absence of disincentive effects.

50. Starting at A, a redistribution of income from Rich to Poor along line ACD0 shows that
    a. strong disincentive effects are present.
    b. it is impossible to equalize incomes.
    c. both persons could be better off with income inequality than equality.
    d. both (a) and (c) are true.

51. According to John Rawls, economic justice would be served if total income were distributed
    a. according to point C in the presence of disincentive effects.
    b. according to point E in the presence of disincentive effects.
    c. according to any point on line 0E, regardless of the presence or absence of disincentive effects.
    d. according to any point on line ACD0, regardless of the presence or absence of disincentive effects.

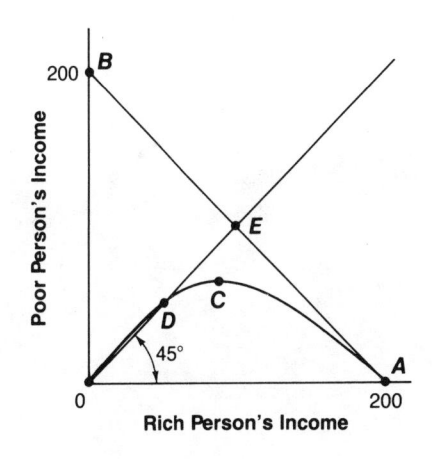

TRUE-FALSE QUESTIONS

In each space below, write a *T* if the statement is true and an *F* if the statement is false.

_____ 1. Economic or allocative efficiency exists whenever it is possible, with a given technology, to produce a larger welfare total within an economy from given stocks of resources.

_____ 2. If one producer used 9 units of capital to make 12 units of output, while another used 21 units of capital to make 98 units of output, economic inefficiency would clearly exist.

_____ 3. Economic efficiency requires that the marginal rate of transformation between any two goods be the same for any two producers producing both goods.

_____ 4. Economic efficiency requires that the marginal rate of transformation between any two goods produced by any producer be equal to the marginal rate of substitution between these two goods for any consumer of both.

_____ 5. A consumer who consumed 9 units of food plus 12 units of clothing in a year, instead of 12 units of food and 6 units of clothing, would be better off—provided the consumer's marginal rate of substitution was 1 unit of food for 3 units of clothing.

_____ 6. Any allocation of goods among consumers that is associated with equal marginal rates of substitution is equally efficient according to the Pareto criterion.

_____ 7. When firms maximize profit and households maximize utility in the context of perfectly competitive markets, a Pareto optimum results.

_____ 8. As long as each consumer is free to equate the marginal rate of substitution between any two goods with the market exchange ratio implied by the goods' prices, any *MRS* of one consumer equals that of any other consumer and the allocation of goods among consumers is efficient.

_____ 9. According to Harberger's original estimate, the deadweight welfare loss from economic inefficiency may be as low as .1 percent of U.S. GNP.

_____ 10. In situations of economic efficiency it is impossible to raise the welfare of anyone any further.

_____ 11. In the Edgeworth-box diagram, the tangency of two consumption-indifference curves denotes a logical absurdity (because indifference curves cannot be tangent to one another, but only to budget lines).

_____ 12. In the Edgeworth-box diagram, contract curve and conflict curve are identical.

_____ 13. In the Edgeworth-box diagram, situations of economic inefficiency can be depicted as lying off a contract curve.

_____ 14. According to Karl Marx, goods will be apportioned equitably (in accordance with people's needs) after the demise of socialism.

_____ 15. The Gini coefficient can range from 0 (denoting perfect inequality) to 1 (denoting perfect equality).

PROBLEMS

1. Consider the following situation: Producer A uses 20 units of labor and produces 50 bicycles per week. Producer B uses 10 units of labor and produces 50 bicycles per week. Each producer, of course, also uses unspecified amounts of other inputs.
   a. Is the allocation of labor resources technically efficient? Explain.
   b. Is the allocation of labor resources economically efficient? Explain.
   c. If you found inefficiency, what remedy would you suggest?

2. Consider the following situation: Producer A is producing locomotives and trucks, and the marginal rate of transformation is $1L$ for $2T$. Producer B produces the same products, but the MRT is $1L$ for $4T$. Prove that it is possible, without changing input totals,
   a. to increase truck output without decreasing that of locomotives.
   b. to increase locomotive output without decreasing that of trucks.
   c. to increase the output of both goods at the same time.

3. Consider the following situation: Producer A is producing sweaters and towels, and the marginal rate of transformation is $3S$ for $10T$. Consumer B is consuming both goods, and the marginal rate of substitution is $3S$ for $5T$. Prove that it is possible to raise consumer welfare without changing the totals of inputs used.

4. Consider the following situation: Consumer A is consuming 5 pounds of meat and 30 pounds of vegetables per month, while Consumer B is consuming 20 pounds of meat and 40 pounds of vegetables per month. How, if at all, could total economic welfare be raised with certainty
   a. if consumer A's MRS was $1M$ for $1V$, while B's was $1M$ for $3V$?
   b. if both consumers had an MRS of $1M$ for $7V$?

5. Consider the following alternatives:
   I. Consumer A is consuming 6 pounds of meat and 30 pounds of vegetables per month, while consumer B is consuming 20 pounds of meat and 40 pounds of vegetables per month. Both consumers have an identical MRS of $1M$ for $4V$.
   II. Consumer A is consuming 13 pounds of meat and 35 pounds of vegetables per month, while Consumer B is consuming identical quantities. Both consumers have an identical MRS as well: $1M$ for $9V$.
   a. Which of the two situations is economically more efficient?
   b. Which is more fair?

6. **a.** Illustrate, with reference to the following graphs, why fertilizer would be allocated efficiently among different farms producing asparagus under perfect competition, if all farms maximized profit.
   b. Illustrate also why a central planner who allocated equal quantities of fertilizer to all farms would create economic inefficiency (unless all producers had identical marginal-value-product curves).

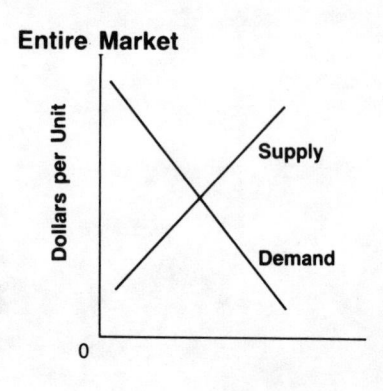

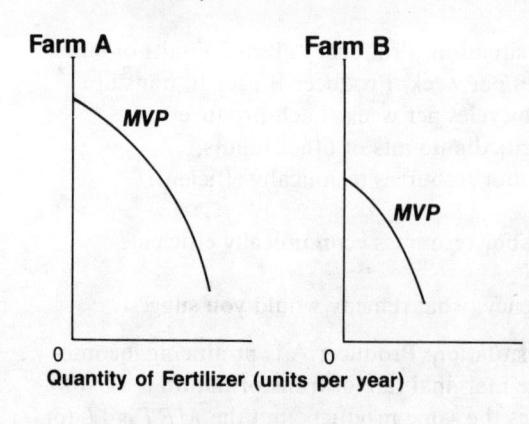

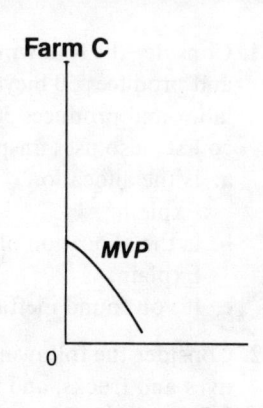

7. **a.** Illustrate, with reference to the following graphs, why production tasks would be allocated efficiently among different producers under perfect competition, if all firms maximized profit.
   b. Illustrate also why a central planner who assigned equal output quotas to all firms would create economic inefficiency (unless all producers had identical marginal-cost curves).

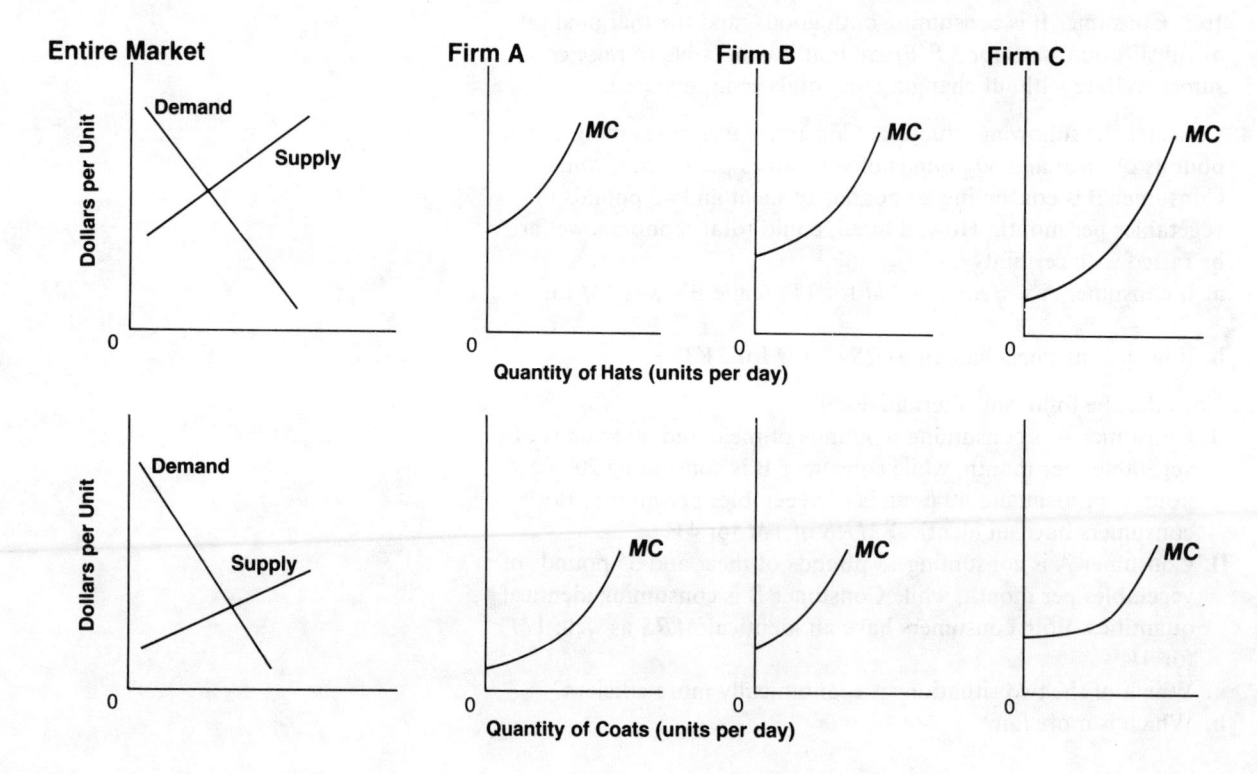

8. Illustrate, with reference to the following graphs, why production tasks might very well be allocated inefficiently among different producers under imperfect competition, if all firms maximized profit.

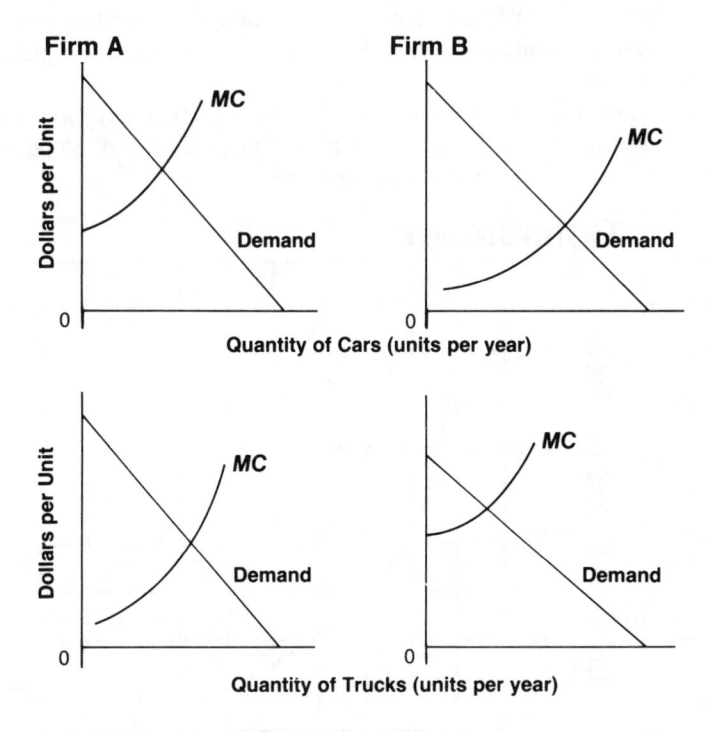

Firm A — MC, Demand; Quantity of Cars (units per year)
Firm B — MC, Demand; Quantity of Cars (units per year)

Firm A — MC, Demand; Quantity of Trucks (units per year)
Firm B — MC, Demand; Quantity of Trucks (units per year)

9. a. Illustrate, with reference to the following graphs, why production tasks might be allocated efficiently among producers even under imperfect competition, if all firms maximized profit.
   b. Why is your answer here different from that to problem 8?

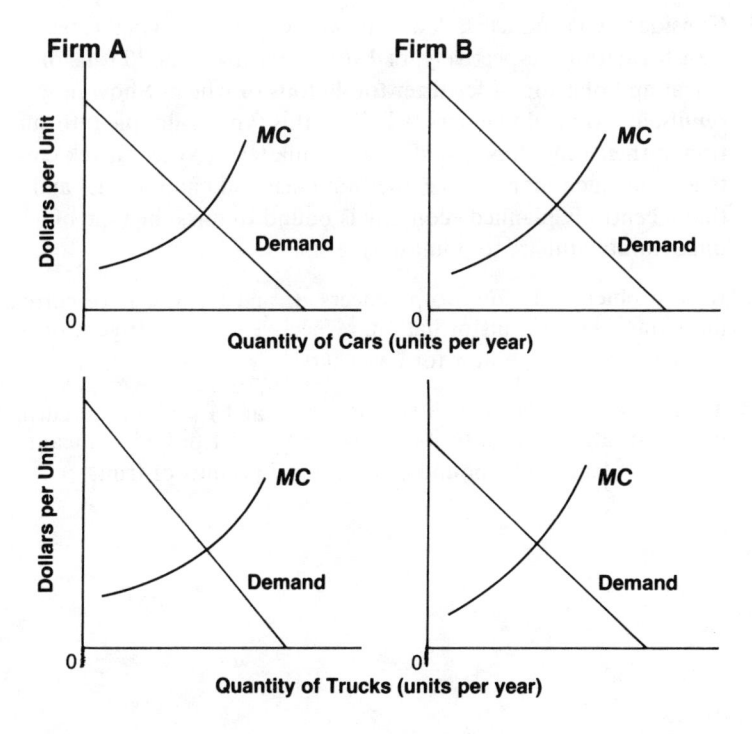

Firm A — MC, Demand; Quantity of Cars (units per year)
Firm B — MC, Demand; Quantity of Cars (units per year)

Firm A — MC, Demand; Quantity of Trucks (units per year)
Firm B — MC, Demand; Quantity of Trucks (units per year)

10. Consider the two sets of indifference curves given in the following graphs. Assume that consumer A had a budget of $100 per week, but consumer B had one of $200.
   a. Prove that consumption goods would be allocated efficiently between the two consumers if both maximized utility on the basis of identical prices of $1 per unit of bread and $10 per unit of meat.
   b. Show that the allocation would become inefficient if price discrimination existed and consumer B faced a price of $5 per unit of meat, all else remaining the same.

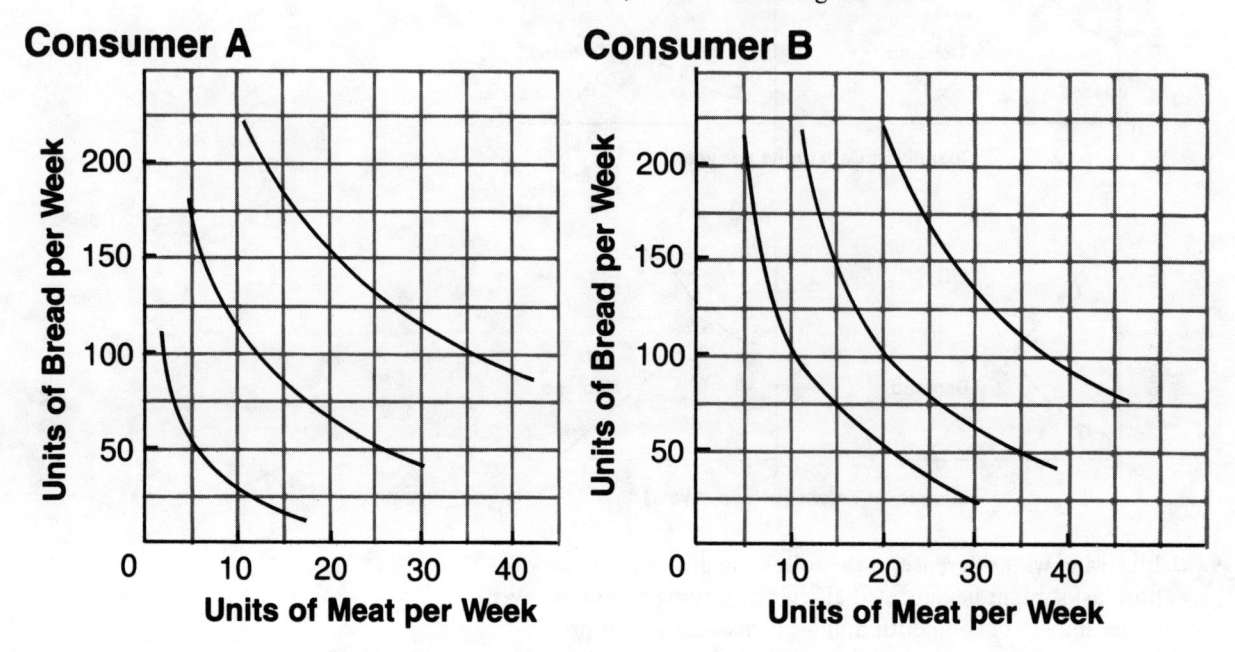

Consumer A

Consumer B

11. Consider two producers, A and B, with current marginal rates of transformation, respectively, of 1 ton of fertilizer for 30 tons of wheat and of 1 ton of fertilizer for 40 tons of wheat. Show how *omniscient* central planners (who had this type of unique information at their fingertips) could improve matters. (Note: Hayek contends that such omniscience does not exist and cannot exist and that a centrally planned economy is bound to miss the type of opportunity you are to point out).

12. Redo problem (11) for two producers, C and D, who have current marginal rates of transformation, respectively, of 1 refrigerator for 1 washer and 1 refrigerator for 2 washers.

13. Redo problem (11) for two consumers, E and F, who have current marginal rates of substitution, respectively, of 1 pound of meat for 2 pounds of fruit and 1 pound of meat for 7 pounds of fruit.

14. Consider the following Edgeworth box.
    a. Draw in the contract curve.
    b. Determine whether the allocation of goods between P (Poor) and R (Rich) is efficient at points *B* and *E*. Explain your answers.
    c. Identify all the possible allocations of goods that would make both persons better off compared to initial allocation *E*.

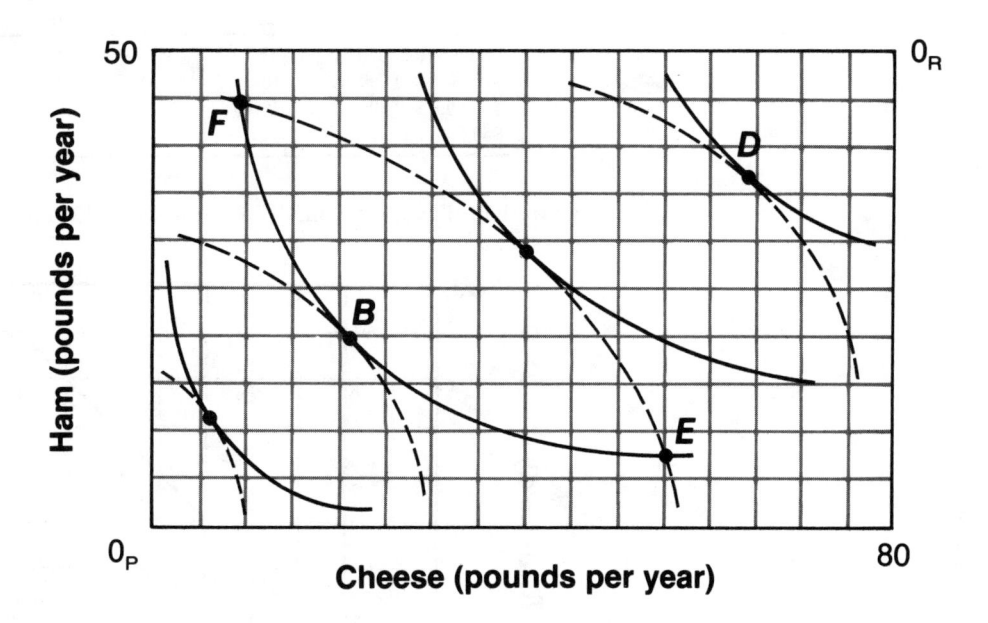

    d. Would it be possible to raise social welfare by changing allocation *B* to *D*? Explain.

15. With the help of the following graph, prove the following:
    a. "If the marginal-utility-of-income functions were known to be identical for all people, only an absolutely equal distribution of income among them could maximize social welfare."
    b. "If the marginal-utility-of-income functions were known to be lower for B-people than A-people, only a distribution of income that gave more to A-people than B-people could maximize social welfare."

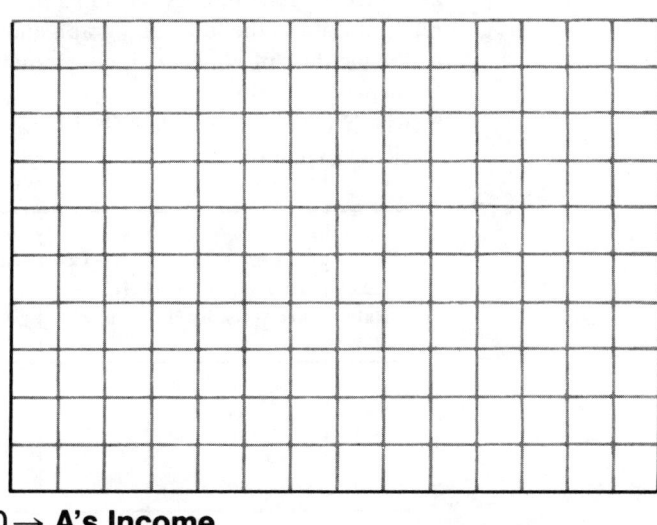

16. Consider the following hypothetical data for lawyers:

Income Class (dollars per year) (1)	Percent of Lawyers in Class (2)	Percent of Total Lawyers' Income Received by Lawyers in Class (3)
under $10,000	3	1
$10,000–$29,999	10	4
$30,000–$49,999	30	20
$50,000 and over	57	75

a. In the following graph, draw the lawyers' Lorenz curve.

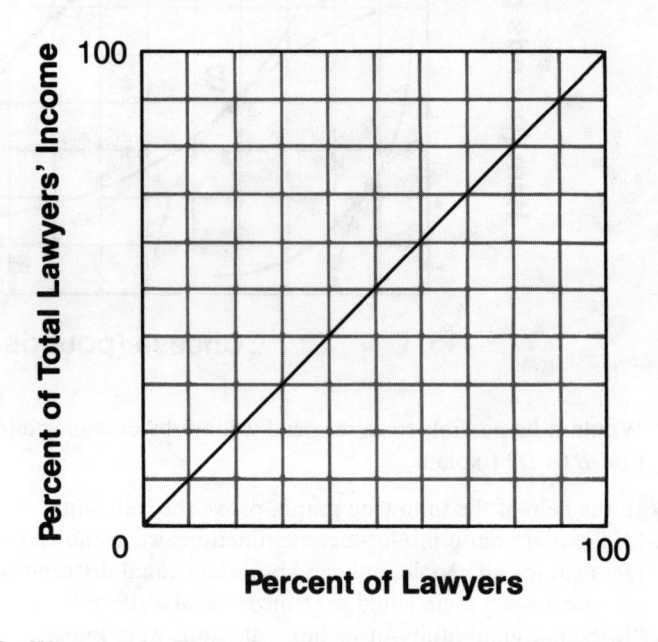

b. What would you have to do to calculate the Gini coefficient?

17. Suppose a high-income person, H, who was earning $50 per hour, were to respond as in the following table to taxes designed to help zero-income person, Z. After filling in columns (3)–(5), plot the relevant data in the following graph and indicate

a. the incomes of H and Z, if government insisted on equalizing them.

b. whether the situation described in (a) would be economically efficient. Explain.

Tax Rate (1)	Hours Worked per Week by H (2)	Pre-Tax Income of H (3)	Tax Revenue and Transfer to Z (4)	After-Tax Income of H (5)
0	60			
25	44			
50	20			
75	10			
100	0			

Z's Income (dollars per week)

[Graph with Z's Income on vertical axis (0 to 500+ dollars per week) and H's Income on horizontal axis (0 to 3000 dollars per week)]

H's Income (dollars per week)

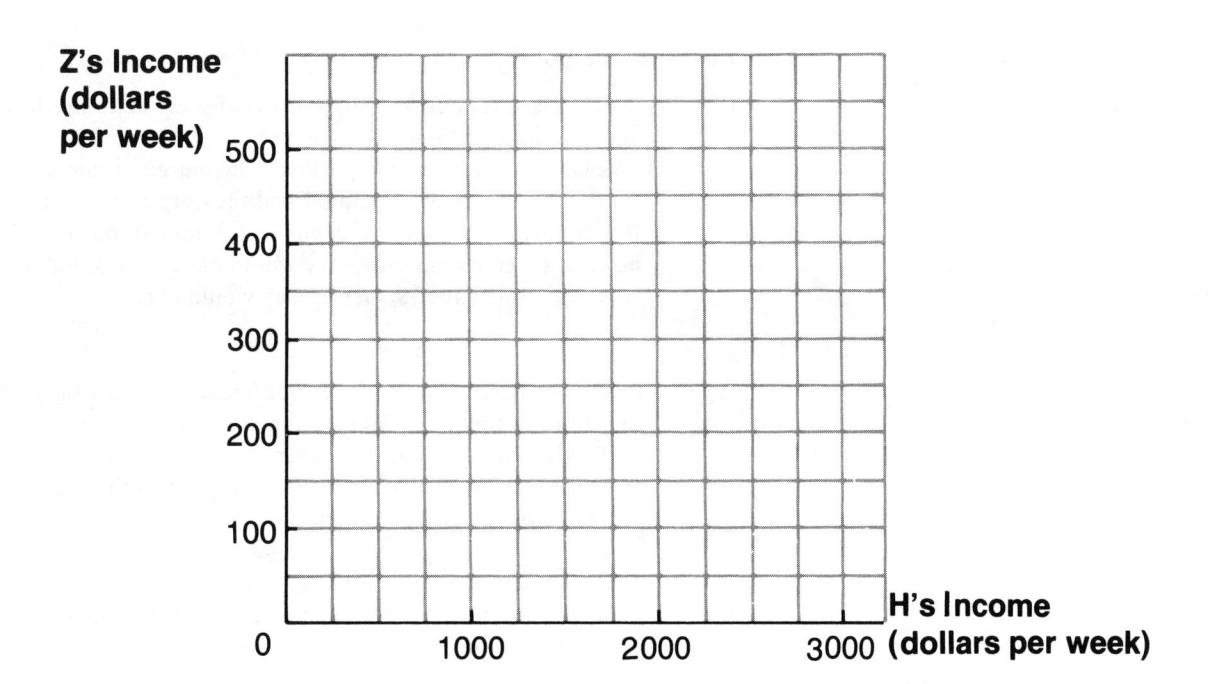

18. Overheard in an economics class: "The Lorenz curve of a country's personal income distribution is in effect a mirror image of that country's production-possibilities frontier." In the following graph, plot the U.S. production-possibilities frontier with the help of text Figure 14.14, "The Lorenz Curve."

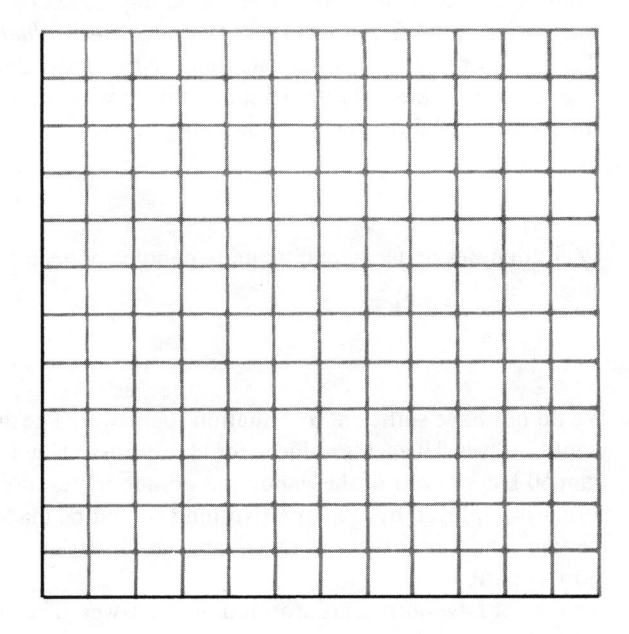

[Blank grid]

ANSWERS

Multiple-Choice Questions

1. c	2. b	3. a	4. c	5. b	6. d	7. d	8. a	9. c
10. d	11. a	12. c	13. d	14. b	15. b	16. c	17. b	18. d
19. a	20. c	21. b	22. c	23. d	24. a	25. a	26. c	27. c
28. b	29. a	30. b	31. c	32. d	33. c	34. b	35. d	36. b
37. a	38. b	39. c	40. d	41. a	42. c	43. c	44. d	45. b
46. a	47. a	48. b	49. d	50. d	51. a			

True-False Questions

1. F (Such efficiency exists whenever it is *impossible* to produce a larger welfare in these circumstances.)
2. F (We cannot tell unless we are given the *marginal* rate of transformation between capital and the output in question. If it equaled, say, 1 unit of capital for 2 units of output in both cases, economic efficiency would exist. If it differed between the producers, inefficiency would exist.)
3. T
4. T
5. F (The consumer would then be indifferent about giving up 1 unit of food for 3 units of clothing—or 2 units of food for 6 units of clothing. In fact, however, 3 units of food were lost in exchange for 6 units of extra clothing. Thus the consumer would be worse off.)
6. T
7. T
8. F (This is true only in the absence of price discrimination. See the answer to problem 10 below.)
9. T
10. F (It is impossible to do so without reducing the welfare of others but it is possible to do so at the expense of others.)
11. F (Different indifference curves of any one individual cannot touch, for that would denote—for the given combination of goods at the tangency point—two different associated utility totals at the same time. Yet there is nothing illogical about tangencies of indifference curves of *different individuals*. These points would only indicate that the two individuals had identical marginal rates of substitution, while consuming identical or different combinations of goods.)
12. T
13. T
14. T
15. F (Zero denotes perfect equality; unity denotes perfect inequality.)

Problems

1. **a.** We do not have sufficient information to answer. The answer would be "yes," if each producer could not possibly get more than 50 bicycles out of the (labor and nonlabor) resources actually employed by the firm. (Nothing should be made of the fact that A uses twice as much labor as B. Perhaps B uses more capital.)
   b. We do not have sufficient information to answer. The answer would be "yes," if the marginal rate of transformation between labor and bicycles were identical in both firms and equal, say, to $1L$ for $6B$. In that case, it would not help to reallocate labor between the firms. The firm losing $1L$ would lose $6B$; the firm gaining $1L$ would gain $6B$. Overall output would still equal 100 bicycles (consisting not of $50B + 50B$, but of $44B + 56B$).
   c. You haven't found any inefficiency.

2. **a.** If A produced 1 extra locomotive, it would have to cut truck production by 2. If B then produced 1 less locomotive, it could raise truck production by 4. Overall, the same locomotive output would be associated with 2 extra trucks.
   b. If A produced 1 extra locomotive, it would have to cut truck production by 2. If B then raised truck production by 2, it would have to cut locomotive production by 1/2. Overall, 1/2 extra locomotive would be associated with the same output of trucks. (Don't be bothered by *half* a locomotive. Production figures are flows per unit of time. Thus 1/2 locomotive per month, say, comes to 6 extra locomotives per year.)
   c. If A produced 1 extra locomotive, it would have to cut truck production by 2. If B then raised truck production by 3, it would have to cut locomotive production by 3/4. Overall, 1/4 extra locomotive would be associated with 1 extra truck.

3. If A cut sweater *production* by 3, 10 extra towels could be produced. Thus B would have to cut sweater *consumption* by 3, but could consume 10 extra towels. Given B's *MRS*, 5 extra towels would leave B equally well off; 10 extra towels, therefore, would make B better off.

4. **a.** By letting B consume more meat and A more vegetables (each being valued relatively more by the person just named). For example, A could give up 1 pound of meat for 1 pound of vegetables (and would then be equally well off, given A's *MRS*). Then B could consume 1 pound of extra meat and would have to give up 1 pound of vegetables (while being indifferent about giving up 3 pounds of vegetables). B would be better off. Note: If B gave up 3*V*, B would be equally well off, but A better off. If B gave up more than 1*V*, but less than 3*V*, both consumers would be better off after the exchange.
   b. In this case, total economic welfare could not be raised by any exchange between A and B. If an exchange occurred at the rate of 1*M* for 7*V*, both consumers would end up equally well off. Any other terms of trade would make one consumer better off, the other one worse off.

5. **a.** Both situations are economically efficient. Period. (Neither is more or less efficient than the other. What matters is the equality of the *MRS*.)
   b. There is no objective way of telling. It all depends on one's favorite criterion of equity.

6. **a.** In the market as a whole, an equilibrium price of 0*A* would be established for fertilizer. This price would become a given for all three farms pictured (and for thousands of others not shown). Note the horizontal line at the level of 0*A*. Each profit-maximizing farm would buy that quantity of fertilizer which equated the falling marginal value product of fertilizer with the fertilizer's price. (Recall the text discussion of Figure 11.1, "A Firm's Demand for Labor: The Simple Case," which can now be applied to fertilizer.) Thus Farm A would buy 0*B*, Farm B would buy 0*C*, and Farm C would buy nothing. Economic efficiency is implied by the fact that *BD* = *CE*: all actual users of fertilizer would end up with the identical mar-

ginal value product. Given an identical output price as well (given, that is, an identical price of asparagus for all perfectly competitive farms), this implies identical marginal physical products as well. (Recall that the marginal value product of any input equals its marginal physical product multiplied by product price.) Identical marginal physical products of an input used by different producers to make the same type of output denotes economic efficiency. (Recall text Figure 14.5, "Perfect Competition and Pareto's First Condition.")

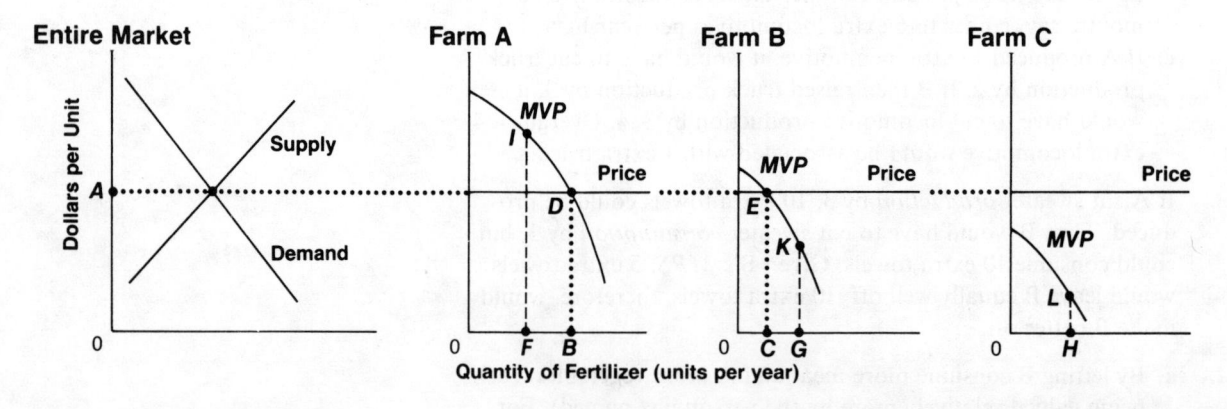

[Figure: Entire Market, Farm A, Farm B, Farm C diagrams showing Supply/Demand and MVP curves with Price line at level A; Farm A shows points I, D at quantities F, B; Farm B shows points E, K at quantities C, G; Farm C shows point L at quantity H. X-axis: Quantity of Fertilizer (units per year); Y-axis: Dollars per Unit]

   b. Suppose a central planner allocated identical amounts of fertilizer to all asparagus farms: $0F$ to Farm A, $0G$ to Farm B, $0H$ to Farm C. Then note how FI would differ from GK and HL. These differences imply economic inefficiency. An omniscient central planner could reallocate a unit of fertilizer from Farms B and C towards Farm A, losing output of lesser value in the former places (GK and HL, respectively) than would be gained in the latter (FI). This process could be continued until the solution noted in (a) above was achieved. Barring omniscience, a central planner would be unlikely to achieve it.

7. a. In the markets as a whole, equilibrium prices of $5 and $100, respectively, would be established for hats and coats. These prices would become givens for all three firms pictured (and for thousands of others not shown). Note the horizontal lines at the levels of $5 and $100. Each profit-maximizing firm would produce that quantity of each good which equated the rising marginal cost of producing the good with the good's price. (Recall, for example, the text discussion of Figure 5.2, "A Profitable Business.") Thus firm A would produce 8 hats

and 14 coats. Firm B would produce 12 hats and 8 coats. Firm C would produce 17 hats and 13 coats. Economic efficiency is implied by the fact that Firm A's marginal cost ratio a/b would equal Firm B's ratio c/d and Firm C's ratio e/f, all, of course, being equal to 5/100. This means that all firms would have a marginal rate of transformation of 20 hats for 1 coat or, if you prefer, 1 hat (costing \$5 of resources at the margin) for 1/20 coat (costing \$5 as well). Recall a similar discussion in text Figure 14.6, "Perfect Competition and Pareto's Second Condition."

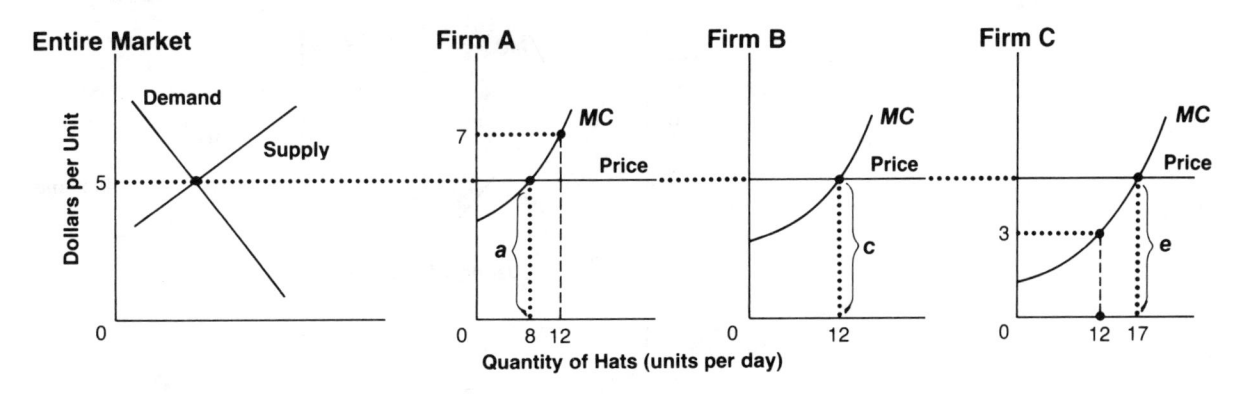

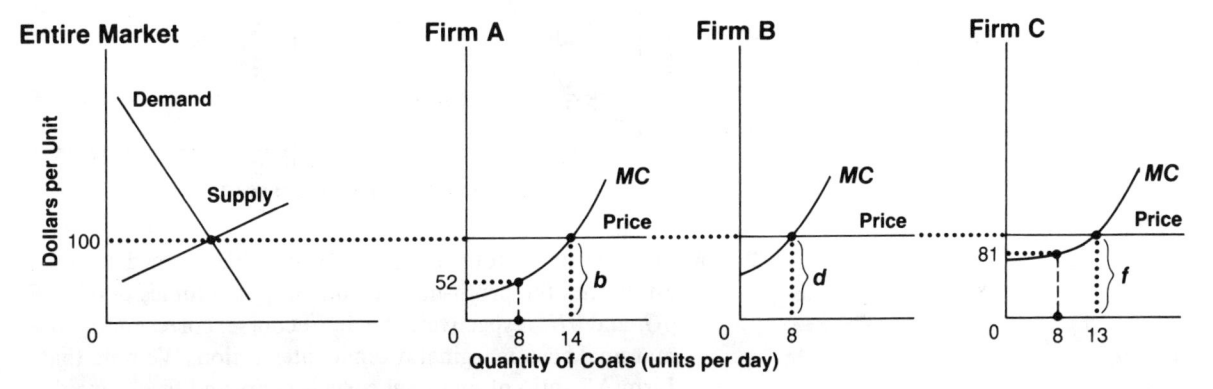

b. Suppose a central planner assigned identical output quotas of 12 hats plus 8 coats per day to all firms. Then note how the firms' marginal costs would differ: For hats, they would equal \$7, \$5, and \$3, respectively. For coats, they would equal \$52, \$100, and \$81, respectively. Thus the three firms' marginal cost ratios would equal 7/52, 5/100, and 3/81; this would mean marginal rates of transformation, respectively, of 7.4 hats or 20 hats or 27 hats for 1 coat. These differences imply economic inefficiency. An omniscient central planner could reassign output quotas in accordance with the solution noted in (a) above, but, barring omniscience, a central planner would be unlikely to achieve such economic efficiency.

8. Drawing in the four marginal-revenue lines immediately establishes the profit-maximizing output volumes of $0A$, $0B$, $0C$, $0D$, respectively. Each, of course, corresponds to a marginal cost-marginal revenue intersection. We note that Firm A's ratio of marginal costs for cars and trucks would equal a/b, clearly different from Firm B's ratio of c/d. Thus the two firms' marginal rates of transformation between cars and trucks would differ as well; economic inefficiency would exist.

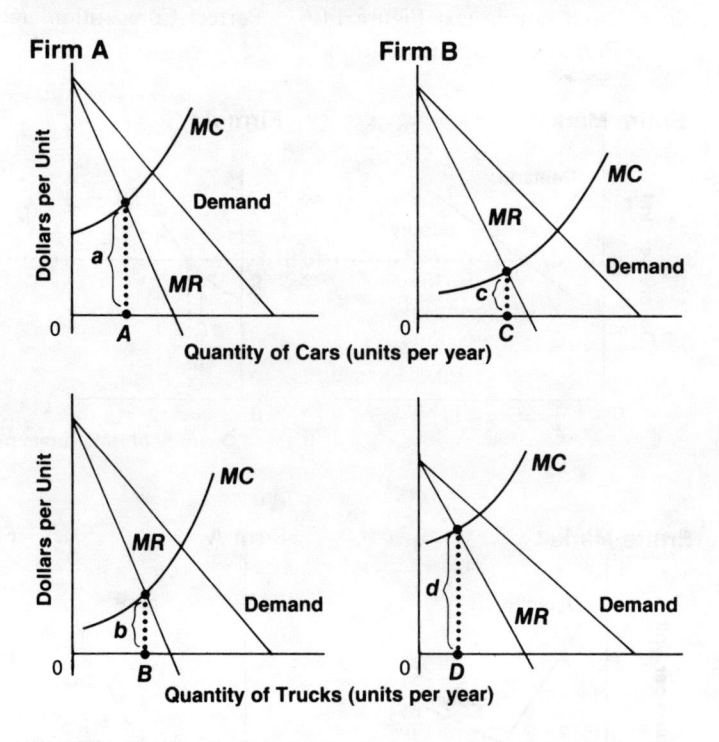

9. a. Drawing in the four marginal-revenue lines immediately establishes the profit-maximizing output volumes of $0E$, $0F$, $0G$, and $0H$, respectively. Each, of course, corresponds to a marginal cost/marginal revenue intersection. We note that Firm A's ratio of marginal costs for cars and trucks would equal e/f, exactly equal to Firm B's ratio of g/h. Thus the two firms' marginal rates of transformation between cars and trucks would be equal as well; economic efficiency would exist.

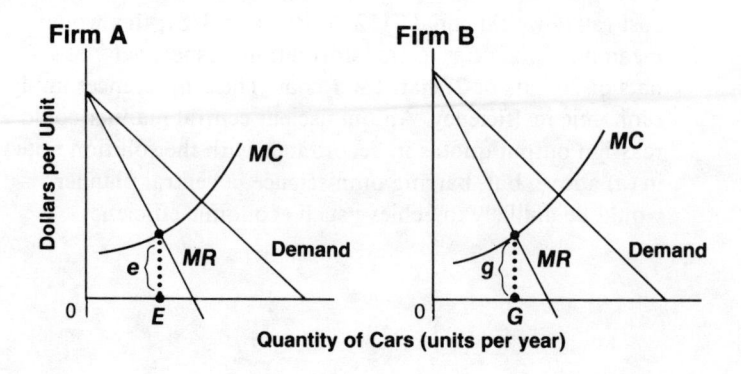

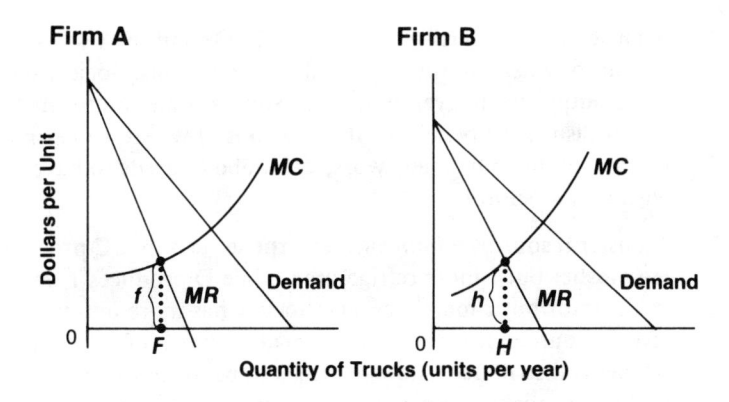

Firm A

[Graph: Dollars per Unit vs Quantity of Trucks (units per year); showing Demand, MR, MC curves with point F and bracket f]

Firm B

[Graph: showing Demand, MR, MC curves with point H and bracket h]

b. The demand and marginal-cost lines are sufficiently different to produce economic efficiency. Yet its achievement under imperfect competition is an accident, not a logical necessity.

10. a. From the information given, it is possible to construct Consumer A's budget line *ab* and Consumer B's line *cd*. Clearly, consumer A would maximize utility by purchasing 50 units of bread plus 5 units of meat per week—the combination corresponding to point *e*. At *e*, the consumer's marginal rate of substitution (equal to the slope of budget line and indifference curve) would be 10 units of bread for 1 unit of meat. Consumer B, on the other hand, would maximize utility by purchasing 100 units of bread plus 10 units of meat per week—the combination corresponding to point *f*. At *f*, the consumer's marginal rate of substitution would also equal 10 units of bread for 1 unit of meat. Hence efficiency would exist in the allocation of consumption goods.

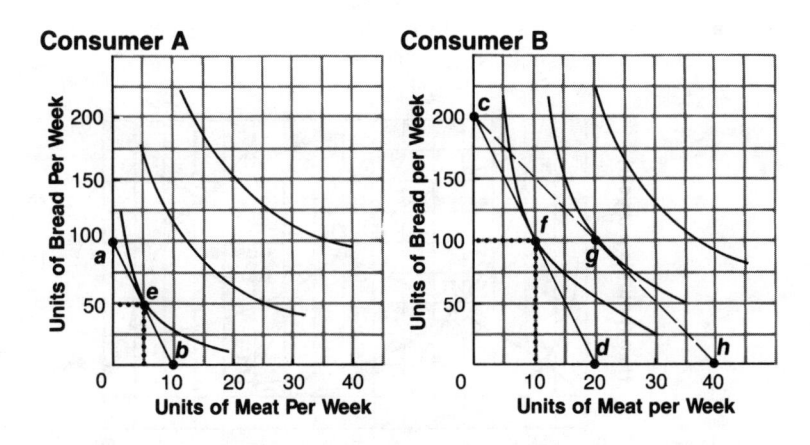

Consumer A

[Graph: Units of Bread Per Week vs Units of Meat Per Week; showing budget line ab with point e at (5 meat, 50 bread) and indifference curves]

Consumer B

[Graph: Units of Bread per Week vs Units of Meat per Week; showing budget line cd with point f at (10 meat, 100 bread), point g, and dashed line ch with point h, and indifference curves]

b. In that case, Consumer B's budget line would be dashed line *ch*. The consumer would purchase 100 units of bread plus 20 units of meat—combination *g*. Significantly, the marginal rate of substitution at *g* would equal 5 units of bread for 1 unit of meat, unlike that of consumer A at *e*. This would be inefficient. (Proof: If A now gave B, say, 7 units of bread for 1 unit of meat, both consumers would be better off. This is so because A could give up as many as 10 units of bread for 1 unit of meat and still be indifferent, while B would be indifferent when being paid as few as 5 units of bread for the sacrifice of 1 unit of meat.)

11. Transfer 1 ton of fertilizer from A to B. Output of A will fall by 30 tons of wheat, output of B will rise by 40 tons. Society has more output from given resources. Some people can be made better off (as the net benefit of 10 extra tons of wheat is distributed); nobody needs to be made worse off (nobody needs to have less wheat than before).

12. Transfer resources *within* each enterprise such that C produces 1 less washer but 1 more refrigerator, while D produces 1 less refrigerator but 2 more washers. Society has more output from given resources. Someone can be made better off (as the extra 1 washer is distributed), nobody needs to be made worse off (nobody needs to receive fewer washers or refrigerators).

13. Let E consume 1 pound less of meat, and let F consume 1 pound more. Let E consume 7 pounds more of fruit and let F consume 7 pounds less. As a result, E is better off (because the *MRS* of 1 pound of meat for 2 pounds of fruit indicates that E is *indifferent* about giving up 1 pound of meat and receiving 2 pounds of fruit. In fact, E receives 7 pounds of fruit). By analogous reasoning, F is equally well off. Society has squeezed greater welfare from given quantities of goods (differently distributed). Can you think of a still different rearrangement that would leave E as well as F better off? Can you think of a rearrangement that would leave E equally well off, and only F better off? (Try moving 1 pound of meat from E to F in exchange for 5 pounds of fruit or 2 pounds of fruit.) *Caution:* All marginal rates of transformation and substitution vary with circumstances. Do not regard them as eternal constants. We cannot assume that any of the reallocations discussed above can be repeated with equally beneficial results.

14. **a.** The contract curve connects the points of tangency of indifference curves, such as *A, B, C,* and *D*.

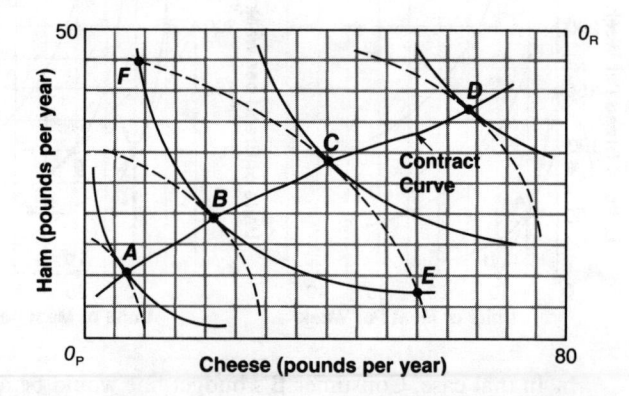

    b. *Efficient at* B: The indifference curves are tangent; thus the marginal rates of substitution (measured by the curves' slopes) are equal for the two customers. *Inefficient at* E: The indifference curves intersect; thus the marginal rates of substitution are not equal for the two consumers. An exchange of ham and cheese that yielded allocation *B* would leave Poor equally well off (*B* and *E* lie on the same indifference curve) but would greatly advance the welfare of Rich (*B* lies on a higher indifference curve than *C* for Rich). An exchange that yielded allocation *C* would leave Rich equally well off but would greatly advance the welfare of Poor. Any new allocation between B and C on the contract curve would make both persons better off at the same time.

c. All allocations within area $EBFC$, but not on the two indifference curves enclosing this area.

d. That is debatable. Poor would become much better off; Rich would become worse off. Lacking a means of measuring and comparing the two persons' utilities, we cannot predict the effect on social welfare.

15. a. Consider marginal-utility-of-income function ab for A-people. Then draw an identical one for B-people, such as cd. The lines intersect at I, indicating that the marginal utilities will be equalized only at identical incomes $0e = 0'e$. No other allocation of total income could yield a greater utility total than $0aIe$ (for A-people) plus $0'cIe$ (for B-people).

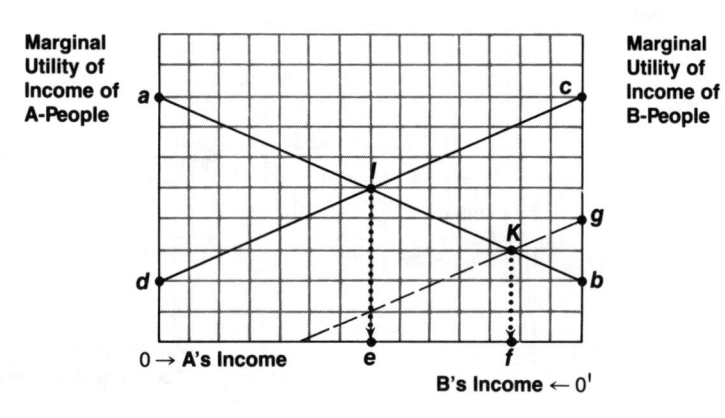

b. All else being equal, consider a new dashed function for B-people that is lower than that of A-people. The lines now intersect at some point to the right of I, such as K, indicating that the marginal utilities will be equalized only if A-people receive higher income $(0f)$ than B-people $(0'f)$. No other allocation of total income could yield a greater utility total than $0aKf$ (for A-people) plus $0'gKf$ (for B-people).

16. a. By cumulating the data of columns (2) and (3), data for the Lorenz curve can be derived as follows in the table. These data can then be graphed:

Percent of Lawyers in Class or Lower Ones	Percent of Total Lawyers' Income Received by Lawyers in Class or Lower Ones
(A) 3	1
(B) 13	5
(C) 43	25
(D) 100	100

b. You would have to measure areas a and b and take the ratio of a to $a + b$.

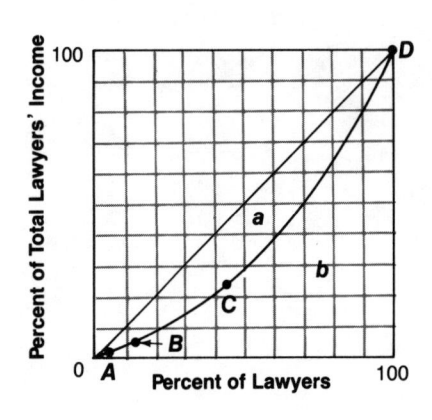

17. a. Both would have an income of $500 per week (point C). Note: The line of equal distribution is not a 45-degree line because the units on the graph's two axes are not identical.

Pre-Tax Income of H (3)	Tax Revenue and Transfer to Z (4)	After-Tax Income of H (5)
$3,000	(A) $ 0	$3,000
2,200	(B) 550	1,650
1,000	(C) 500	500
500	(D) 375	125
0	(E) 0	0

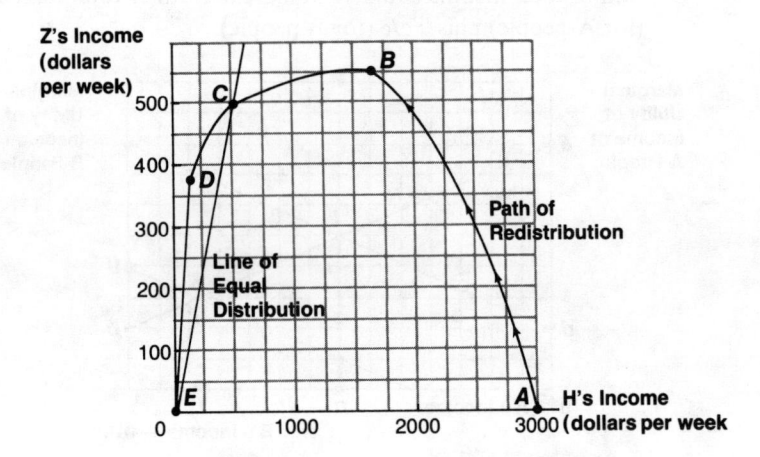

b. The situation would be inefficient. By lowering the tax rate from 50 percent to 25 percent, the total pie would expand such that both persons could be better off under inequality (point B).

18. You didn't try, did you? The quoted sentence is utter nonsense. You may wish to review the Glossary definition of *production-possiblities frontier* and *Lorenz curve*.

BIOGRAPHY 14.1

Vilfredo Pareto

Vilfredo Pareto (1848–1923) was born in Paris to an Italian father and a French mother. He studied in Italy, received a doctoral degree in engineering, and eventually became manager general of the Italian Iron Works. After years of industrial practice, he turned to economics. In 1893, he was appointed to succeed Léon Walras (see Biography 13.2) at the University of Lausanne, but he resigned that chair in 1906 and retired to Céligny on Lake Geneva. He devoted most of his later years to the study of sociology.

His major books on economics include *Cours d'Économie Politique,* vols. 1 and 2 (1896–97) and *Manuel d'Économie Politique* (1910). An Italian version of the latter work had appeared in 1906, and an English translation was published in 1971. Although he devoted no more than two decades of his long life to economics, Pareto has become one of the patron saints of the discipline. From its early beginnings, thinkers in the field had attempted to specify the meaning of social welfare and the kind of allocation of resources and goods that would maximize it. Pareto broke away decisively from traditional practice, most recently exemplified by Bentham (see Biography 3.1). Pareto rejected the cardinal measurement of utility and any interpersonal comparisons thereof. He restricted economic science to welfare comparisons that require only intrapersonal comparisons (in which affected individuals themselves testified to the direction of change in their welfare). He insisted that economic science should make pronouncements only about unambiguous changes in social welfare. If a reallocation of resources or goods left some individuals, in their own estimation, equally well off but others better off, social welfare had increased. If some felt equally well off but others worse off, social welfare had decreased. If some were better off and others worse off, the situation could not be evaluated by economic science—unless, that is, the gainers actually compensated the losers to the losers' full satisfaction and were still better off. Such a case would, of course, be indistinguishable from one in which no one was worse off, but some people were better off.

Additional Readings

Amoroso, Luigi. "Vilfredo Pareto." *Econometrica,* January 1938, pp. 1-21.
   An expanded story of Pareto's life and work.

Schumpeter, Joseph A. *Ten Great Economists: From Marx to Keynes.* New York: Oxford University Press, 1951, chap. 5 on Vilfredo Pareto.

Seligman, Ben B. *Main Currents in Modern Economics.* Chicago: Quadrangle, 1962, pp. 386-403.
   A critical review of the work of Pareto.

BIOGRAPHY 14.2

Friedrich A. von Hayek

Friedrich August von Hayek (1899-   ) was born and educated in Vienna. He began his career as director of the Austrian Institute for Economic Research and lecturer in economics at the University of Vienna. Starting in 1931 he served as professor first at the London School of Economics, then at the University of Chicago, and finally at the University of Freiburg in Germany. In 1974, while serving as visiting professor at the University of Salzburg in Austria, von Hayek was awarded the Nobel Memorial Prize in Economic Science (jointly with Sweden's Gunnar Myrdal). Von Hayek's greatest insight, perhaps, is that markets, above all else, are mechanisms for utilizing knowledge. He considered the question of what institutional arrangement could best enable large numbers of people—each possessing only bits of knowledge—to cooperate with each other so as to achieve the best use of resources. He rejected the notion that one could put at the disposal of some center all the knowledge that ought to be used but that was initially dispersed among many. The relevant knowledge is made up of elements of such number, diversity, and variety, he argued, that its explicit, conscious combination in a single mind is impossible. Yet the spontaneous interaction of people in free markets can bring about that which could be achieved by deliberate action only by someone possessing the combined knowledge of all. Consider his own words:[1]

> It is worth contemplating for a moment a very simple and commonplace instance of the action of the price system to see what precisely it accomplishes. Assume that somewhere in the world a new opportunity for the use of some raw material, say tin, has arisen, or that one of the sources of supply of tin has been eliminated. It does not matter for our purpose—and it is very significant that it does not matter—which of these two causes has made tin more scarce. All that the users of tin need to know is that some of the tin they used to consume is now more profitably employed elsewhere. . . . There is no need for the great majority of them even to know where the more urgent need has arisen. . . . If only some of them know directly of the new demand, and switch resources over to it, and if the people who are aware of the new gap thus created in turn fill it from still other sources, the effect will rapidly spread throughout the whole economic system and influence not only all the uses of tin, but also those of its substitutes and the substitutes of these substitutes, the supply of all the things made of tin, and their substitutes. . . .
>
> The most significant fact about this system is the economy of knowledge with which it operates. . . . In abbreviated form, by a kind of symbol, only the most essential information is passed on, and passed on only to those concerned. . . . The marvel is that in a case like that of a scarcity of one raw material, without an order being issued, without more than perhaps a handful of people knowing the cause, tens of thousands of people whose identity could not be ascertained by months of investigation, are made to use the material or its products more sparingly. . . .
>
> I have deliberately used the word "marvel" to shock the reader out of the complacency with which we often take the working of this mechanism for granted. I am convinced that if it were the result of deliberate human design, and if the people guided by the price changes understood that their decisions have significance far beyond their immediate aim, this mechanism would have been acclaimed as one of the greatest triumphs of the human mind. . . . But those who clamor for "conscious direction"—and who cannot believe that anything which has evolved without design (and even without our understanding it) should solve problems which we should not be able to solve consciously—should remember this: The problem is precisely how to extend the span of our utilization of resources beyond the span of the control of any one mind; and, therefore, how to dispense with the need of conscious control and how to provide inducements which will make the individuals do the desirable things without anyone having to tell them what to do.

Von Hayek is more than an economist. He is also an eminent political and legal theorist. He is convinced that markets do the best job of solving the problem of resource allocation, but only if they are free from any distortions introduced by ill-advised government. In a best-selling book, *The Road to Serfdom* (1944), von Hayek warns that the enthusiasm of governments for intervening in the market leads us

---

[1] Friedrich A. von Hayek, "The Use of Knowledge in Society," *The American Economic Review,* September 1945, pp. 519–30.

down a path that ends in central planning and totalitarianism. Government intervention will thus cause the end of the free society, humanity's highest social achievement.

Von Hayek's most recent books from the University of Chicago Press are magnificent statements of all of these themes: *The Constitution of Liberty* (1960); and *Law, Legislation, and Liberty,* vol. I, *Rules and Order* (1973), vol. II, *The Mirage of Social Justice* (1976), and vol. III, *The Political Order of a Free Society* (1979).

Additional Readings

Hayek, Friedrich A. *The Road to Serfdom.* Chicago: University of Chicago Press, 1944; *The Counter-Revolution of Science.* Glencoe: Free Press, 1952; *The Constitution of Liberty.* Chicago: University of Chicago Press, 1960; *Law, Legislation, and Liberty.* vols. 1–3. Chicago: University of Chicago Press, 1973–1979. *The Fatal Conceit: The Errors of Socialism.* Chicago: University of Chicago Press, 1989.

> Magnificent works, by the 1974 Nobel-Prize winner in economics, on the market economy, the centrally planned economy, and their implications for human freedom.

BIOGRAPHY 14.3

Karl Marx

Karl Marx (1818–1883) was born in Trier, Germany. He studied history, law, and philosophy at the universities of Bonn, Berlin, and Jena. He earned a doctoral degree in philosophy, then worked as a newspaper editor. His newspaper work brought trouble with the Prussian authorities and forced him to emigrate to France, then Belgium, and finally, England. In London, he was part-time correspondent for *The New York Daily Tribune* for a decade; his meager earnings were supplemented by Friedrich Engels, who owned factories in Germany and England and was a lifelong friend and collaborator. Marx spent much time in the British Museum studying and writing. Through his writings, he came to influence the thought of generations; today, over a third of the world's population lives in countries calling themselves Marxist. Many espouse his teachings with religious fervor.

The major works of Marx include: *Economic and Philosophical Manuscripts* (1844), *The Communist Manifesto* (1848), *The Grundrisse* (1857–58), *Theories of Surplus-Value* (1861–63), and *Capital: A Critique of Political Economy* (vols. 1–3, 1867–80, edited by Engels 1883–94).

Marx provided a grandiose vision of historical evolution. As he saw it, *economic conditions* (the ways in which resources are owned and used and newly produced goods are apportioned) shape people's attitudes and actions and, ultimately, history. Capitalism, for example, is characterized by the crucial fact that natural and capital resources are owned by a small minority of the population—the capitalist class, or *bourgeoisie*. The vast majority of people own only their bodies, and have only their labor to sell. They are the working class, or *proletariat*. By virtue of their economic position, and independent of individual volition, argued Marx, these classes are antagonistic to each other.

Inevitably, they struggle over the *economic surplus,* the difference between the total of goods produced and the portion needed to maintain and reproduce the capital and human resources who helped produce that total. To the extent that the bourgeoisie keeps the economic surplus, said Marx, it *exploits* the proletariat. This exploitation does not arise from individual circumstances, occasionally and accidentally, but from the logic of the capitalist system—unavoidably and independently of individual intention.

Equally unavoidable is revolution. Workers will expropriate the bourgeois expropriators and seize political power. A new era of *socialism* will be ushered in; workers will enjoy ownership of nonhuman resources and will be the masters of the productive process rather than its slaves. "Let the ruling classes tremble.... The proletarians have nothing to lose but their chains. They have a world to win. WORKING MEN OF ALL COUNTRIES, UNITE!"[1]

But socialism will still have a defect, said Marx: the attitudes of workers will still be influenced by their experiences in the old society. Thus workers cannot be expected to work without material incentives or harsh commands. Because they will still have to be rewarded according to their contribution to production (such as hours worked) there will still be income inequality. Eventually, though, socialism will turn into *communism,* which will have no defects at all. Communism will be an industrialized, classless, and nonexploitative society. Above all, the socialist transition stage will have produced a dramatic change in the outlook of people. A "new person" unlike the present one will emerge, who will contribute freely and gladly to the well-being of all, being neither coaxed by material incentives nor by bureaucratic commands. The abundance created by economic growth and the lack of egoism exhibited by the new type of human being will make possible a new principle of production and distribution: "From each according to his ability, to each according to his need."[2]

Additional Readings

Schumpeter, Joseph A. *Ten Great Economists: From Marx to Keynes.* New York: Oxford University Press, 1951, chap. 1.

A discussion of Marx and his work.

---

[1] Karl Marx and Friedrich Engels, *Manifesto of the Communist Party,* in Robert C. Tucker, ed., *The Marx-Engels Reader,* 2nd ed. (New York: W. W. Norton, 1978), p. 500.

[2] Karl Marx, *Critique of the Gotha Program,* in Robert C. Tucker, ed., *The Marx-Engels Reader,* 2nd ed. (New York: W. W. Norton, 1978), p. 531.

CHAPTER 15

Property Rights, Antitrust, and Regulation

MULTIPLE-CHOICE QUESTIONS

Circle the letter of the *one* answer that you think is correct or closest to correct.

1. Governmental policies in favor of competition in the market economy are carried out in order to
   a. disperse economic power.
   b. provide equal opportunities for all.
   c. escape the economic inefficiency and inequity of imperfect competition.
   d. achieve all of the above.

2. All of the following are devices used by private firms to replace competition with monopoly *except:*
   a. the horizontal merger.
   b. the natural monopoly.
   c. the holding company.
   d. the cartel.

3. Assume a corporation acquired all of its $100 million worth of assets by issuing $9 worth of nonvoting preferred stock for every $1 worth of common stock, while selling bonds worth $99 for every $1 worth of stock. An investor could control all of the corporation's assets by buying just over
   a. $50,000 worth of common stock.
   b. $100,000 worth of common stock.
   c. $450,000 worth of common stock.
   d. $50 million worth of the bonds.

4. The Sherman Act
   a. clearly outlawed certain market structures (such as monopoly), but not necessarily attempts to create them.
   b. clearly outlawed certain types of business conduct (such as attempts to establish monopoly), but not necessarily already existing monopolies.
   c. gave wide latitude of interpretation to the courts.
   d. is correctly described by all of the above.

5. The Rule of Reason
   a. was a U.S. Supreme Court interpretation of the intent of the Sherman Act.
   b. was a section of the Sherman Act itself (which outlawed unreasonable restraint of trade).
   c. was a section of the Sherman Act itself (which focused on a firm's *intent* to exercise monopoly power).
   d. is not correctly described by any of the above.

6. Among the first firms to be convicted of Sherman Act violations was
   a. Alcoa.
   b. American Tobacco.
   c. du Pont.
   d. Eastman Kodak.

7. The Clayton Act of 1914 is an antitrust law that
   a. outlaws consent agreements.
   b. outlaws vertical mergers.
   c. focuses on conduct rather than market structure.
   d. focuses on market structure rather than conduct.

8. An agreement by which sellers agree to lease or make a sale of commodities, but only on the condition that the lessee or purchaser thereof shall not use or deal in the commodity of a competitor, is called
   a. an exclusive contract.
   b. a requirements contract.
   c. a tying contract.
   d. a consent agreement.

9. The antitrust law that is most concerned with mergers is
   a. the Celler-Kefauver Act of 1950.
   b. the Wheeler-Lea Act of 1938.
   c. the Clayton Act of 1914.
   d. the Sherman Act of 1890.

10. The typically low fines handed down upon convicton in U.S. antitrust cases do not tell the whole story about punishment because the firms involved typically incur other and more substantial costs, such as
    a. lawyers fees.
    b. consumer brand switching due to injury to the firm's "image."
    c. sanctions by administrative government agencies.
    d. all of the above.

When answering questions 11-16, refer to the graph, on page 265, about a natural monopoly.

11. In the absence of government regulation, and in order to maximize profit, this telephone company would choose to produce an output
    a. equal to 0A.
    b. in excess of 0A.
    c. equal to 0B.
    d. in excess of 0B.

12. In the absence of government regulation, this firm would set a profit-maximizing price
    a. of 0F.
    b. above 0E.
    c. of 0E.
    d. of 0D.

13. In the absence of government regulation, this firm would contribute to a deadweight efficiency loss equal to
    a. GHK.
    b. GHCA.
    c. KHCA.
    d. none of the above.

14. In principle, the deadweight efficiency loss could be eliminated by forcing this firm to charge a price that is
    a. equal to 0D.
    b. higher than 0D, but lower than 0E.
    c. lower than 0D.
    d. none of the above.

15. Setting a price so as to eliminate the deadweight efficiency loss would
    a. raise the firm's profit.
    b. reduce the firm's profit to zero.
    c. drive this firm into bankruptcy.
    d. not affect this firm's profit.

16. If government regulators set a price so as to eliminate profit precisely, they would
    a. fail to eliminate the deadweight efficiency loss.
    b. also and precisely eliminate the deadweight efficiency loss.
    c. increase the deadweight efficiency loss.
    d. achieve both (a) and (c).

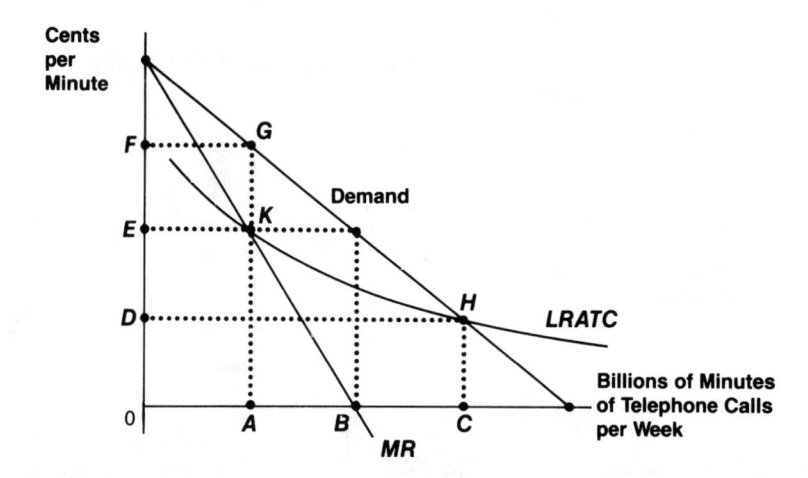

17. Which of the following comes closest to explaining why government regulators of a natural monopoly may permit it to engage in price discrimination?
    a. Such a firm's marginal cost lies below average total cost.
    b. Such a firm is bound to make losses at all output levels if it charges a uniform price.
    c. Such a firm would make excessive profit otherwise.
    d. The price elasticity of demand for such a firm's output is less than |1|.

18. Government regulators of natural monopolies can be relied upon to determine the rate base of such firms as
    a. the dollars ever received by the firms through the sale of common stock.
    b. the original cost of the firms' assets minus depreciation thereon.
    c. the current replacement cost of the firms' assets, given their present conditions.
    d. none of the above.

19. Under regulatory lag,
    a. the X-inefficiency of regulated firms is likely to be enhanced.
    b. the dynamic efficiency of regulated firms is likely to be enhanced.

c. the owners of well-managed regulated firms can easily make more than a "fair" return on their investment in times of inflation.
d. all of the above are true.

20. The owners of regulated firms may well undertake new investments in the firm, even if these investments do not increase output, as long as the "fair" rate of return guaranteed by regulators
a. falls short of the cost of capital.
b. falls short of the rate base.
c. exceeds the cost of capital.
d. exceeds the rate base.

When answering questions 21–22, refer to the following graph of a competitive market.

21. If government regulators set a below-equilibrium price of 0A, there would occur
a. a fall in supply.
b. a rise in demand.
c. a shortage.
d. all of the above.

22. The action noted in the previous question would also produce a deadweight welfare loss of
a. *EHI*.
b. *EDI*.
c. *DIKF*.
d. *CED*.

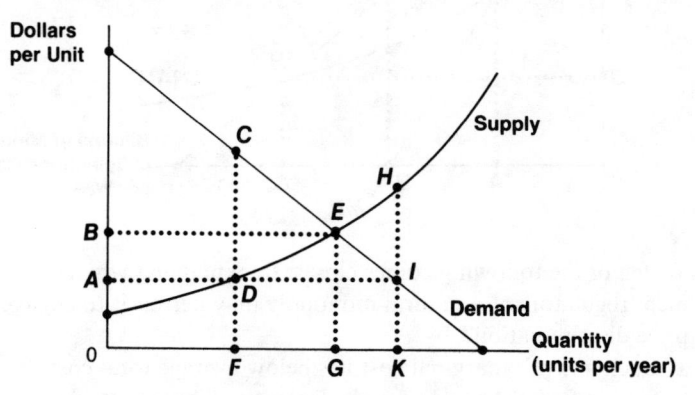

23. The list of government regulatory agencies that have sprung up since 1970 includes
a. the Food and Drug Administration.
b. the Consumer Products Safety Commission.
c. the Federal Communications Commission.
d. the Federal Power Commission.

24. In recent years, some economists have
a. seriously questioned the worth of many types of government regulation (because their benefits are low or nonexistent, while their costs are high or unnecessary).
b. had second thoughts on theoretical grounds about the desirability of regulating natural monopolies (the theory of the second best).
c. shown that, in situations in which one or more Pareto conditions cannot be satisfied, economic efficiency is not served by satisfying as many of the other ones as possible.
d. done all of the above.

25. The high costs of government regulation include
   a. the direct costs of administering the regulatory agencies.
   b. the direct costs of compliance by the affected parties.
   c. such indirect costs as reductions in X-efficiency, economic efficiency, and dynamic efficiency.
   d. all of the above.

TRUE-FALSE QUESTIONS

In each space below, write a *T* if the statement is true and an *F* if the statement is false.

_____ **1.** Holding companies typically hold no productive assets at all.

_____ **2.** U.S. courts have consistently held that pricing behavior that is consciously parallel, but involves no direct communication among the firms in question, is illegal *per se* under the Sherman Act.

_____ **3.** The Clayton Act outlaws price discrimination *per se;* that is, regardless of the reasons for it.

_____ **4.** Critics of U.S. antitrust policy would like to see more stress placed on controlling business conduct rather than market structure.

_____ **5.** A combination of regulatory lag plus inflation is likely to spell high economic profits for regulated firms.

_____ **6.** The theory of the second best deals a fatal blow to any policy that attempts to approach economic efficiency by bringing price in line with marginal cost for as many goods as possible and doing so in a piecemeal fashion.

_____ **7.** The theory of the third best favors general policies (such as promoting free market entry and exit) over specific policies (such as bringing prices in line with marginal costs).

_____ **8.** Redistributing income through regulation usually involves the taking of several dollars from some people in order to give one dollar to favored recipients.

_____ **9.** In recent years, the total direct cost of government regulation came to about $500 per year for every person in the United States.

_____ **10.** Deregulation during the early 1980s has drastically reduced the number of regulatory agencies (as well as the size of regulatory budgets.)

PROBLEMS

1. Consider the demand and long-run average-total-cost conditions of the natural monopoly depicted in the following graph.

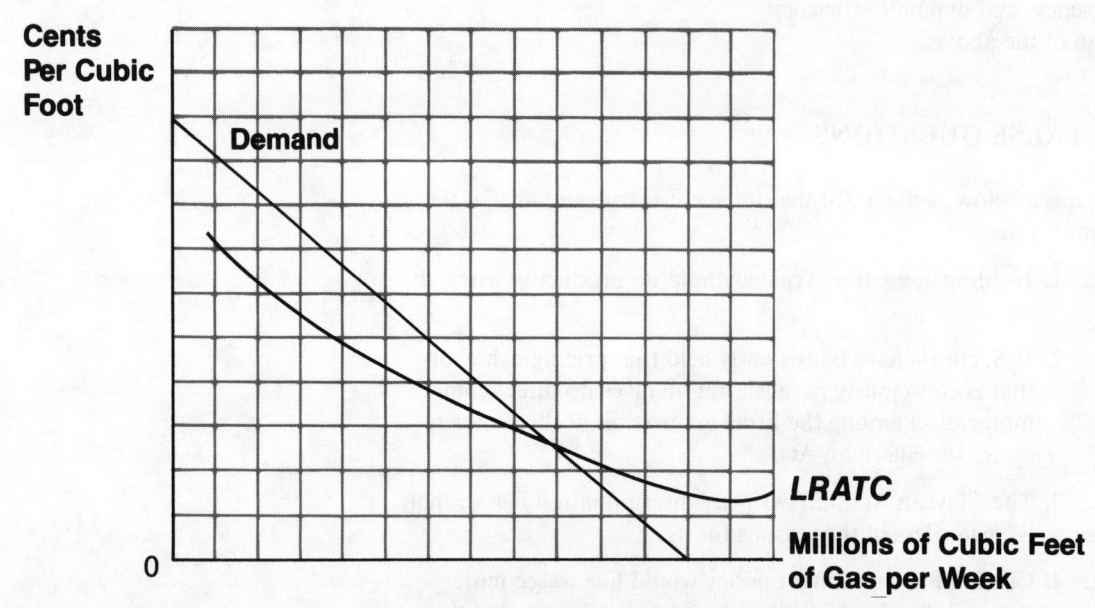

a. Draw in logically possible lines of long-run marginal cost and marginal revenue.
b. Determine how much the firm would produce if it were unregulated.
c. Which price would it then charge?
d. Which would be the associated deadweight efficiency loss?
e. Which would be its economic profit?
f. What would happen to (d) and (e) in the short run, if a government regulator insisted on the production of output salable at a price equal to marginal cost? In the long run?

2. If you noticed any long-run problem in 1(f) above, name at least three ways to escape it.

3. Consider the demand and cost conditions of the competitive industry depicted in the following graph.

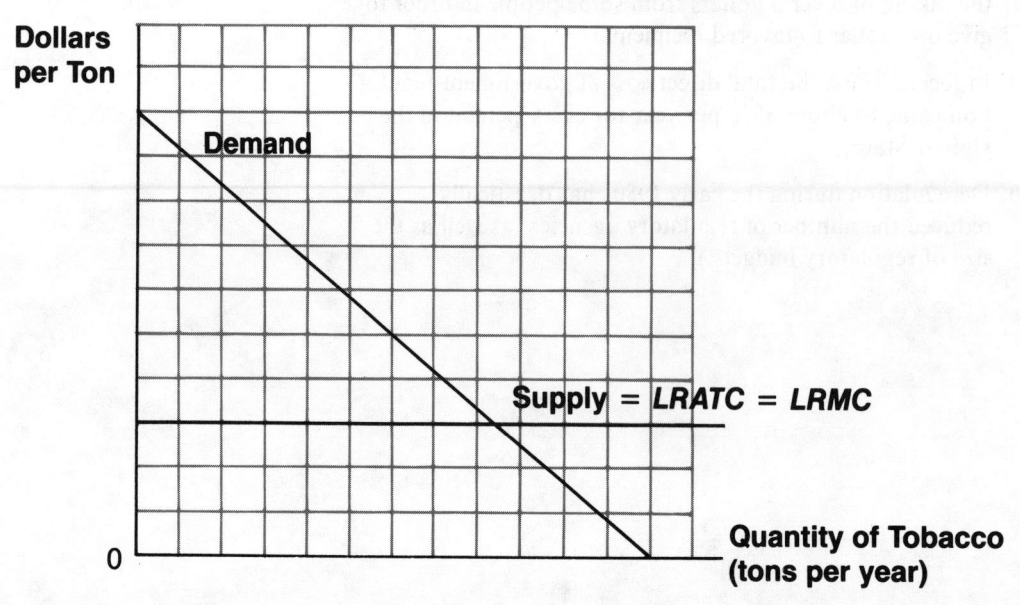

a. Imagine you were a government regulator. In the graph, draw a minimum above-equilibrium price that you might set and indicate the maximum quantity you could let firms supply in order to make the price stick (while avoiding surpluses or mandated purchases).
b. Indicate the extent of the resultant deadweight welfare loss.
c. Indicate the monopoly profit you would have created.
d. Indicate who would eventually get the monopoly profit.

ANSWERS

Multiple-Choice Questions

1. d   2. b   3. a   4. c   5. a   6. b   7. c   8. a   9. a
10. d   11. b   12. b   13. d   14. c   15. c   16. a   17. a   18. d
19. b   20. c   21. c   22. d   23. b   24. d   25. d

True-False Questions

1. T
2. F (The courts have been quite inconsistent.)
3. F (There is no *per se* rule; the text indicates many exceptions.)
4. F (The opposite is the case.)
5. F (It is likely to spell a lower than "fair" rate of return on the owners' investment, because unregulated input prices and costs rise, while regulated selling prices and revenues lag.)
6. T
7. T
8. T
9. T
10. F (The number of agencies and their budgets have continued to rise.)

Problems

1. a. As long as long-run average total cost is falling, long-run marginal cost must lie below it. *LRMC* must equal *LRATC* at the latter's minimum, then rise above it. Marginal revenue for the straight demand line given here must lie halfway between demand and the vertical axis at any given price.

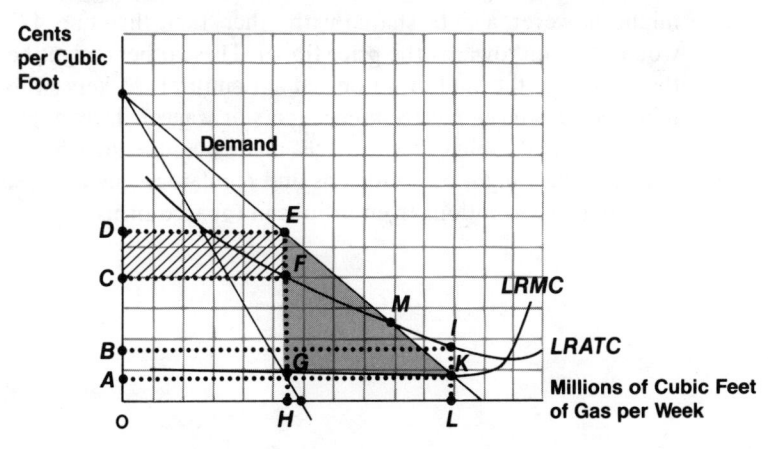

b. The firm would produce $0H$ million cubic feet of gas per week, because this is the quantity corresponding to the marginal cost and marginal revenue intersection at G.
   c. The firm would charge a price of $HE = 0D$ cents per cubic foot, because only at this price would people demand the profit-maximizing output quantity $0H$.
   d. The associated deadweight efficiency loss would equal EKG (shaded).
   e. The firm's economic profit would equal $CDEF$ (crosshatched).
   f. Output would then equal $0L$, corresponding to intersection K. In the short run, the deadweight welfare loss would be zero; the profit would turn into a loss equal to $ABIK$. In the long run, the firm would go bankrupt.

2. Bankruptcy could be escaped by
   a. not regulating the natural monopoly at all.
   b. regulating it but allowing it to produce the output salable at a price equal to average total cost (point M on the graph).
   c. regulating it but allowing it to engage in price discrimination so as to make total revenue equal the total cost $0BIL$ of producing $0L$.

3. **a.** If you set a minimum price of $0B$, firms could not be allowed to supply more than $0E$ tons of tobacco per year.

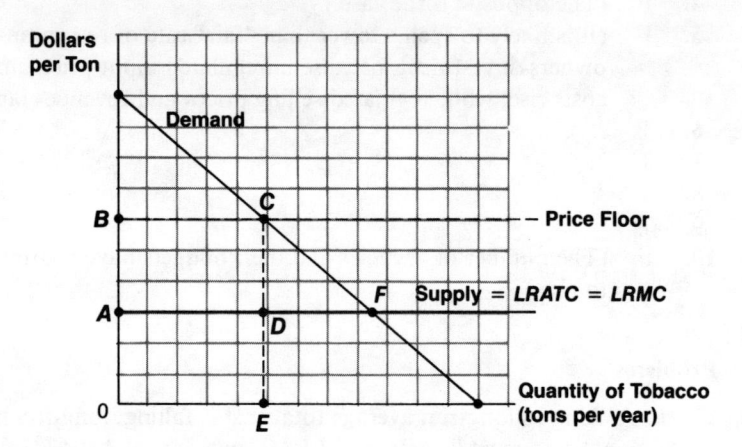

   b. There would exist a deadweight welfare loss of CDF.
   c. There would exist a monopoly profit of $ABCD$.
   d. The profit might go to the owners of tobacco-growing firms. It might, however, also be shared with others (and then line AF would drift up towards the price floor). These others might be the owners of the land or unionized agricultural workers (who might, respectively, receive higher rents or wages). They might be lawyers and lobbyists whose fees made possible the whole scheme. They might be politicians and regulators who received campaign contributions from or jobs in tobacco firms.

BIOGRAPHY 15.1

George J. Stigler

George Joseph Stigler (1911–   ) was born in Renton, Washington. He studied at the University of Washington, Northwestern, and the University of Chicago. After teaching at Iowa State, the University of Minnesota, and Columbia, he returned to a distinguished career at the University of Chicago. He served in many professional capacities, including the Attorney General's Commission for the Study of Antitrust Laws, the presidency of the American Economic Association, and the editorship of the *Journal of Political Economy*.

Among his many writings are *Production and Distribution Theories: The Formative Period* (1941), *Domestic Servants in the United States: 1900–1940* (1946), *Five Lectures on Economic Problems* (1950), *The Demand and Supply of Scientific Personnel* (1957), *The Theory of Price* (1962), *Capital and Rates of Return in Manufacturing Industries* (1963), *The Intellectual and the Market Place, and Other Essays* (1963), *Essays in the History of Economics* (1965), *The Organization of Industry* (1968), *The Behavior of Industrial Prices* (1970, with James Kindahl), and *The Citizen and the State: Essays on Regulation* (1975).

In 1982, "for his studies of industrial structures, functioning of markets, and the effects of public regulation," Stigler was awarded the Nobel Memorial Prize for Economic Science.

One theme pervades all his work: his defense of competitive markets. Stigler's research indicates that competitive forces are strong even in our present economy, that they can be kept strong by a moderate amount of antitrust action, and that government regulation, procured by industries wishing to escape competition, is a serious threat to the market economy. He has, therefore, long argued for a reduction in the power of such agencies as the CAB and ICC and against the notion that government can do almost anything, if it really tries. According to Stigler:

> Our faith in the power of the state is a matter of desire rather than demonstration. When the state undertakes to achieve a goal, and fails, we cannot bring ourselves to abandon the goal, nor do we seek alternative means of achieving it, for who is more powerful than a sovereign state? We demand, then, increased efforts of the state, tacitly assuming that where there is a will, there is a governmental way. Yet . . . the sovereign state is not omnipotent. . . .
>
>   1. The state cannot do anything quickly . . . : (Deliberation is intrinsic to large organizations: not only does absolute power corrupt absolutely, it delays fantastically) . . .
>
>   2. When the national state performs detailed economic tasks, the responsible political authorities cannot possibly control the manner in which they are performed, whether directly by governmental agencies or indirectly by regulation of private enterprise. (The lack of control is due to the impossiblity of the central authority either to know or to alter the details of a large enterprise) . . .

3. The democratic state strives to treat all citizens in the same manner; individual differences are ignored if remotely possible. (The striving for uniformity is partly due to a desire for equality of treatment, but much more to a desire for administrative simplicity. Thus men with a salary of $100,000 must belong to the Social Security System; professors . . . must take a literacy test to vote; . . . the same subsidy per bale of cotton must be given to the hillbilly with two acres and the river valley baron with 5,000 acres. We ought to call him Uncle Same.) . . .

4. The ideal public policy, from the viewpoint of the state, is one with identifiable beneficiaries, each of whom is helped appreciably, at the cost of many unidentifiable persons, none of whom is hurt much. (The preference for a well-defined set of beneficiaries has a solid basis in the desire for votes.) . . .

5. The state never knows when to quit. (One great invention of a private enterprise system is bankruptcy, an institution for putting an eventual stop to costly failure. No such institution has yet been conceived of in the political process, and an unsuccessful policy has no inherent termination. Indeed, political rewards are more closely proportioned to failure than to success, for failure demonstrates the need for larger appropriations and more power.)[1]

---

[1] George J. Stigler, *A Dialogue on the Proper Economic Role of the State*, Selected Paper No. 7 (University of Chicago: Graduate School of Business, 1963).

CHAPTER 16

Externalities and Public Goods

MULTIPLE-CHOICE QUESTIONS

Circle the letter of the *one* answer that you think is correct or closest to correct.

1. Real externalities are commonly divided into
   a. negative vs. positive ones.
   b. pecuniary vs. nonpecuniary ones.
   c. spillovers vs. neighborhood effects.
   d. all of the above.

2. Beneficial real externalities are the benefit some people provide for bystanders who are not being charged for this favor; these benefits can take the form of
   a. higher prices for things sold.
   b. lower prices for things bought.
   c. increased utility or output.
   d. any of the above.

3. Examples of negative real externalities include
   a. air and water pollution.
   b. crowding on freeways.
   c. crowding on ocean fisheries.
   d. all of the above.

When answering questions 4–10, refer to the graph, on page 274, of a perfectly competitive paint industry. Assume that the demand line correctly reflects the marginal social benefit of paint production.

4. In the absence of government intervention, this industry would
   a. produce $0F$ and charge a price of $0B$.
   b. produce $0I$ and charge a price of $0A$.
   c. produce $0I$ and charge a price of IG.
   d. do none of the above.

5. In the absence of government intervention, the marginal external cost of paint production
   a. cannot be determined from the above information.
   b. would equal CE.
   c. would equal GI.
   d. would equal GH.

6. In the absence of government intervention, the social net benefit from paint production and consumption would be smaller than its possible maximum by the amount of
   a. CGH.
   b. CEH.
   c. $CHIF$.
   d. $EHIF$.

7. Given the situation described by the correct answers to questions 4–6, Pigou would recommend
   a. a per-unit tax of *GH* on paint producers.
   b. a per-unit tax of *CE* on paint producers.
   c. a lump-sum tax of *CGH* on paint producers.
   d. none of the above.

8. According to Pigou, one consequence of his recommended policy (described in question 7) would be
   a. a higher quantity produced and a lower price for consumers.
   b. a lower quantity produced and a lower price for consumers.
   c. a lower quantity produced and a higher price for consumers.
   d. none of the above.

9. If it were easily possible for paint producers and consumers to get together with outsiders who were injured by paint production, they might negotiate a mutually advantageous reduction in paint output from the actual to the socially optimal level, because such a reduction would decrease damages to outsiders by
   a. *FCGI*, while lowering the paint industry producer surplus by only *FEHI* and lowering the consumer surplus by only *ECH*.
   b. *ECGH*, while lowering the paint industry producer surplus by only *EDH* and lowering the consumer surplus by only *DCH*.
   c. *FEHI*, while raising the paint industry producer surplus by *ECH* and raising the consumer surplus by *CGH*.
   d. *FDHI*, while lowering the paint industry producer and consumer surplus by less than *FDHI*.

10. The private deal described in question 9 would not take place, of course, if the transactions costs involved exceeded the potential net gain of
    a. *CEH*.
    b. *DCH*.
    c. *CGH*.
    d. *FCGI*.

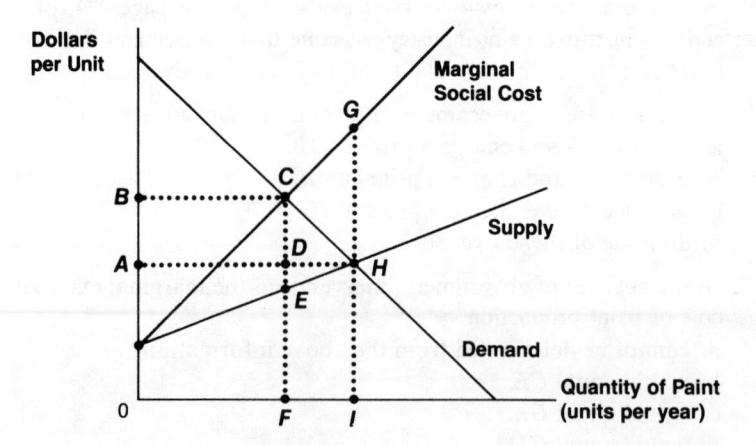

When answering questions 11–17, refer to the graph, on page 275, of a perfectly competitive flower-seed industry. Assume that the supply line correctly reflects the marginal social cost of flower-seed production.

11. In the absence of government intervention, this industry would ignore the obviously
    a. positive externality in production.
    b. positive externality in consumption.
    c. negative externality in production.
    d. negative externality in consumption.

12. In the absence of government intervention, there would exist
    a. a marginal external benefit of flower-seed production of *IK*.
    b. a marginal external benefit of flower-seed consumption of *GI*.
    c. a marginal external benefit of flower-seed consumption of *DE*.
    d. none of the above.

13. In the absence of government intervention, the social net benefit from flower-seed production and consumption would be smaller than its possible maximum by the amount of
    a. *ACE*.
    b. *CDE*.
    c. *EGI*.
    d. *DEG*.

14. Given the situation described by the correct answers to questions 11–13, Pigou would recommend
    a. a per-unit subsidy of *GI* for flower-seed consumers.
    b. a per-unit subsidy of *DE* for flower-seed producers.
    c. a per-unit tax of *DE* on flower-seed producers.
    d. none of the above.

15. According to Pigou, one consequence of his recommended policy (described in question 14) would be
    a. a larger quantity produced and a lower price for producers.
    b. a larger quantity produced and a higher price for producers.
    c. a lower quantity produced and a lower price for producers.
    d. an equality between marginal internal benefit and marginal external benefit.

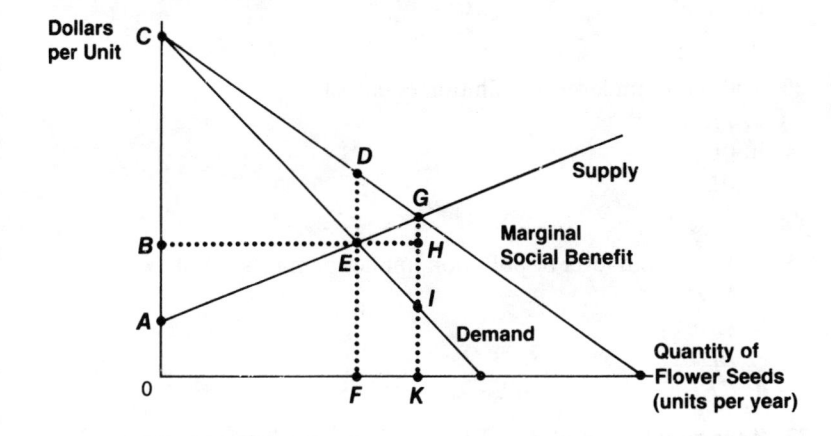

16. If it were easily possible for flower-seed producers and consumers to get together with outsiders who were benefited by flower growing, they might negotiate a mutually advantageous increase in flower-seed output from the actual to the socially optimal level, because such an increase would provide benefits to outsiders of
    a. 0BEF, while damaging flower-seed producers and consumers by less than 0BEF.
    b. FEIK, while providing additional benefits to flower-seed producers of EHI and additional benefits to flower-seed consumers of EGH.
    c. FEGK, while damaging flower-seed producers by only EGH and damaging flower-seed consumers by only EHI.
    d. EDGI, while damaging flower-seed producers by only EGH and damaging flower-seed consumers by only EHI.

17. The private deal described in the previous question would not take place, of course, if the transactions costs involved exceeded the potential net gain of
    a. DEG.
    b. FEIK.
    c. FEGK.
    d. ABE.

When answering questions 18–24, refer to the graph on page 277.

18. The line labeled *MSB* clearly represents the marginal social benefit of
    a. producing goods.
    b. abating wastes.
    c. dumping wastes.
    d. none of the above.

19. The total benefit (in terms of other goods not sacrificed for the sake of treating wastes) of dumping 0E wastes equals
    a. 0BDE.
    b. 0BDC.
    c. EDH.
    d. CDE.

20. The optimum level of pollution equals
    a. zero.
    b. 0C.
    c. 0E.
    d. 0H.

21. The optimum level of pollution implies damages equal to
    a. CDE.
    b. zero.
    c. EDH.
    d. CFH.

22. If the supply of pollution opportunities coincided with the horizontal axis, society's net benefit from dumping wastes would equal
    a. 0BDE minus EDFH.
    b. 0BDC minus DFH.
    c. 0BDC.
    d. DFH.

23. Government could induce a reduction of unlimited dumping to the optimum amount by imposing
    a. a lump-sum tax of DFH on polluters.
    b. a lump-sum tax of $EDFH$ on polluters.
    c. a tax of HF per unit of wastes dumped.
    d. a tax of $0A$ per unit of wastes dumped.

24. The taxing scheme referred to in the previous question would produce the optimum amount of dumping because
    a. $0BDE$ exceeds $0ADE$.
    b. $EDGH$ exceeds EDH.
    c. of neither (a) nor (b).
    d. of both (a) and (b).

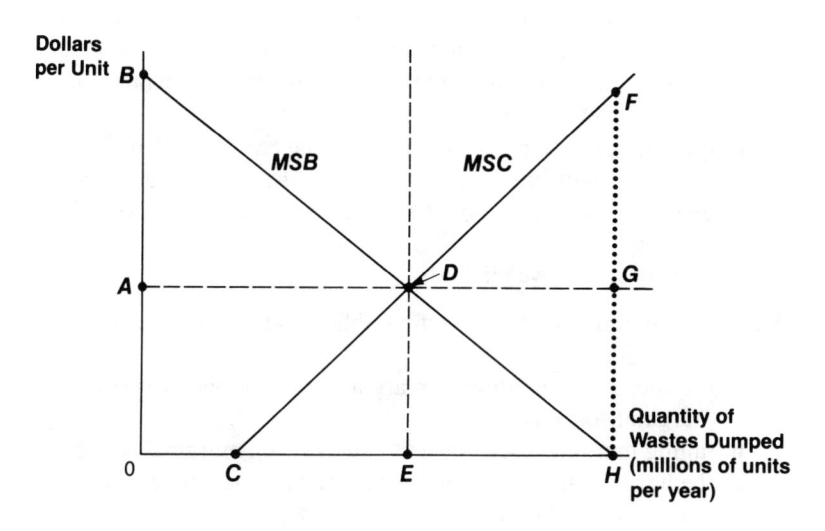

25. The concept of a pure public good is most closely associated with the name of
    a. Ronald Coase.
    b. Alfred Marshall.
    c. A. C. Pigou.
    d. Paul Samuelson.

26. A pure public good is
    a. any good the production or consumption of which imposes unwanted costs on outside parties or enables them to snatch free benefits.
    b. any good that provides nonexcludable and nonrival benefits to all people in a given society.
    c. any good that imposes nonexcludable and nonrival costs on all people in a given society.
    d. all of the above.

27. Private producers are unlikely to produce pure public goods because
    a. their benefits will be instantly and equally available to all.
    b. it will be next to impossible to collect payment from the benefited parties.
    c. of (a) and (b).
    d. any goods they produce are private goods by definition.

28. Examples of pure public goods include
    a. national defense.
    b. public education.
    c. public sanitation services.
    d. all of the above.

29. The marginal social cost of a good
    a. reflects the opportunity cost of producing it.
    b. reflects the minimum value of other goods that must be forgone to produce it.
    c. is correctly described by (a) and (b).
    d. always equals the good's marginal social benefit.

30. The market-demand line for a pure private good is derived by
    a. the vertical summation of individual consumers' marginal private benefit lines at each possible quantity.
    b. the horizontal summation of individual consumers' marginal private benefit lines at each possible price.
    c. the horizontal summation of individual consumers' marginal private benefit lines at each price plus the vertical addition of other consumers' marginal external benefits at each possible quantity.
    d. none of the above methods.

31. The optimum amount of a pure public good is that quantity at which
    a. the vertical summation of marginal external benefits equals marginal social cost.
    b. falling marginal social benefit equals rising marginal social cost.
    c. the horizontal summation of marginal internal benefits equals marginal social cost.
    d. all of the above are true.

When answering questions 32–36, refer to the graph on page 279.

32. The optimum amount of the good in question
    a. would be zero, if it were a private good.
    b. would be $0F$, if it were a public good.
    c. would be $0E$, regardless of whether it was a private or public good.
    d. cannot be determined from the given information.

33. If the good were a pure public good, the net benefit derived from its production and consumption in the absence of government intervention would be
    a. ACD.
    b. BCD.
    c. ABD.
    d. zero.

34. An omniscient and benevolent government, which sought to maximize society's net benefit from this good, would see to it that
    a. the total benefit was $0CF$.
    b. the total benefit was $0CDE$.
    c. the total cost was $0BDE$.
    d. the total cost was zero.

35. If the good were a pure private good and its optimum quantity were produced, we would know that each consumer's marginal private benefit equaled
    a. ABD.
    b. 0E.
    c. 0B.
    d. zero.

36. If the good were a pure public good and its optimum quantity were produced, we would know that each consumer's marginal private benefit equaled
    a. BCD.
    b. ABD.
    c. 0B.
    d. less than 0B.

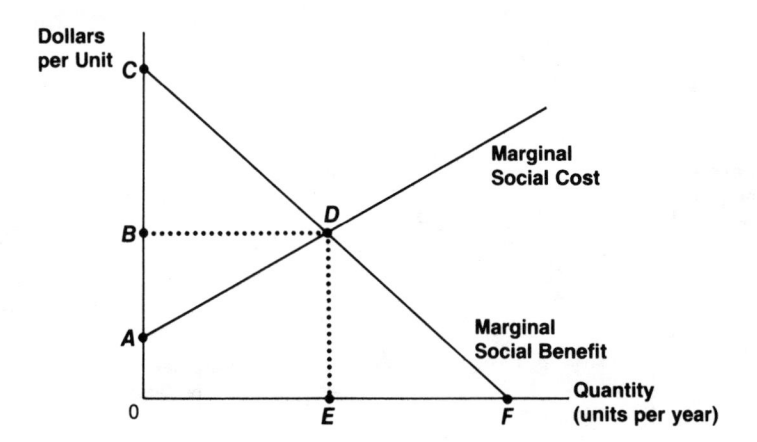

37. The desire to tax consumers of pure public goods in accordance with the marginal private benefits derived by them from the consumption of such goods necessitates
    a. the imposition of different taxes on different consumers.
    b. the provision of different amounts of such goods for different consumers.
    c. both (a) and (b).
    d. neither (a) nor (b).

When answering questions 38–42, refer to the graph on page 280.

38. Government officials who were omniscient and benevolent (who had the information depicted by this graph and wanted to maximize society's net benefit from pure public goods) would assure the provision of public goods in the amount of
    a. 0L.
    b. 0K.
    c. 0F.
    d. less than 0F.

39. The private market, in contrast, would provide pure public goods equal to
    a. 0K.
    b. 0F.
    c. less than 0F, but greater than zero.
    d. zero.

40. In the presence of internalities, it would not be surprising if government officials provided public goods equal to
    a. 0K.
    b. 0F.
    c. less than 0F, but greater than zero.
    d. zero.

41. In that case, society's net benefit from public goods would equal
    a. 0CEF minus 0AEF.
    b. ACE minus EHI.
    c. BCD.
    d. zero.

42. If X-inefficiency raised marginal social cost to dashed line BDG, government officials who provided society with 0K of pure public goods would be giving people a net benefit of
    a. 0CIK minus 0BGK.
    b. BDC minus DGI.
    c. either (a) or (b).
    d. EDGH.

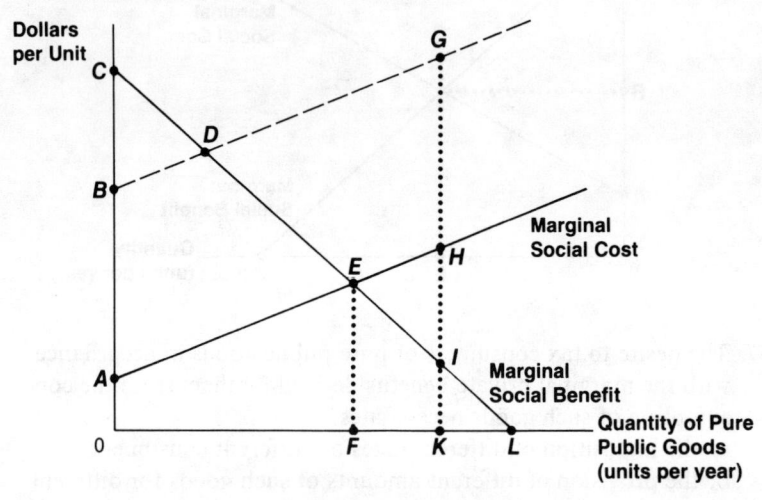

43. The performance of government bureaus might be improved by all of the following *except*
    a. benefit-cost analysis.
    b. transitivity analysis.
    c. competition among government bureaus.
    d. replacing government production with government provision.

44. When benefit-cost analysis is applied to governmental projects,
    a. benefits and costs that are spread out over time are discounted to the present.
    b. the future values of streams of benefits and costs are compared.
    c. projects with positive net present values are rejected.
    d. all of the above occur.

45. When a benefit-cost ratio of 1.827 has been calculated for a public project, we know
    a. that the public project provides public benefits that exceed the forgone private ones by 82.7 percent.
    b. very little, unless we can trust the underlying estimates of benefits and costs.
    c. very little, unless we can trust the selection of the discount rate used in the calculations.
    d. (a), given the conditions noted in (b) and (c).

46. X-inefficiency in government may be effectively countered by
   a. government decentralization.
   b. government consolidation.
   c. government production (as well as provision) of public goods.
   d. all of the above.

TRUE-FALSE QUESTIONS

In each space below, write a *T* if the statement is true and an *F* if the statement is false.

_____ 1. Nonpecuniary externalities are externalities involving households rather than firms.

_____ 2. Positive real externalities are likely to occur when people plant trees, keep bees, and bury high-tension wires.

_____ 3. Marginal social benefits minus marginal private benefit equals marginal internal benefit.

_____ 4. According to Pigou, maximum economic welfare in society requires an equality of all activities' marginal internal benefits and marginal external costs.

_____ 5. In cases where the marginal social cost of production and consumption exceeded the marginal private cost, Pigou recommended a per-unit tax on producers equal to the *MSC-MPC* gap at the optimal output volume.

_____ 6. In cases where the marginal social benefit of production and consumption fell short of the marginal private benefit, Pigou recommended a per-unit tax on consumers equal to the *MSB-MPB* gap at the pretax output volume.

_____ 7. In cases where the marginal social cost of production and consumption fell short of the marginal private cost, Pigou recommended a per-unit subsidy for producers equal to the *MSC-MPC* gap at the optimal output volume.

_____ 8. In cases where the marginal social benefit of production and consumption exceeded the marginal private benefit, Pigou recommended a per-unit subsidy for consumers equal to the *MSB-MPB* gap at the optimal output volume.

_____ 9. When producers or consumers provide free benefits to others, competitive markets allow production and consumption to fall short of the amount at which marginal social cost just equals marginal social benefit.

_____ 10. If a producer had to make damage payments equal to the marginal external costs imposed on outside parties, the producer would view marginal social cost in the same way as marginal private benefit was viewed before.

_____ 11. When the producer or consumer of a good alone bears all of the cost and enjoys all of the benefit associated with it, the good is called a *pure private good*.

_____ 12. A *local* public good is a self-contradictory term.

_____ 13. It is technically impossible or extremely costly to exclude any individual from the enjoyment of a pure public good.

_____ 14. The unwillingness of individuals voluntarily to help cover the cost of a pure public good (and their eagerness to let others produce the good so they can enjoy its benefits at a zero cost) is called the *indivisibility problem.*

_____ 15. While we know that when two consumers consume 3 and 6 units, respectively, of a pure private good, 9 units of it are being consumed, such a conclusion would be unthinkable for a pure public good.

_____ 16. The equation $MSC = MPB_1 + MPB_2 + \ldots MPB_n$ identifies the optimum quantity of a pure private good.

_____ 17. Economists are fairly confident that government can overcome the bad effects of market failure (such as the failure to produce optimal amounts of pure public goods).

_____ 18. Government failure refers to the fact that attempts by real-world, less-than-omniscient governments to raise economic welfare may well lower it instead.

_____ 19. The "iron triangle" tends to promote insufficient budgets for pure public goods.

PROBLEMS

1. Consider the four perfectly competitive markets pictured in the following graphs and use them to answer questions (a)–(h) graphically.

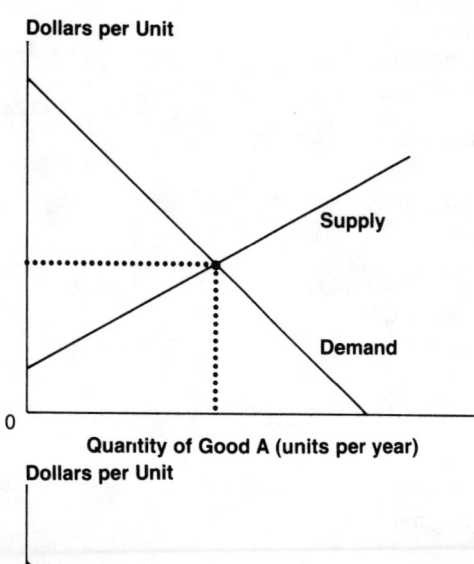

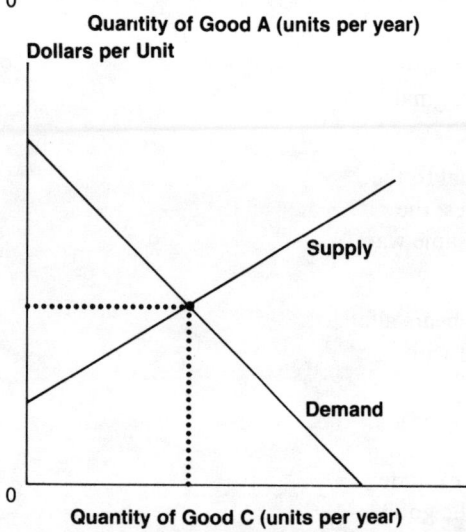

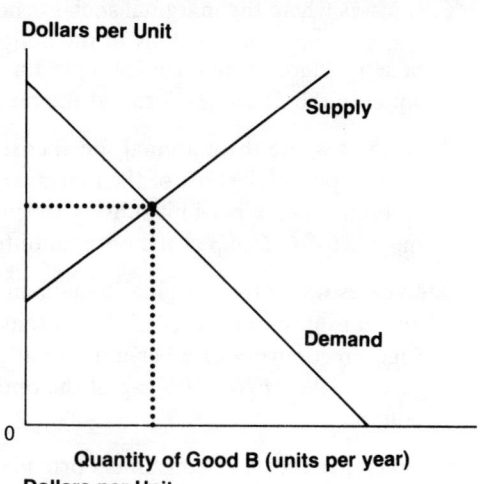

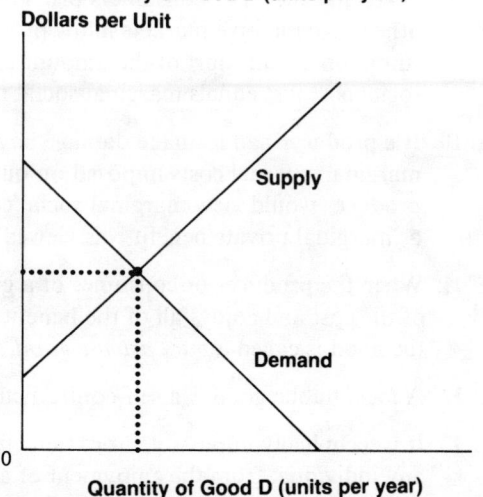

a. Show why social welfare would not be maximized if the producers of good A imposed a negative externality on other producers.
   b. Show how Pigou would have us deal with problem (a).
   c. Show why social welfare would not be maximized if the producers of good B bestowed a positive externality on other producers.
   d. Show how Pigou would have us deal with problem (c).
   e. Show why social welfare would not be maximized if the consumers of good C imposed a negative externality on other consumers.
   f. Show how Pigou would have us deal with problem (e).
   g. Show why social welfare would not be maximized if the consumers of good D bestowed a positive externality on other consumers.
   h. Show how Pigou would have us deal with problem (g).

2. Reconsider your answers to 1(b), (d), (f), and (h) and determine in each case the solution Coase would predict (in the absence of excessive transactions costs).

3. Consider the following graph and determine:
   a. the optimum level of pollution.
   b. society's total benefit, total cost, and net benefit from polluting at the optimal level.
   c. society's total benefit, total cost, and net benefit from polluting at the actual level likely to occur if dumping were unrestricted.
   d. the best taxing scheme that would convert (c) to (b). How much tax would the government collect?
   e. the best subsidy scheme that would convert (c) to (b). How much subsidy would the government have to pay?
   f. the best pollution-rights-market scheme that would convert (c) to (b). How much revenue would the government collect from selling pollution rights?

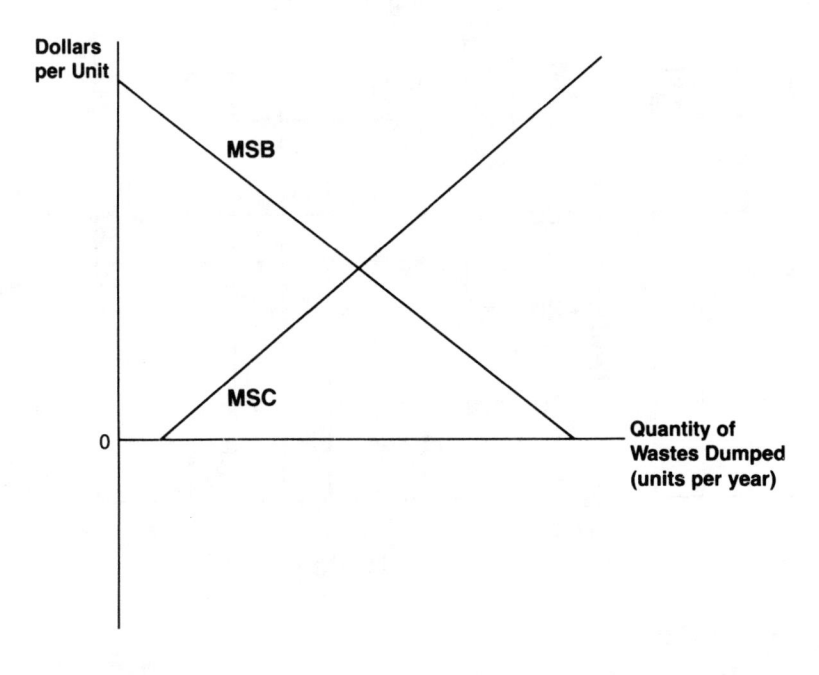

*4. Imagine three firms emitted the following wastes per year to avoid the following (constant) waste treatment costs: Firm A: 50 units ($1 per unit); Firm B: 500 units ($10 per unit); and Firm C: 450 units ($5 per unit).
   a. If the government wanted to cut waste emissions in half and issued uniform emission standards for all polluters, what would be the cost of achieving the cleaner environment?
   b. What would be the lowest possible cost, and how might it be achieved by emission standards?
   c. How might this lowest cost be achieved with the help of a pollution-rights market? What would be the equilibrium price for a pollution right?

*5. Consider the information in the table.
   a. If it cost 80 pounds of fish to run a boat for a day (fuel, wages and the like being paid in fish), how many vessels would go out fishing? Explain.
   b. What would the optimum number of boats, assuming society wanted to maximize the total daily catch *regardless of cost*? Explain the discrepancy to (a) above.
   c. What would be the optimum number of boats for a profit-maximizing private owner of the Lake Superior fishery? (*Hint:* It would be the same number that would maximize the daily *net* catch.)
   d. Show how a "congestion tax" on fishing boats under free access to the fishery could bring about the result noted in (c).

Fishing Vessels on Lake Superior	Average Catch (in pounds per day)
1	200
2	180
3	160
4	140
5	120
6	100
7	80
8	60

6. Consider the marginal private benefits received by Consumer A and B for various units of a good that are depicted in the following graphs.

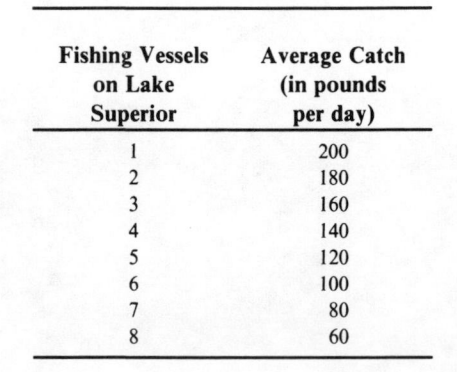

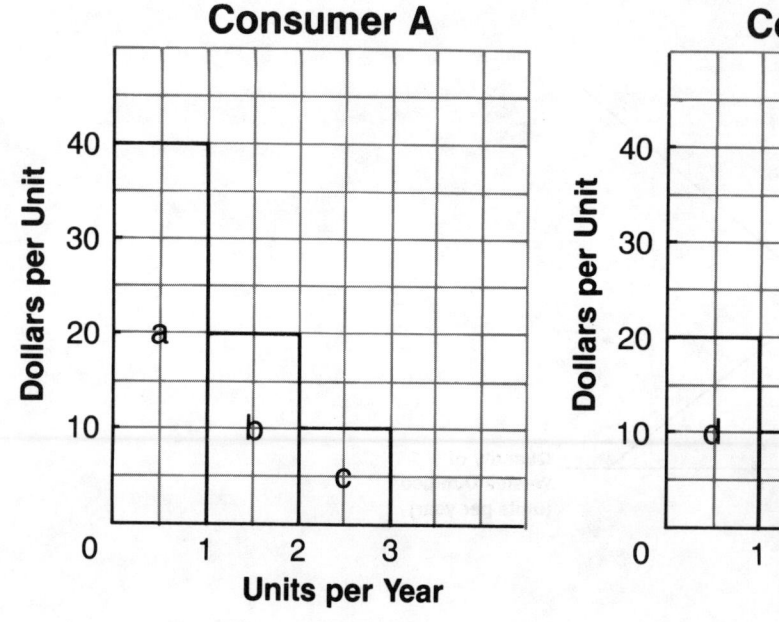

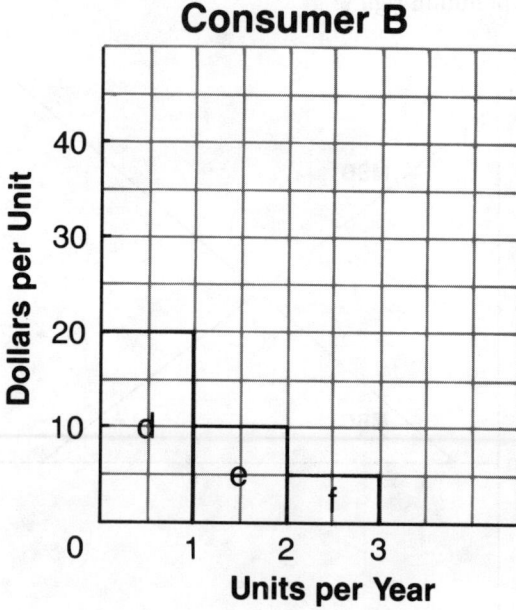

In the space provided, plot:
   a. the market demand line for the good in question if it were a pure private good.

---

*These *are* difficult questions; they cover material not directly explained in the text.

b. the pseudo-market-demand line for the good in question if it were a pure public good (and explain why the term *pseudo-market-demand* is used).

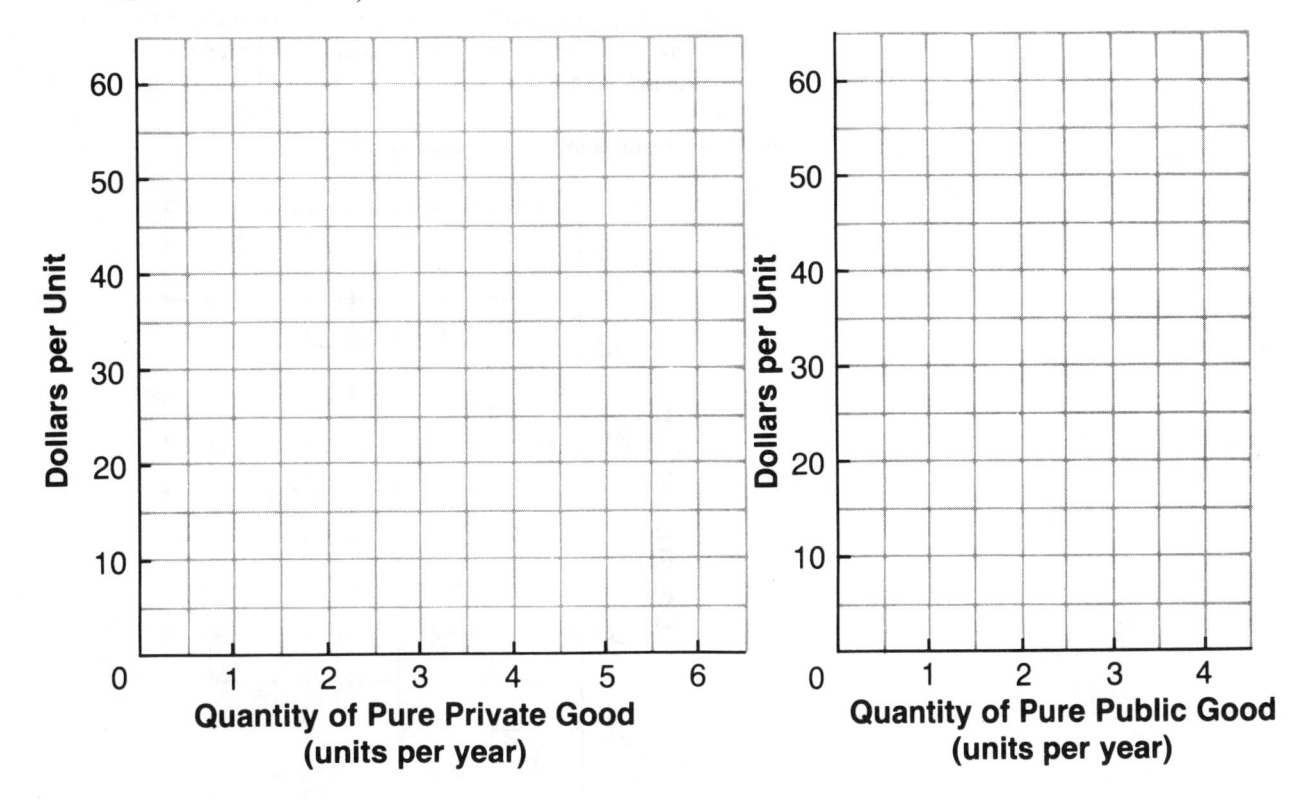

7. Plot your pseudo-market-demand line from 6(b) above as the marginal social benefit line in the following graph.
   a. Determine the optimum quantity of the pure public good.
   b. Identify the society's net benefit from the optimal quantity.

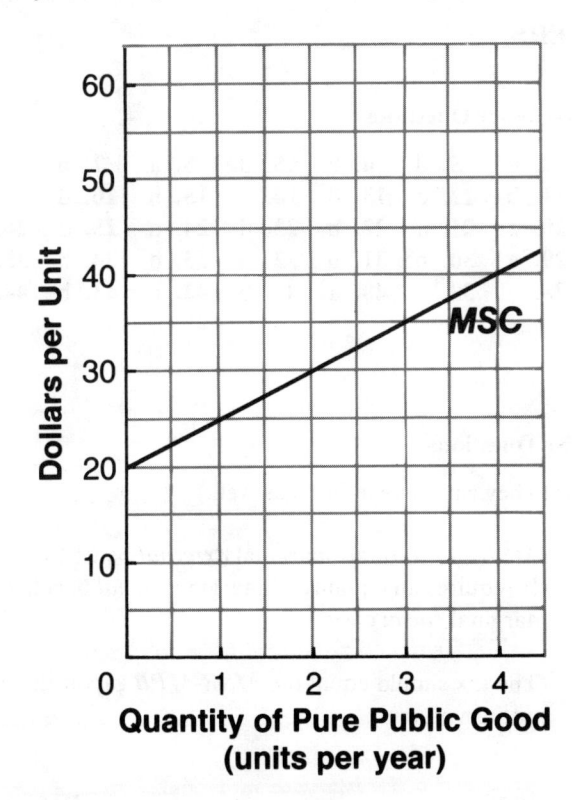

8. Consider the accompanying graph. Once more plot your pseudo-market-demand line from 6(b) as the line of marginal social benefit.
   a. Assume that government officials 1. produce an excess amount of the pure public good 2. at an excessive cost (above the *MSC* line printed in the graph). Show graphically that such government action and failure may still be preferable to no government action.
   b. Is this result a logically necessary one?

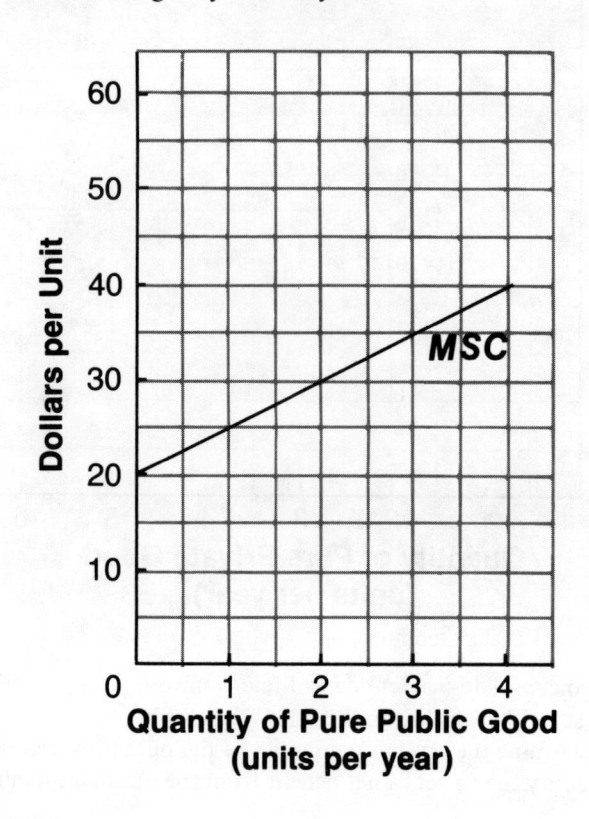

ANSWERS

Multiple-Choice Questions

1. a	2. c	3. d	4. b	5. d	6. a	7. b	8. c	9. b
10. c	11. b	12. c	13. d	14. a	15. b	16. d	17. a	18. c
19. a	20. c	21. a	22. b	23. d	24. d	25. d	26. b	27. c
28. a	29. a	30. b	31. b	32. c	33. d	34. b	35. c	36. d
37. a	38. c	39. d	40. a	41. b	42. c	43. b	44. a	45. d
46. a								

True-False Questions

1. F (They can involve firms as well.)
2. T
3. F (*MSB-MPB* equals marginal *external* benefit.)
4. F (It requires an equality of marginal *social* benefits and marginal *social* costs.)
5. T
6. F (The tax should equal the *MSB-MPB* gap at the *optimal* output volume.)

7. T
8. T
9. T
10. F (The producer would view it as marginal private *cost* was viewed before.)
11. T
12. F (The size of the society to which the term *public good* applies is arbitrary.)
13. T
14. F (This unwillingness is called the *freerider* problem.)
15. T
16. F (The equation identifies the optimum quantity of a pure *public* good.)
17. F (While the private market would produce zero amounts of pure public goods, government may well produce excessive amounts of them and may do so at an excessive cost. As a result, society may be burdened with a *negative* net benefit, as in panel (c) of text Figure 19.3, "Internalities," which is worse than the zero net benefit under market failure.)
18. T
19. F (It tends to promote *overgenerous* budgets for pure public goods.)

Problems

1. **a.** In that case, marginal social cost, *MSC,* would lie above the supply line (representing marginal private cost). Assuming demand reflected marginal social benefit as well as marginal private benefit, the ideal output quantity would correspond to intersection *b,* not *a.* The ideal quantity would maximize the social net benefit from producing and consuming good A at *cbd.* By producing too much, the perfectly competitive market would reduce this maximum potential net benefit by *abe* (shaded).

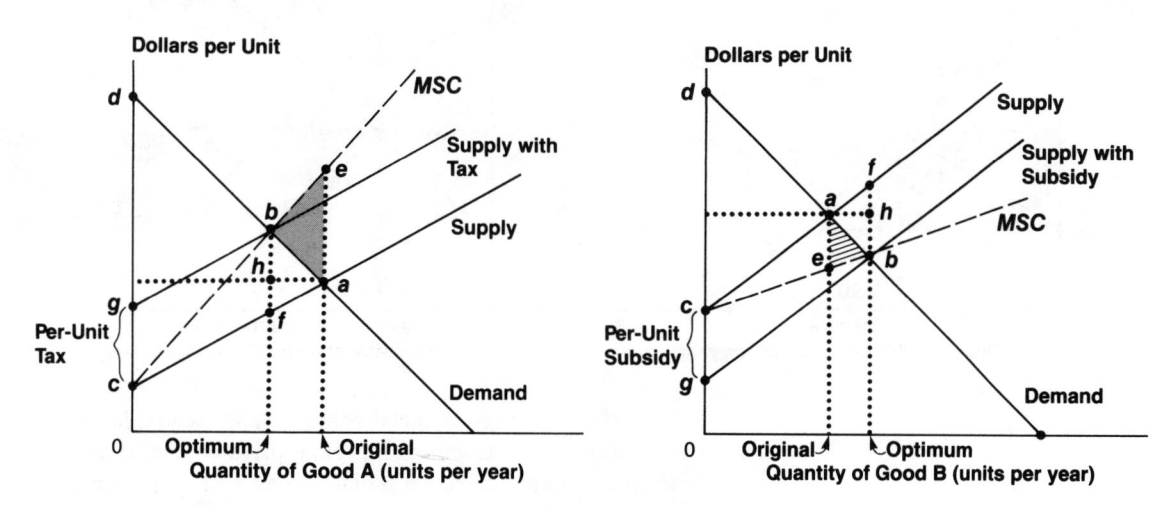

b. Pigou would have us impose a per-unit tax of *bf* = *cg* on producers (equal to the divergence between marginal social cost and marginal private cost at the optimal output volume). This would effectively cause a parallel upward shift of the supply line until it passed through *b* and the optimal quantity was produced.

c. In that case, marginal social cost, *MSC,* would lie below the supply line (representing marginal private cost). Assuming demand reflected marginal social benefit as well as marginal private benefit, the ideal output quantity would correspond to intersection *b,* not *a.* The ideal quantity would maximize the social net benefit from producing and consuming good B at *cbd.* By producing too little, the perfectly competitive market would fail to capture the crosshatched portion *(abe)* of this maximum potential net benefit.

d. Pigou would have us offer a per-unit subsidy of *bf = cg* to producers (equal to the divergence between marginal social cost and marginal private cost at the optimal output volume). This would effectively cause a parallel downward shift of the supply line until it passed through *b* and the optimal quantity was produced.

e. In that case, marginal social benefit *MSB,* would lie below the demand line (representing marginal private benefit). Assuming supply reflected marginal social cost as well as marginal private cost, the ideal output quantity would correspond to intersection *b,* not *a.* The ideal quantity would maximize the social net benefit from producing and consuming good C at *cbd.* By producing too much, the perfectly competitive market would reduce this maximum potential net benefit by *abe* (shaded).

f. Pigou would have us impose a per-unit tax of *bf = dg* on consumers (equal to the divergence between marginal social benefit and marginal private benefit at the optimal output level). This would effectively cause a parallel downward shift of the demand line until it passed through *b* and the optimal quantity was produced (and consumed).

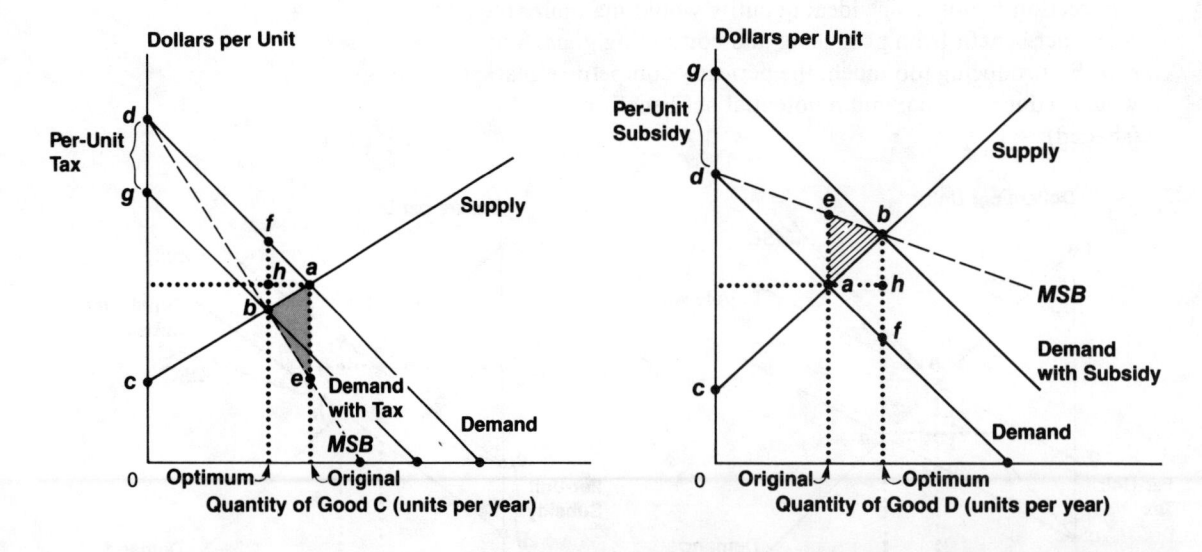

g. In that case, marginal social benefit, *MSB,* would lie above the demand line (representing marginal private benefit). Assuming supply reflected marginal social cost as well as marginal private cost, the ideal output quantity would correspond to intersection *b,* not *a.* The ideal quantity would maximize the social net benefit from producing and consuming good D at *cbd.* By producing too little, the perfectly competitive market would fail to capture the crosshatched portion *(abe)* of this maximum potential net benefit.

h. Pigou would have us offer a per-unit subsidy of *bf* = *dg* to consumers (equal to the divergence between marginal social benefit and marginal private benefit at the optimal output volume). This would effectively cause a parallel upward shift of the demand line until it passed through *b* and the optimal quantity was produced (and consumed).

2. The Coase solutions would be as follows:

   a. *Good A:* Coase would notice that the damage to outsiders from excess production would equal *fbea*, which would be the maximum amount they would pay the producers and consumers of A to cut production to the optimal level. Such a cut would cause producers to lose producer surplus of *fha* and would cause consumers to lose consumer surplus of *hba*, the sum of which would be smaller than the potential bribe. Thus the parties could get together, agree on cutting the production and consumption of A to the optimum, and jointly reap the social net gain of this action equal to *abe*.

   b. *Good B:* Coase would notice that the benefit to outsiders from extra production would equal *eafb*, which would be the maximum amount they would pay the producers and consumers of B to raise production to the optimal level. Such an increase would cause producers to lose producer surplus of *afh* and would cause consumers to lose consumer surplus of *ahb*, the sum of which would be smaller than the potential bribe. Thus the parties could get together, agree on raising the production and consumption of B to the optimum, and jointly reap the social net gain of this action equal to *abe*.

   c. *Good C:* Coase would notice that the damage to outsiders from excess production and consumption would equal *bfae*, which would be the maximum amount they would pay the producers and consumers of C to cut production to the optimal level. Such a cut would cause producers to lose producer surplus of *bha* and would cause consumers to lose consumer surplus of *hfa*, the sum of which would be smaller than the potential bribe. Thus the parties could get together, agree on cutting the production and consumption of C to the optimum, and jointly reap the social net gain of this action equal to *abe*.

   d. *Good D:* Coase would notice that the benefit to outsiders from extra production and consumption would equal *aebf*, which would be the maximum amount they would pay the producers and consumers of D to raise production to the optimal level. Such an increase would cause producers to lose producer surplus of *abh*, and would cause consumers to lose consumer surplus of *ahf*, the sum of which would be smaller than the potential bribe. Thus the parties could get together, agree on raising the production and consumption of D to the optimum, and jointly reap the social net gain of this action equal to *abe*.

3. a. Quantity 0*A*
   b. Total benefit: 0*BCA*
      Total cost: *DCA*
      Net benefit: 0*BCD*
   c. The level would then equal 0*E*.
      Total benefit: 0*BE*
      Total cost: *DFE*.
      Net benefit: 0*BCD* minus *CFE*

d. A tax of AC per unit dumped. The government would collect taxes of $0GCA$.
e. A subsidy of $AC = EH$ per unit abated. The government would have to pay a subsidy of $ACEH$.
f. Selling $0A$ rights per year at a price of $0G$ each. Thus the revenue would equal $0GCA$.

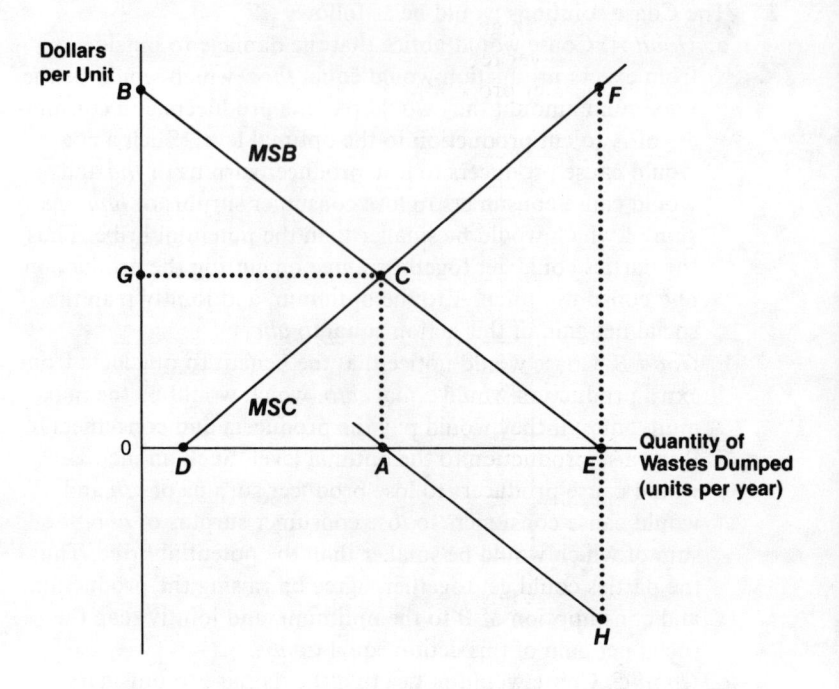

4. a. $25(\$1) + 250(\$10) + 225(\$5) = \$3,650$.
   b. $50(\$1) + 450(\$5) = \$2,300$. It could be achieved by forcing A and C to treat all wastes, while B treated none.
   c. The government could prohibit all waste dumping, except with pollution rights. It could then offer to sell 500 such rights per year. The equilibrium price would lie between $5 and $10 per right. At any price below $5, firms B and C would demand 950 rights rather than treat wastes for more. The shortage would drive the price up. At any price above $10, nobody would demand rights, because all firms would find waste treatment to be cheaper. The surplus would drive the price down. At $10 precisely, firm B would be indifferent between treating and dumping; at $5 precisely, the same would hold for Firm C.

5. a. As many as 7 vessels, because a boat's daily revenue (the average catch) would exceed or be equal to the daily cost of 80 pounds of fish as long as 7 or fewer vessels were on the lake.
   b. The optimum would be 5–6 boats. This answer differs from (a) because under the circumstances one would want to reduce to zero the marginal social benefit shown in the accompanying table. This would occur at a lower number of boats than 7, because the marginal private benefit (the average catch another vessel can expect) here always exceeds the marginal social benefit. Thus the latter reaches zero before the former does.
   c. The optimum would be 3–4 boats, because this would equate (at 80 pounds per day) the marginal social benefit with the marginal social cost.

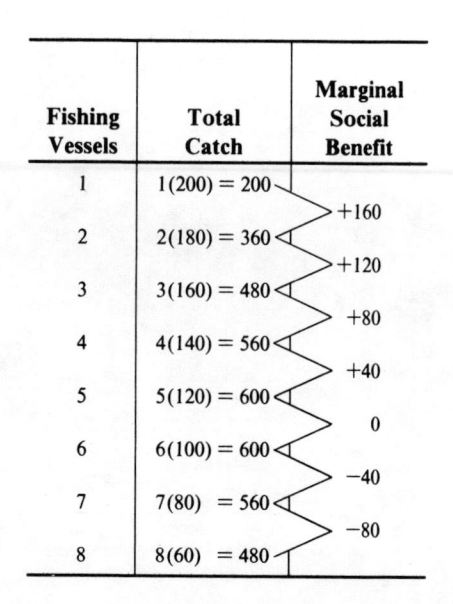

Fishing Vessels	Total Catch	Marginal Social Benefit
1	1(200) = 200	
2	2(180) = 360	+160
3	3(160) = 480	+120
4	4(140) = 560	+80
5	5(120) = 600	+40
6	6(100) = 600	0
7	7(80) = 560	−40
8	8(60) = 480	−80

d. Let a tax be imposed equal to the divergence (at the optimal number of 3.5 boats) between marginal private benefit (150 pounds) and marginal social benefit (80 pounds). This 70-pound tax, when added to the 80-pound cost of running a boat, would discourage all but 3.5 boats from the fishery (given the average catch data of this example).

6. The demand line in graph (b) is a pseudo demand line because people would never reveal these preferences in a private market, given the freerider problem.

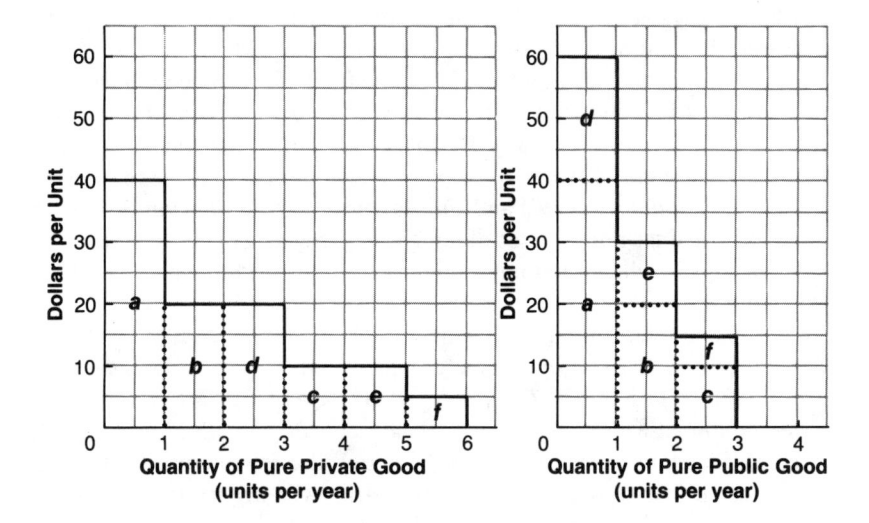

7. a. The optimum quantity of the pure public good is 2 units per year (corresponding to intersection A).
   b. Society's net benefit from the optimal quantity equals area $ABCDE$.

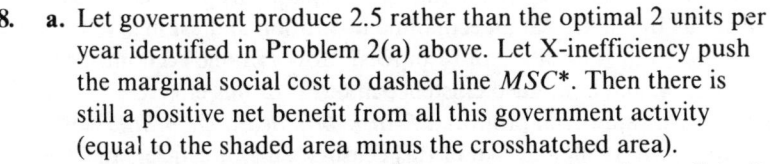

8. a. Let government produce 2.5 rather than the optimal 2 units per year identified in Problem 2(a) above. Let X-inefficiency push the marginal social cost to dashed line MSC^*. Then there is still a positive net benefit from all this government activity (equal to the shaded area minus the crosshatched area).
   b. Of course not. See panel (c) of text Figure 16.9, "Internalities," for an example to the contrary.

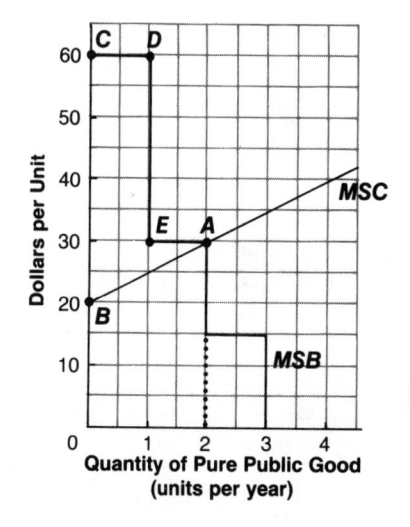

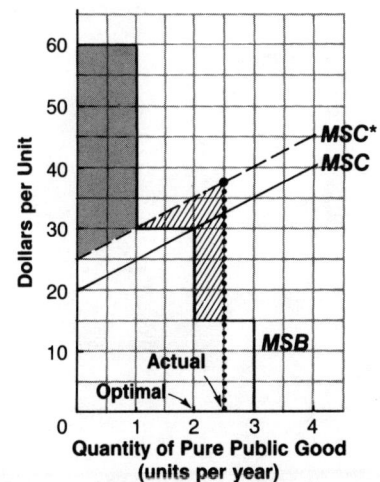

BIOGRAPHY 16.1

Arthur C. Pigou

Arthur Cecil Pigou (1877–1959) was born in Ryde on England's Isle of Wight. He studied at Cambridge, where he became a prize pupil and then the successor of Alfred Marshall (see Biography 6.1). Pigou revered his great teacher and firmly carried on his tradition. But Pigou was also a pioneer in welfare economics. He introduced the distinction (explained in this chapter) between private and social costs and benefits and stressed the need for corrective taxes or subsidies to achieve a Pareto optimum.

Until challenged by Keynes (in the field of macroeconomics), Pigou was Britain's leading economist. His works ranged widely, from his classic *The Economics of Welfare* (which first appeared as *Wealth and Welfare* in 1912) and *A Study in Public Finance* (1928) to *Socialism Vs. Capitalism* (1937) and *Employment and Equilibrium* 1941). All his writings reflected his abiding concern with practical matters, with ways in which economics might help improve humanity's condition.

"The social enthusiasm which revolts from the sordidness of mean streets and the joylessness of withered lives . . . ," Pigou says in the 1920 edition of *The Economics of Welfare* (p. 5), "is the beginning of economic science." Pigou was always concerned with improving conditions *now* so that life might be better also for the future. He demurred from those of his contemporaries who argued that poverty, after all, could not be inherited: "My reply is that the environment of one generation *can* produce a lasting result, because it can affect the environment of future generations. Environments, in short, as well as people, have children" (p. 98).

BIOGRAPHY 16.2

Ronald H. Coase

Ronald Harry Coase (1910–   ) was born in Willesden, England. He studied and later taught at the London School of Economics. In 1951, he came to the United States, teaching at the Universities of Buffalo, Virginia, and, finally, Chicago. There Coase became the founder of the "new institutional economics," which unifies analytical and institutional economics—two branches that have often been antagonistic toward one another. Again and again, Coase has produced great insights by looking at an economic institution and asking: What are its effects on the allocation of resources? How can one account for its evolution and continued existence in terms of this effect?

This approach produced *British Broadcasting: A Study in Monopoly* (1950) and two of this century's major articles. In "The Nature of the Firm" (1937), Coase examined what determines which transactions are conducted within firms and which are conducted in the marketplace.[1] In "The Problem of Social Cost," he challenged Pigou in

---

[1] Ronald H. Coase, "The Nature of the Firm," *Economica*, New Series 4 (November 1937): 386–405. Reprinted in American Economic Association, *Readings in Price Theory* (Chicago: Irwin, 1952).

a way noted at length in this chapter.[2] Through his editorship of *The Journal of Law and Economics*, Coase has rapidly advanced studies in these overlapping fields. He has summed up one result by saying: "It is now generally accepted by all students of the subject that most (perhaps almost all) government regulation is anti-competitive and harmful in its effects." In 1980, the American Economic Association honored Coase by making him one of its Distinguished Fellows.

BIOGRAPHY 16.3

Paul A. Samuelson

Paul Anthony Samuelson (1915–   ) was born in Gary, Indiana. He studied at the University of Chicago and Harvard before joining the faculty at the Massachusetts Institute of Technology.

While still a graduate student, he wrote 11 major articles, including the now classic analysis of interaction between accelerator and multiplier. At age 23, he prepared a pathbreaking dissertation, which he called nothing less than *Foundations of Economic Analysis*. Using such a title was a rather bold move for such a young man and quite a contrast to the narrow titles of typical dissertations. Samuelson's goal was to cast all of economics in mathematical terms. Thus he showed that the concept of maximization under constraints pervaded all of economic analysis. (Consumers maximized utility, given preferences, incomes, and the prices of goods; firms maximized profit, given technology and the prices of inputs and outputs; governments maximized net social benefits.) He described mathematically the state of an economic system in equilibrium and the process of adjustment from one state to another. He thereby showed that the comparison of equilibrium positions in static states was meaningful only if a dynamic analysis of stability was conducted. Unlike Alfred Marshall (see Biography 6.1), who buried his mathematics in footnotes and appendices, Samuelson was not bashful about his mathematical approach. "The laborious literary working over of essentially mathematical concepts such as is characteristic of much modern economic theory," he says, "is not only unrewarding from the standpoint of advancing the science, but involves as well gymnastics of a peculiarly depraved type" (p. 6).

Samuelson's sophisticated and oftentimes highly mathematical treatment of advanced economics (ranging from consumer behavior and international trade to business cycles, public finance, and welfare economics) can be sampled in his four-volume *Collected Scientific Papers* (1966–78).

His work has always been supplemented by innovative approaches at the level of the novice. His introductory *Economics* went through 12 editions between 1948 and 1985 and was used by millions. The book presented a "grand neoclassical synthesis," which showed that the time-honored principles of the founding fathers (Smith, Ricardo, and the like) remained valid in a world of full employment brought about with the help of modern national income analysis (as developed by Keynes). Many millions more have followed

---

[2] Ronald H. Coase, "The Problem of Social Cost," *The Journal of Law and Economics* 3 (October 1960): 1–44.

Samuelson's columns in *Newsweek,* which frequently advocate actively interventionist government policies (in contrast to Milton Friedman—Biography 1.1—the magazine's other economic columnist). Not surprisingly, Samuelson has been showered with many honors. The American Economic Association awarded him its first John Bates Clark medal in 1947 and made him its president in 1961. In 1970, he became the first American to receive the Nobel Memorial Prize for Economic Science. The award committee cited his "outstanding ability to derive important new theorems and to find new applications for existing ones. By his contributions Samuelson has done more than any other contemporary economist to raise the level of scientific analysis in economic theory."

Samuelson resisted urgent requests by Presidents Kennedy and Johnson to join the Council of Economic Advisers. He preferred to remain an economist's economist. As he put it in his presidential address to the American Economic Association, "In the long run, the economic scholar works for the only coin worth having—our own applause.... This is not a plea for 'Art for its own sake,' 'Logical elegance for the sake of elegance.' It is not a plea for leaving the real-world problems of political economy to noneconomists. It is not a plea for short-run popularity with members of a narrow in-group. Rather it is a plea for calling shots as they really appear to be (on reflection and after weighing all evidences), even when this means losing popularity with the great audience of men and running against 'the spirit of the times.'"[1]

BIOGRAPHY 16.4

James M. Buchanan

James McGill Buchanan (1919–   ) was born at Murfreesboro, Tennessee. He studied at the Universities of Tennessee and Chicago, then taught at Florida State, Virginia, UCLA, and Virginia Polytechnic Institute. He now directs the Center for the Study of Public Choice at George Mason University. He has served as president of the Southern Economic Association and, in 1983, the American Economic Association made him one of its Distinguished Fellows. In 1986, "for his pioneering development of new methods for analyzing economic and political decision-making," he was awarded the Nobel Memorial Prize in Economic Science.

Buchanan is a leader among the new institutional economists; his work has been devoted to exploring the common ground between economics and political science. More than anyone, he has been responsible for the "public choice revolution" of the past two decades, which has taken a radically new look at government. Instead of characterizing government as a superindividual—omniscient, benevolent, and always correcting market failures with precision—scholars influenced by Buchanan are now more apt to look at government from the perspective of individual officials who seek to serve their own purposes. Buchanan developed these ideas in *The Calculus of Consent* (1962, with Gordon Tullock), *Demand and Supply of Public Goods* (1968),

---

[1] Paul A. Samuelson, "Economists and the History of Ideas," *The American Economic Review,* March 1962, p. 18.

The Limits of Liberty (1975), *Freedom in Constitutional Contract* (1978), *What Should Economists Do?* (1980), *Liberty, Market, and State* (1986), and many other books and articles. Some of his major articles have been republished as *Economics: Between Predictive Science and Moral Philosophy* (1988).

Buchanan has been highly critical of attempts (such as those by Samuelson) to build a bridge from theories of maximization (of utility or profit) by individuals to one of maximization by society. Society, Buchanan has argued, is not a thinking and acting being that could possibly maximize its well-being on the basis of a social welfare function that somehow relates different states of the world to social welfare. Hence, it is absurd to expect that social choices (which in fact are not the choice of a superbeing, but are simply the aggregate result of individual choices) will be internally consistent. Such an expectation should only be held for individual choice. According to Buchanan:

> The object for economists' research is "the economy," which is, by definition, a *social organization,* an interaction among separate choosing entities. . . . "The economy" does not maximize . . . there exists no one person, no single chooser, who maximizes *for* the economy. . . . Paul Samuelson . . . in his *Foundations of Economic Analysis* . . . extended the maximizing construction to welfare economics, extolling the virtues of [a] social welfare function. . . . I have no quarrel with the elaboration and refinements of the maximizing models for individual and firm behavior. . . . My strictures are directed exclusively at the extension of this basic maximizing paradigm to social organization where it does not belong. This is the bridge which economists should never have crossed, and which has created major intellectual confusion. "That which emerges" from the . . . exchange process . . . is not the solution to a maximizing problem. . . . "That which emerges" is "that which emerges" and that is that. . . .
>
> . . . What should be the role for the economist . . . ? He should neither revert to nihilism nor seek the escapism of social welfare functions. His productivity lies in his ability to search out and to invent social rearrangements which will embody Pareto-superior moves. If an observed position is inefficient, there must be ways of securing agreement on change, agreement which signals mutuality of expected benefits. . . . Yet how many economists do we observe working out such schemes? . . . As persons, both from the streets and the ivory towers, observe modern governmental failures, they can scarcely fail to be turned off by those constructions which require beneficent wisdom on the part of political man. And they can hardly place much credence in the economist consultant whose policy guidelines apply only within institutions that embody such wisdom. Something is amiss, and economists are necessarily being forced to take stock of the social productivity of their efforts. When, as, and if they do, they will, I think, come increasingly to share what I have called the contractarian paradigm. . . .
>
> Economics comes closer to being a "science of contract" than a "science of choice." And with this, the "scientist," as political economist, must assume a different role. The maximizer must be replaced by the arbitrator, the outsider who tries to work out compromises among conflicting claims. The Edgeworth . . . box becomes the first diagram in our elementary textbooks. . . . The unifying principle becomes *gains-from-trade,* not maximization.[2]

---

[2]James M. Buchanan, "A Contractarian Paradigm for Applying Economic Theory," *The American Economic Review,* May 1975, pp. 225–30.